C0-AKZ-595

SPRINGER HANDBOOK OF AUDITORY RESEARCH

Series Editors: Richard R. Fay and Arthur N. Popper

Springer
New York
Berlin
Heidelberg
Hong Kong
London
Milan
Paris
Tokyo

Springer Handbook of Auditory Research

Andrea Megela Simmons
Arthur N. Popper
Richard R. Fay
Editors

Acoustic Communication

With 55 Illustrations

Andrea Megela Simmons
Departments of Psychology and Neuroscience
Brown University
Box 1853
Providence RI 02912, USA
Andrea_Simmons@brown.edu

Arthur N. Popper
Department of Biology and Neuroscience and Cognitive Science Program
University of Maryland
College Park, MD 20742-4415, USA
AP17@umail.umd.edu

Richard R. Fay
Department of Psychology and Parmly Hearing Institute
Loyola University of Chicago
Chicago, IL 60626, USA
rfay@wpo.it.luc.edu

Series Editors: Richard R. Fay and Arthur N. Popper

Cover illustration: Courtesy of Daniel Baleckaitis and Daniel Margoliash.

Library of Congress Cataloging-in-Publication Data
Acoustic communication / editors, Andrea Megela Simmons, Arthur N. Popper, Richard R. Fay.
p. cm.—(Springer handbook of auditory research; 16)
Includes bibliographical references.
ISBN 0-387-98661-8 (alk. paper)
1. Animal sounds. 2. Sound production by animals. 3. Animal communication. I. Simmons, Andrea Megela. II. Popper, Arthur N. III. Fay, Richard R. IV. Springer handbook of auditory research; v. 16.
QL765 .A32 2002
591.59′4—dc21 2001057680

ISBN 0-387-98661-8 Printed on acid-free paper.

Printed in the United States of America.

9 8 7 6 5 4 3 2 1 SPIN 10697639

www.springer-ny.com

Springer-Verlag New York Berlin Heidelberg
A member of BertelsmannSpringer Science+Business Media GmbH

Photograph courtesy of Dong Lin.

This volume of the *Springer Handbook of Auditory Research* is dedicated to Peter Marler in recognition of his many incisive and influential conceptual and technical contributions to the understanding of animal communication. Throughout his career, Peter has served as an inspiration and a mentor to numerous students and scholars in this field. We are honored to have the opportunity to dedicate this volume to him.

Series Preface

The *Springer Handbook of Auditory Research* presents a series of comprehensive and synthetic reviews of the fundamental topics in modern auditory research. The volumes are aimed at all individuals with interests in hearing research, including advanced graduate students, postdoctoral researchers, and clinical investigators. The volumes are intended to introduce new investigators to important aspects of hearing science and to help established investigators to better understand the fundamental theories and data in fields of hearing that they may not normally follow closely.

Each volume is intended to present a particular topic comprehensively, and each chapter will serve as a synthetic overview and guide to the literature. As such, the chapters present neither exhaustive data reviews nor original research that has not yet appeared in peer-reviewed journals. The volumes focus on topics that have developed a solid data and conceptual foundation rather than on those for which a literature is only beginning to develop. New research areas will be covered on a timely basis in the series as they begin to mature.

Each volume in the series consists of five to eight substantial chapters on a particular topic. In some cases, the topics will be ones of traditional interest for which there is a substantial body of data and theory, such as auditory neuroanatomy (Vol. 1) and neurophysiology (Vol. 2). Other volumes in the series will deal with topics that have begun to mature more recently, such as development, plasticity, and computational models of neural processing. In many cases, the series editors will be joined by a coeditor having special expertise in the topic of the volume.

Richard R. Fay, Chicago, Illinois
Arthur N. Popper, College Park, Maryland

Preface

Previous volumes in the *Springer Handbook of Auditory Research* have focused on mechanisms of hearing and sound processing using laboratory experiments to study the detection capabilities of the auditory system and neurophysiological experiments to study the underlying neural bases of these capabilities. Although the series has included a number of chapters dealing with various aspects of acoustic behavior, few, if any, chapters have examined the concept of acoustic communication and the related issue of how sound is used in a natural behavioral context.

At the same time, it is important to remember that the auditory system has probably evolved for two reasons. Its earliest evolution is likely to have involved detection and recognition of environmental signals that enabled animals to obtain information about the nonbiological and biological components of their environment. Because sound detection and recognition probably evolved in the aquatic environment, sound processing was highly useful because visual senses (if they existed at all) have a very limited range in the water. With sensitivity to sound, the earliest vertebrates could have extended the range over which they could detect and recognize objects, including potential predators or prey that might have made sounds as they moved.

The second, and later, reason for further evolution of the auditory system was certainly for communication. There is little doubt that the process of acoustic communication has shaped some aspects of the evolution of hearing, the ear, and central processing mechanisms, and that is one of the major themes of this volume.

Analyses of acoustic communication and behavior in this volume are based primarily on observations and experiments on animals in their natural habitats, where social, ecological, and environmental factors all influence sound detection and perception. All chapters in this volume share as a common theme the neuroethological approach to acoustic communication. The neuroethological approach attempts to understand the underlying neural and hormonal bases of complex, species-typical behavior in both an evolutionary and an ecological context. Thus, multiple levels of analysis are explored, from mechanistic to phylogenetic. Moreover, the chapters are not limited to discussing data from only one group of animals but attempt to derive common principles from a wide range of vertebrate species.

In the first chapter, Simmons provides an overview of acoustic communication within a neuroethological perspective, emphasizing both the progress that has been made and the fascinating questions yet to be explored. Bass and Clark, in Chapter 2, tackle the difficult problem of sound communication by fishes in shallow and deep water. In Chapter 3, Fitch and Hauser explore the interface between vocal production mechanisms and the meanings of vocal signals. Boughman and Moss, in Chapter 4, explore how acoustic signals are shaped over ontogeny by psychological processes of learning. The physical acoustics of sound transmission in air are considered by Ryan and Kime in Chapter 5. These authors address the issue of how signal structure and sensory processing mechanisms might have evolved to take advantage of the physical constraints of sound transmission pathways. Hormonal mechanisms mediating both sensory perception and vocal production are discussed by Yamaguchi and Kelley in Chapter 6. Gentner and Margoliash, in Chapter 7, explore the variety of neural processing mechanisms underlying perception and cognition of complex vocal signals.

Past *Springer Handbook of Auditory Research* volumes do have a number of chapters related to acoustic communication in various species. In particular, fish and amphibian communication were discussed by Zelick, Mann, and Popper in Vol. 11 of this series (*Comparative Hearing: Fish and Amphibians*), and environmental constraints on communication in insects were considered by Römer in Vol. 10 (*Comparative Hearing: Insects*). Acoustic behavior and communication by dolphins and whales were the theme of the chapter by Tyack and Clark, and communication by wild dolphins was discussed by Herzing, in Vol. 12 (*Hearing by Whales and Dolphins*).

Andrea Megela Simmons, Providence, Rhode Island
Arthur N. Popper, College Park, Maryland
Richard R. Fay, Chicago, Illinois

Contents

Contributors

ANDREW H. BASS
Department of Neurobiology and Behavior, Cornell University, Ithaca, NY 14850, USA, and UC Bodega Marine Laboratory, Bodega Bay, CA 94923, USA

Janette Wenrick Boughman
Department of Zoology, University of Wisconsin, Madison, WI 53706, USA

Christopher W. Clark
Bioacoustics Research Program, Laboratory of Ornithology, Cornell University, Ithaca, NY 14853, USA

W. Tecumseh Fitch
Department of Psychology, Harvard University, 33 Kirkland Street, Cambridge, MA 02138, USA

Timothy Q. Gentner
Department of Organismal Biology and Anatomy, University of Chicago, 1025 E. 57th Street, Chicago, IL 60637, USA

Marc D. Hauser
Department of Psychology, Harvard University, 33 Kirkland Street, Cambridge, MA 02138, USA

Darcy B. Kelley
Department of Biological Sciences, Sherman Fairchild Center for Life Sciences, Mail Code 2432, Columbia University, New York, NY 10027, USA

Nicole M. Kime
Section of Integrative Biology C0930, University of Texas, Austin, TX 78712, USA

Daniel Margoliash
Department of Organismal Biology and Anatomy, University of Chicago, 1025 E. 57th Street, Chicago, IL 60637, USA

Cynthia F. Moss
Department of Psychology, University of Maryland, College Park, MD 20742, USA

Michael J. Ryan
Section of Integrative Biology C0930, University of Texas, Austin, TX 78712, USA

Andrea Megela Simmons
Departments of Psychology and Neuroscience, Brown University, Providence, RI 02912, USA

Ayako Yamaguchi
Department of Biological Sciences, Sherman Fairchild Center for Life Sciences, Mail Code 2432, Columbia University, New York, NY 10027, USA

1 Perspectives and Progress in Animal Acoustic Communication

Andrea Megela Simmons

1. Introduction

Several hours after twilight, on a hot summer night in June, a female bullfrog, *Rana catesbeiana*, swims in the shallow waters of a Rhode Island pond. Periodically, she raises her head above the water level and appears to listen to a cacophony of sounds around her, including the advertisement calls of male bullfrogs, advertisement calls of male green frogs (*Rana clamitans*), the buzzing of insects, the prowling of raccoons and cats around the pond's edges, shouts and laughter of humans nearby, music, traffic noise, wind noise, and the low rumble of airplanes flying overhead. Through the water come additional sounds, including those of fish, turtles and insects. After a time of seemingly random swimming, the female bullfrog approaches a calling male bullfrog and mates with him. After amplexus is complete, the female deposits her eggs, and the male resumes calling.

This seemingly simple and straightforward instance of acoustic communication requires a great deal of sophisticated biological and psychological processing by the animals involved. Temperate-zone frogs in their natural habitats only breed during the summer, and then only when rainfall and temperature are appropriate. Because mating is seasonal, these animals must have a mechanism for detecting environmental conditions. In addition, mating only occurs when both the female and the male are hormonally ready. Their hormonal control systems must thus have access to environmental sensors in order to coordinate internal events with the external milieu. The female's task is to identify and choose among males of her own species using acoustic cues embedded in a complex, noisy environment. She must be able to identify and separate sounds conveying the message of "bullfrog advertisement calls" from those conveying irrelevant messages such as "a motorcycle is going by." Not only must the female discriminate between conspecific bullfrog males and heterospecific green frog males, but she also needs to distinguish among the different types of calls (advertisement, aggressive, territorial) within the male bullfrog's repertoire. Moreover, the female may target particular individual males within the larger chorus of vocalizing conspecifics by focusing on particular acoustic features of their calls. These acoustic features may provide an indication of an individual male's vigor or quality as a mate.

Male bullfrogs emit advertisement calls not only to attract females for mating but also to announce their occupation of a territory to neighboring males. Because calling is such an energetically costly activity, males might develop certain rules regulating their vocal output and interactions with vocally competing males (Boatright-Horowitz et al. 2000). These rules might involve psychological processes of habituation and learning. Production of calls also requires an underlying hormonally sensitive mechanism of laryngeal control. The auditory systems of both the male sender and the female receiver must process the male's vocal signals in order for detection, discrimination, and identification to occur. These mechanisms might involve specialized "mating call detectors" (Capranica and Moffat 1983) or they might be an adaptation of some more general neural processing strategy not unique to frogs or their particular communication system.

The use of sounds for communication, from mate calling in frogs to speech in humans, is widespread among vertebrate animals. As the example above shows, even the frog's acoustic communication is a complex process both behaviorally and physiologically. The focus of this book is to analyze the communicative use of audition in nonhuman vertebrate animals from the interdisciplinary perspective of neuroethology. Neuroethology attempts to unravel the physiological underpinnings of complex, species-typical behavior within a broad evolutionary framework. All chapters in this volume share as their common theme such a neuroethological perspective on acoustic communication. In doing so, they span multiple levels of inquiry, from single-unit recordings to phylogenetic analysis. Moreover, they are not limited to discussing data from only one group of animals but seek to derive common principles of communication from comparisons of a wide range of vertebrate species. Even though the focus of this volume is on vertebrates, several chapters include relevant comparisons with invertebrate animals. Due to space limitations, acoustic communication by speech in humans is not discussed.

This volume complements and extends the comparative approach to hearing outlined in previous volumes in this series (Fay and Popper 1994, 1999; Hoy et al. 1998; Dooling et al. 2001b). Chapters in these earlier volumes examined the acoustic capabilities of animals primarily in terms of psychophysical limits on auditory performance (Brown 1994; Long 1994; Stebbins and Moody 1994; Moss and Schnitzler 1995; Simmons et al. 1995; Fay and Simmons 1999; Dooling et al. 2001a). In laboratory psychophysical experiments, social and environmental influences on perception are necessarily excluded from analysis, and detection of simplified stimuli with well-defined acoustic parameters is emphasized. The psychophysical approach contributes crucial information about the sensory limitations that constrain perception of sounds but does not directly address the interactions between individual members of a species that form the basis of communicative behavior. The neuroethological approach presented here builds on the psychophysical perspective by extending analysis to the perception of natural, complex sounds (or simulations thereof) in addition to sounds embodying variations of particular acoustic parameters. In addition, the influence of social, ecological and environmental factors on auditory perception and discrimination is explicitly addressed. Similar neuroethological perspectives can be found in chapters by Zelick et al. (1999), which focuses on field studies of sound communication in anurans, and Römer (1998), which examines ecological factors constraining insect communication. Clearly, both psychophysical limits and real-world manifestations of performance are influenced by the evolutionary history of the species, and neuroethological analyses attempt to integrate both types of experimental and theoretical approaches.

1.1. The Neuroethological Perspective

According to Tinbergen (1963), a useful way to answer the question of "what makes behavior happen the way it does" is to dissect this question into different levels of conceptual and experimental analysis. On one level, behavior can be analyzed mechanistically in terms of its underlying physiological and hormonal bases. A second level of analysis is ontogenetic, which considers the differing influences of genetic and environmental factors on the expression of behavior during an individual's lifetime. Tinbergen called the mechanistic and ontogenetic levels the "proximate" causes of behavior because they concern events within an individual organism. Behavior also can be understood phylogenetically, from its evolutionary history, and in terms of its fitness consequences and adaptive significance. Tinbergen called these two additional levels of analysis the "ultimate" causes of behavior because they concern events taking place across multiple individuals on a time scale spanning numerous generations. Neuroethological studies attempt to integrate all four of these levels of analysis into a unified view of behavior. Although some chapters in this volume focus more closely on ultimate (Bass and Clark, Chapter 2; Fitch and Hauser, Chapter 3; Ryan and Kime, Chapter 5) or proximate (Boughman and Moss, Chapter 4; Yamaguchi and Kelley, Chapter 6; Gentner and Margoliash, Chapter 7) explanations, all of them examine acoustic communication within Tinbergen's theoretical framework. For example, Fitch and Hauser creatively place the issue of the evolution of honest and dishonest signals in the context of the mechanistic constraints imposed by the signaler's vocal apparatus. This direct inclusion of proximate mechanisms in ultimate explanations distinguishes chapters in this volume from those in earlier, classic volumes on animal acoustic communication (e.g., Lanyon and Tavolga 1960; Busnel 1963), which focused primarily on evolutionary origins and functional descriptions of vocal signals.

Neuroethologists have commonly followed a "model-systems" approach to behavior by focusing on species-specific behavior patterns typical of particular groups of animals. This has yielded cohesive "stories" about prominent examples, such as advertisement calling in frogs, singing in birds, and echolocation in bats. The advantage of the model- systems approach derives from the fact that, for any particular species, the acoustic stimuli that have to be considered are well-defined and do not range widely over all possible sounds. An integrated approach to behavior need not be limited to highly specialized systems, however. The neuroethological perspective allows insights into both more general and more species-specific physiological mechanisms and evolutionary adaptations. For example, both songbirds and humans can selectively discriminate and attend to specific sounds of interest within a complex acoustic scene, and they may use similar acoustic cues to achieve this segregation (Hulse et al. 1997). Neural mechanisms of sound segregation obviously are much more amenable to study in birds than in humans, even with the advent of noninvasive imaging techniques. Results from neuroethological studies have ranged far beyond descriptions of unusual or bizarre behaviors in strange organisms with no relevance to understanding "less specialized" animals. Crucially, some widespread principles of neural organization (combination-sensitive neurons, distorted tonotopic maps, parallel cortical representations) were first quantified in the brains of highly specialized model systems and then later observed to operate in the nervous systems of other animals (Suga 1988; Gentner and Margoliash, Chapter 7). Then, too, in an evolutionary sense, all communication systems are specialized in that each has evolved in relation to specific ecological niches and selection pressures to emphasize particular types of signals and variability in their parameters.

2. What Is Communication?

Acoustic communication, using speech, is something humans do every day. Acoustic communication using both simple and complex sounds is also very common among vertebrate and invertebrate animals. Operational definitions of what we mean by communication are many (see, for example Green and Marler 1979; Hauser 1997; Bradbury and Vehrencamp 1998). Each set of authors in this volume provides a short definition of acoustic communication as it has guided their choice of material.

Most broadly, we can define communication as the transfer of information (a message) between a sender and a receiver. Although communication is often thought of as an intragroup interaction among members of the same species, it need not be restricted in this way. Acoustic signals used for communication can have widely varying information content and thus convey either very explicit or very vague messages. For example, signals might convey a broad message, such as species identity between groups, or a more narrow message, such as information about the identity of a specific individual within a group. Sounds might be used in such contexts as mate selection, predator avoidance, parental behavior, foraging, and coordinating group activities. Within these different behavioral contexts, sounds may communicate the sender's motivational or cognitive state. Transmission of the message of species identity in the form of mating or advertisement signals has been a major topic of study both in the ethological and in the neuroethological literature (Ryan and Kime, Chapter 5). Fitch and Hauser (Chapter 3), Boughman and Moss (Chapter 4), and Yamaguchi and Kelley (Chapter 6) all add to this literature by their neuroethological analyses of the evolution, function and/or hormonal mechanisms underlying other types of communication sounds, such as courtship calls, alarm calls, contact calls, and isolation calls. In particular, Boughman and Moss ask whether certain kinds of acoustic signals used by animals can be understood as individual signatures, much like names.

The transfer of information in animal communication systems is sometimes presumed to be intentional and of mutual benefit both to the sender and the receiver. An example of this mutual benefit is the male túngara frog, *Physalaemus pustulosus*, vocalizing to attract females to him for mating. However, sound by its very nature moves in all directions from its source and, because it is broadcast, can be detected by any receiver within range. Thus, information transfer either to conspecific animals (competitors) or to heterospecific animals (predators) can be an unwanted by-product of the communicative process. Ryan and Kime (Chapter 5) discuss, among other examples, how the fringe-lipped bat, *Trachops cirrhosus*, exploits the male túngara frog's calling behavior to locate him as prey. These unintended consequences of acoustic communication play an important role in the evolution of signal structure.

The issues of intent and mutual benefit in communicative interactions are further analyzed by Fitch and Hauser (Chapter 3) in the context of the role of honesty in communication. Dawkins and Krebs (1978) first proposed that, from an evolutionary point of view, communication signals are designed to manipulate the behavior of receivers to the advantage of signalers and thus do not necessarily convey an accurate (honest) picture of the signaler. Fitch and Hauser analyze this proposition in terms of the mechanisms of signal production, arguing that morphological features of the signaler (such as body size and length of the vocal tract) constrain the form and the variability of communication sounds and thus their ability to deceive a receiver. These physiological constraints may maintain signal honesty in the absence of any particular selective pressure for honesty.

Communication is also described as a process by which one individual's (the sender) behavior effects some change in a second individual (the receiver). These changes can be behavioral, hormonal, or morphological. Most often, we focus on behavioral changes because these are readily observable. A good example is mate calling by male anurans. Here, we say that communication has occurred because the female has changed her motor behavior in response to the male's call by approaching the source of the call (phonotaxis). These and similar examples (see Boughman and Moss, Chapter 4) presuppose that the signal produces an observable or measurable response in the receiver. Long-term, subtle effects of a signal on a receiver's behavior or physiology are equally important but more difficult to unravel. A classic example of how communication signals produce more subtle changes in the receiver is the pioneering work of Lehrman (1964) on the reproductive behavior of ring doves (*Streptopelia risoria*). During pair-bond formation, both acoustic and visual signals emitted by one member of the pair effect subsequent hormonal changes in the other. The behavior of the receiver thus has an important impact on the physiology of the sender as well as on its behavior. This interplay between the male and the female emphasizes the interactive and bidirectional nature of communication. Yamaguchi and Kelley (Chapter 6) provide further examples of these interactive effects in the context of hormonal mediation of communication.

2.1. Deciphering the Message

Deciphering the messages conveyed by communication sounds can be a difficult task both for the receiver of the signals and for the experimenter. Part of this difficulty arises from the process of propagation of the sound through the environment. Both in air (Ryan and Kime, Chapter 5) and underwater (Bass and Clark, Chapter 2), the process of propagation alters both the spectral and the temporal characteristics of the signal. The environment can introduce distortion or noise (either biotic or abiotic) or it can selectively attenuate or accentuate certain sound frequencies in relation to others. In either case, the result would be confusion of the message and interference with the transmission of information. Because of these factors, the structure of a signal at its source (the sender) is not necessarily identical to its structure at the receiver, so the receiver's job in deciphering the signal is made more difficult unless propagation is taken into account during reception. Subtle variations in signal meaning may be missed by the receiver because of interference by the environment. Some behavioral and neural strategies available to the receiver to circumvent these obstacles are discussed by Ryan and Kime (Chapter 5) and by Gentner and Margoliash (Chapter 7).

Examination of the physical acoustics of transmission has led to two fundamental insights into the function of acoustic communication systems. As pointed out in a classic and highly influential paper by Marler (1955), the structural features of acoustic signals, reflecting the physics of sound transmission, can provide many clues about their functions. For example, mating or advertisement calls usually contain frequencies that can be localized easily, whereas certain kinds of alarm calls contain frequencies that can be localized only with great difficulty. Signals used for species recognition are often highly stereotyped, with minimal variation among individuals. On the other hand, signals used to convey aggressive or sexual motivation are often more variable or graded, so that the acoustic variation in the signal could provide to the receiver an indication of the motivational level of the sender. Marler's (1955) analysis inspired a great deal of research on quantitative analysis of acoustic signals and on the formulation of "motivational-structural rules" for predicting signal form and function (Morton 1975; Marten and Marler 1977; see Ryan and Kime, Chapter 5).

Since these early studies, evidence has accumulated that the structure of long-distance communication signals reflects the acoustic habitat of the species. This "acoustic adaptation hypothesis" is discussed by Ryan and Kime (Chapter 5) for transmission of sounds in air. As they point out, the structure of signals is also influenced by morphological constraints (such as body and larynx size; see also Fitch and Hauser, Chapter 3) and the evolutionary history of the species under consideration. In contrast, the physical acoustics of signal production cannot predict the structure of vocal signals used by teleost fishes living in shallow water (Bass and Clark, Chapter 2). This means that animals cannot always fully exploit the acoustic features that would theoretically provide maximal transmission in a given environment. Sensory and motor adaptations used by species to circumvent or exploit these constraints are discussed by Bass and Clark (Chapter 2) and by Fitch and Hauser (Chapter 3) for underwater and in-air communication, respectively.

Finally, deciphering the message depends on detecting individual acoustic features in the signals. However, even in animals, such as frogs, with a restricted number of relatively well-defined and stereotyped signals, the receiver's responses are often guided by a combination of individual acoustic features rather than by a particular feature alone (Gerhardt 1992). Careful quantitative analysis of acoustic parameters, both in isolation and in combination with one another, are needed to determine the relative biological importance of particular signals (Boughman and Moss, Chapter 4).

2.2. *Ontogeny and Learning*

Debates on the relative roles of genetics and learning in mediating behavior have had a long history in both psychology and biology. In the neuroethological literature, these issues have been discussed primarily in two contexts: the perception and production of mating songs by orthopteran insects, where genetic factors are primary (Bentley and Hoy 1972), and the perception and production of song in oscine birds, where learning is the predominant mechanism (Marler 1970; Hauser 1997). Detailed analysis of these issues in other animals is relatively less complete. In social animals with parental care or overlap of generations, experiments to tease out the relative roles of genetic mechanisms and learning in the development of call comprehension and call production can be difficult to design and difficult to interpret. Boughman and Moss (Chapter 4) examine the role of vocal learning (both learned acquisition and social modification) in the communication systems of birds and mammals, citing examples from a wide variety of species and a wide variety of different types of acoustic signals. They provide experimental guidelines to help distinguish between the roles of maturation and learning in the development of communication behavior.

2.3. Sensory Mechanisms

Sensory mechanisms constrain both the structure and the function of vocalizations. An influential model of a signal-processing strategy for effective communication in a noisy environment was proposed by Capranica and Moffat (1983). According to this formulation, an ideal auditory system should be selectively biased, in terms of either peripheral sensitivity or of central recognition, to emphasize distinguishing relevant characteristics of communication sounds. Such a receiver would appear to be specialized for reception of those sounds to the exclusion of others. Because many examples of animal acoustic communication involve transmission of specific frequencies of sound, it is important to consider whether selective reception of just these frequencies is an adequate strategy for the receiver to interpret the message. Using the relationship between the frequency content of advertisement calls and the tuning of eighth-nerve fibers in several different species of anuran amphibians, Capranica and Moffat argued in favor of a "matched-filter" model of receiver processing in the frequency domain. In this formulation of an ideal receiver, the frequency response of the auditory system exactly mirrors the energy spectrum of the sender's vocalizations. Such matching would maximize the signal-to-noise ratio for reception by excluding masking due to frequencies not actually present in the calls. This would make the communication channel more private in the sense that sounds at other frequencies would be selectively rejected at the moment of reception. At more central levels, matched filtering would be manifested by the presence of "mating call detectors" or auditory "grandmother cells" capable of strongly responding only to sounds containing the correct combination of frequencies. Gentner and Margoliash (Chapter 7) discuss recent advances in the search for matched filters and call detectors. As they point out, combination-sensitive cells having properties consistent with these have been found in central auditory nuclei of different species of frogs, birds, and mammals. In a different approach, Ryan and Kime (Chapter 5) present a neural network that allows sensory biasing to a conspecific signal without the explicit need for dedicated mating-call detectors.

2.4. Production Mechanisms

Because perception and production of sounds have coevolved (Ryan and Kime, Chapter 5), selective biasing in sensory perception should be reflected in the motor output system as well. Many studies of motor output have focused on how the morphology of the vocal apparatus limits the types of signals that can be produced (Bass and Clark, Chapter 2; Fitch and Hauser, Chapter 3). Fitch and Hauser describe how the morphological diversity of vocal tracts in nonaquatic vertebrates is related to the types of signals used in different environments and for different communicative functions. Using an ontogenetic perspective, Boughman and Moss (Chapter 4) describe how body morphology influences the structure and development of vocalizations over an animal's lifetime. Yamaguchi and Kelley (Chapter 6) analyze vocal production from a mechanistic perspective, focusing on the influence of gonadal hormones on the vocal organs. Finally, responses of motor output cells in the central auditory and vocal systems of the songbird, and how these are related to sensory processing mechanisms, are examined by Gentner and Margoliash (Chapter 7).

3. Future Directions

Neuroethological studies of acoustic communication have generated insights into behavior on proximate and ultimate levels. However, much remains to be done before a complete, integrated analysis of sound communication is achieved in any species. We know a great deal about certain aspects of communication in particular species, but very little is known in others. For example, links between production and perception are best understood in some species of orthopteran insects and oscine birds but are much less well-studied in other vocalizing species. This is partially due to the neuroethological reliance on popular model systems (songbirds, frogs, bats) while ignoring other animals (most mammals) that also use sounds for communication. This limits the generalizability of the neuroethological approach and opens the work to criticisms that it applies only to highly specialized or unusual animals.

3.1. Neural Processing Mechanisms

On the proximate level, much of our knowledge of perceptual mechanisms remains at the level of the receptor organs and their direct neural innervation. The basic structure and response properties of the auditory periphery in relation to sound communication have been described in a variety of species. With several exceptions (Gentner and Margoliash, Chapter 7), much less is known about responses of central auditory neurons to complex communication signals. A great deal of our knowledge of central response properties in the vertebrate brain is based on experiments in which spike activity is recorded to pure tones presented in isolation or in two-tone pairs. These stimulus conditions, although essential for parametric control and basic descriptions of response properties, do not mimic the conditions under which communication typically occurs. It has been appreciated for some time that responses to natural vocalizations often cannot be predicted from responses to simpler stimuli. Nevertheless, there are many fewer experiments in which neuronal responses are recorded to complex sounds presented in the kinds of time-varying, biologically realistic sequences animals encounter while communicating.

Much research on sensory mechanisms has been guided by the matched spectral filter or "call-detector" models. Although these models are conceptually appealing and have received experimental support (Gentner and Margoliash, Chapter 7), they are insufficient to provide a full view of the mechanisms underlying acoustic communication. Two problems that have emerged on a behavioral level are that animals listen to and respond to sounds other than mating calls and that they tolerate a great deal of individual variation in call parameters. In some instances, animals prefer "supernormal" signals, with features such as duration, rate, and intensity that are outside the normal range of variation found in natural conspecific signals (Ryan 1990; Ryan and Keddy-Hector 1992). A neural response system based on sharply tuned mating-call detectors cannot easily account for such behavioral biases.

Original formulations of call-detector models emphasized detection of spectral features of sounds to the exclusion of temporal processing mechanisms. Temporal codes based on phase-locked responding by peripheral neurons or autocorrelation-type mechanisms at more central levels play a significant role in extracting behaviorally important features of complex sounds (Schwartz and Simmons 1990; Simmons et al. 2000). Temporal "template" models based on cross-correlation have been examined most extensively in bat echolocation (Simmons et al. 1995). Unfortunately, these models have not been used extensively in analysis of other communication systems, even though temporal cues play crucial roles in guiding behavior. In addition, spectral and temporal features can interact in complex, nonlinear ways (Gerhardt 1992). Understanding the neural coding of signals that covary in both spectral and temporal properties is still incomplete, but it presently seems likely that a joint time-and-frequency mechanism may be common to a variety of different auditory tasks. Computational techniques for estimating complex stimulus–response functions, such as the spectro-temporal receptive field, have been available for some time (Aertsen and Johannesma 1981) but are not widely used in the neuroethological literature (Theunissen et al. 2000). More widespread use of this and other computational techniques should yield additional insights into the neural code underlying perception of complex sounds. Another challenge for neuroethology is to derive realistic computational models of central auditory processing based on these complex neuronal response properties.

Call production and perception are inextricably linked in theories of ultimate causation of communication (Ryan and Kime, Chapter 5). On the proximate level, the link between production and perception (sensorimotor integration) has been a major theme of research in the study of birdsong but, except for some work on echolocation (Valentine and Moss 1997), has been relatively neglected in studies of other communication systems. Indeed, our knowledge of central motor mechanisms controlling vocal output is incomplete in many species, and, in spite of the theoretical importance of production/perception linkage, it is still unclear how the outputs of the different cell types described in central auditory nuclei interact to regulate vocal output.

3.2. Behavioral Analysis

The variability or plasticity, rather than the stereotypy, of signals and messages in communication has recently been emphasized in both theoretical and experimental work on acoustic communication (Ryan 1990; Ryan and Keddy-Hector 1992; Fitch and Hauser, Chapter 3). More detailed analysis of variability in both signals and responses will be important in formulating a coherent phylogenetic view of the evolution of communication (Ryan and Rand 1990).

Some of the variability seen in both signaler and receiver may be due to the influence of the social and the sensory environments to which animals are exposed in their natural habitats. For example, Marler and colleagues (Marler et al. 1986) have shown that the composition of the social group in domestic chickens (*Gallus gallus*) can affect the type and rate of acoustic signaling. Little quantitative work has examined audience effects in other vocalizing species. In addition, the influence of other sensory cues (vision, olfaction, tactile sensation) on the perception or production of acoustic signals in natural environments has been relatively neglected. Although experimental designs for teasing out the relative importance of multimodal signals are available (Partan and Marler 1999), there has to date been little progress in implementing these designs. Modern video technology should be exploited to record and quantify the entire sensory environment in which an animal communicates in order to assess the roles of these other cues in mediating acoustic behavior (Evans and Marler 1991).

3.3. Ontogeny and Phylogeny

Compared with the extensive literature on songbirds, much less is known about the development and maturation of acoustic communication in vertebrates that do not appear to learn their vocalizations, even in animals such as frogs that have been popular models for the study of acoustic communication. Developmental studies in fish and frogs as well as in mammals can help elucidate mechanisms underlying vocal learning. A study of sensorimotor integration in species that are dependent on vocal learning compared with those showing more genetic control of vocalizations would be interesting.

Selective-breeding experiments have produced interesting results regarding the evolution of female preferences for certain features of communication sounds (Welch et al. 1998). Additional research using these and other genetic techniques will be important in increasing our understanding of the evolution of signal structure and function.

4. Summary

As detailed in the chapters in this volume, much progress has been made on proximate and ultimate levels in understanding acoustic communication in animals. The neuroethological approach has contributed a richness and diversity to both theory and experiment that would not have emerged from either psychophysical or neurophysiological approaches by themselves, and it promises to continue to contribute to our understanding of this fascinating topic.

References

Aertsen A, Johannesma P (1980) Spectro-temporal receptive fields of auditory neurons in the grassfrog. I. Characterization of tonal and natural stimuli. Biol Cybern 38:223–234.

Bentley DR, Hoy RR (1972) Genetic control of neuronal network generating cricket (*Teleogryllus*, *Gryllus*) song patterns. Anim Behav 28:230–255.

Boatright-Horowitz SL, Horowitz SS, Simmons AM (2000) Patterns of vocal interactions in a bullfrog (*Rana catesbeiana*) chorus: Preferential responding to far neighbors. Ethology 106:701–712.

Bradbury JW, Vehrencamp SL (1998) Principles of Animal Communication. Sunderland, MA: Sinauer.

Brown CH (1994) Sound localization. In: Fay RR, Popper AN (eds) Comparative Hearing: Mammals. New York: Springer-Verlag, pp. 57–96.

Busnel RG (1963) Acoustic Behavior of Animals. New York: Elsevier.

Capranica RR, Moffat AJM (1983) Neurobehavioral correlates of sound communication in anurans. In: Ewert JP, Capranica RR, Ingle DJ (eds) Advances in Vertebrate Neuroethology. New York: Plenum, pp. 701–730.

Dawkins R, Krebs JR (1978) Animal signals: Information or manipulation. In: Krebs JR, Davies NB (eds) Behavioural Ecology. Oxford: Blackwell Scientific, pp. 282–309.

Dooling RJ, Lohr B, Dent ML (2001a) Hearing in birds and reptiles. In: Dooling RJ, Popper AN, Fay RR (eds) Comparative Hearing: Birds and Reptiles. New York: Springer-Verlag, pp. 308–360.

Dooling RJ, Popper AN, Fay RR (2001b) Comparative Hearing: Birds and Reptiles. New York: Springer-Verlag.

Evans CS, Marler P (1991) On the use of video images as social stimuli in birds: Audience effects on alarm calling. Anim Behav 41:177–226.

Fay RR, Popper AN (1994) Comparative Hearing: Mammals. New York: Springer-Verlag.

Fay RR, Popper AN (1999) Comparative Hearing: Fish and Amphibians. New York: Springer-Verlag.
Fay RR, Simmons AM (1999) The sense of hearing of fishes and amphibians. In: Fay RR, Popper AN (eds) Comparative Hearing: Fish and Amphibians. New York: Springer-Verlag, pp. 269–318.
Gerhardt HC (1992) Multiple messages in acoustic signals. Sem in Neurosci 4:391–400.
Green S, Marler PM (1979) The analysis of animal communication. In: Marler P, Vandenbergh JG (eds) Handbook of Behavioral Neurobiology: Vol 3. Social Behavior and Communication. New York: Plenum Press, pp. 73–158.
Hauser MD (1997) The Evolution of Communication. Cambridge, MA: MIT Press.
Hoy RR, Popper AN, Fay RR (1998) Comparative Hearing: Insects. New York: Springer-Verlag.
Hulse SH, MacDougall-Shackleton SA, Wisniewski AB (1997) Auditory scene analysis by songbirds: Stream segregation of birdsong by European starlings (*Sturnus vulgaris*). J Comp Psychol 111:3–13.
Lanyon WE, Tavolga WN (1960) Animal Sounds and Communication. Washington, DC: American Institute of Biological Sciences.
Lehrman DS (1964) The reproductive behavior of ring doves. Sci Am 211:48–54.
Long GR (1994) Psychoacoustics. In: Fay RR, Popper AN (eds) Comparative Hearing: Mammals. New York: Springer-Verlag, pp. 18–56.
Marler P (1955) Characteristics of some animal calls. Nature 176:6–8.
Marler P (1970) Birdsong and speech development: Could there be parallels? Am Sci 58:669–673.
Marler P, Dufty A, Pickert R (1986) Vocal communication in the domestic chicken. II. Is a sender sensitive to the presence and nature of a receiver? Anim Behav 34:194–198.
Marten K, Marler P (1977) Sound transmission and its significance for animal vocalization. Behav Ecol Sociobiol 2:271–290.
Morton ES (1975) Ecological sources of selection on avian sounds. Amer Natur 109:17–34.
Moss CF, Schnitzler H-U (1995) Behavioral studies of auditory information processing. In: Popper AN, Fay RR (eds) Hearing by Bats. New York: Springer-Verlag, pp. 87–145.
Partan S, Marler P (1999) Communication goes multimodal. Science 283:1272–1273.
Römer H (1998) The sensory ecology of acoustic communication in insects. In: Hoy RR, Popper AN, Fay RR (eds) Comparative Hearing: Insects. New York: Springer-Verlag, pp. 63–96.
Ryan MJ (1990) Sexual selection, sensory systems, and sensory exploitation. Oxford Surv Evol Biol 7:157–195.
Ryan MJ, Keddy-Hector A (1992) Directional patterns of female mate choice and the role of sensory biases. Am Nat 139:S4–S35.
Ryan MJ, Rand AS (1999) Phylogenetic influence on mating call preferences in female túngara frogs, *Physalaemus pustulosus*. Anim Behav 57:945–956.
Schwartz JJ, Simmons AM (1990) Encoding of a spectrally-complex communication sound in the bullfrog's auditory nerve. J Comp Physiol A 166:489–500.
Simmons AM, Sanderson MI, Garabedian CE (2000) Representation of waveform periodicity in the auditory midbrain of the bullfrog, *Rana catesbeiana*. J Assoc Res Otolaryngol 1:2–24.

Simmons JA, Ferragamo MJ, Saillant PA, Haresign T, Wotton JM, Dear SP, Lee DN (1995) Auditory dimensions of acoustic images in echolocation. In: Popper AN, Fay RR (eds) Hearing by Bats. New York: Springer-Verlag, pp. 146–190.
Stebbins WC, Moody DB (1994) How monkeys hear the world: Auditory perception in nonhuman primates. In: Fay RR, Popper AN (eds) Comparative Hearing: Mammals. New York: Springer-Verlag, pp. 97–133.
Suga N (1988) Auditory neuroethology and speech processing: complex sound processing by combination-sensitive neurons. In: Edelman GM, Gall WE, Cowan WM (eds) Functions of the Auditory System. New York: John Wiley and Sons, pp. 679–720.
Theunissen FE, Sen K, Doupe AJ (2000) Spectral-temporal receptive fields of nonlinear auditory neurons obtained using natural sounds. J Neurosci 20:2315–2331.
Tinbergen N (1963) On aims and methods in ethology. Z Tierpsychol 20:410–433.
Valentine DE, Moss CF (1997) Spatially-selective auditory responses in the superior colliculus of the echolocating bat. J Neurosci 17:1720–1733.
Welch AM, Semlitsch RD, Gerhardt HC (1998) Call duration as an indicator of genetic quality in male gray tree frogs. Science 280:1928–1930.
Zelick R, Mann DA, Popper AN (1999) Acoustic communication in fishes and frogs. In: Fay RR, Popper AN (eds) Comparative Hearing: Fish and Amphibians. New York: Springer-Verlag, pp. 363–411.

2
The Physical Acoustics of Underwater Sound Communication

ANDREW H. BASS and CHRISTOPHER W. CLARK

1. Overview

The physical attributes of a vocal/acoustic phenotype for any one species or individual are shaped through the process of natural selection by an ecological environment that has biotic and abiotic components (e.g., Morton 1975, 1982; Wiley 1983; McCracken and Sheldon 1997; Bass 1998). The biotic environment mainly includes the influences of other organisms, whereas the abiotic environment includes the physical parameters (e.g., temperature, sound velocity of the transmission medium, geometry of reflective surfaces, composition of boundaries) that can affect acoustic transmission in both the temporal and spatial domains. Here, we focus on the possible influences of the abiotic environment on the vocal communication signals of aquatic vertebrates in part because there is a dearth of information and studies for this group, unlike the case for terrestrial vertebrates and invertebrates (e.g., see Michelsen 1978; Wiley and Richards 1982; Gerhardt 1983; Wiley 1983; Waser and Brown 1986; Bailey 1991; also see Chapter 5 by Ryan and Kime). First, we provide a basic review of the physical attributes of sound and the general principles governing the relationships between underwater acoustic propagation and the physical environment, which can also apply more generally to sound transmission in any medium. Next, we discuss more realistic expressions of sound propagation in terms of two different environments, shallow water and deep water, and consider studies that provide an empirical assessment of sound propagation for these two environments as it impacts acoustic communication. We will focus on two distantly related vertebrate groups, cetaceans and teleost fishes, which are the foci of our own research programs and in some sense represent two extremes relative to the abiotic influence of sound propagation on acoustic signaling in an aquatic environment.

There are several textbooks that we recommend as bioacoustic primers. Speaks' (1992) *Introduction to Sound* provides a comprehensive treatment of sound transmission in air, which is generally applicable to the principles underlying sound propagation in any medium. The introductory chapters in

Yost's (2000) *Fundamentals of Hearing* provide a concise presentation of the principles of acoustics for airborne sounds. Jensen et al.'s (1994) *Computational Ocean Acoustics* offers the most recent review of the general field of underwater acoustics and is a wonderful companion to Urick's (1983) *Principles of Underwater Sound*. Lastly, Bradbury and Vehrencamp's (1998) *Principles of Animal Communication* offers an extensive biologically oriented discussion of acoustic signaling more specifically in the context of animal communication, behavioral ecology, and the functional design of sound-producing organs and receivers.

2. Sound

According to Speaks (1992, p. 2), "there is one principal prerequisite for a body to be a source of sound—it must be able to vibrate. If a body is set into vibratory motion, it must have the physical properties of mass and elasticity and all bodies in nature possess both of these properties to some degree. Hence, sound is a mechanical disturbance that displaces the molecules of a medium whether the medium is a solid, liquid, or gas. A vibration results in the distortion or acceleration of an object and is established by the opposing forces of inertia (determined by an object's mass) and elasticity (an object's ability to recover its original form following distortion). "Because all molecular structures have some finite mass and elasticity, all are capable of being both a source of sound and a medium for its transmission" (Speaks 1992, p. 2). Sound may therefore also be defined either as "the propagation of density changes through an elastic medium . . . [or] the transfer of energy through an elastic medium" (Speaks 1992, p. 41).

Within a sound field, one molecule collides with its neighbor, returns back toward its initial position in the opposite direction, and collides with another molecule. The cycle of collisions and returns through the initial position is determined by the medium's elasticity, and this oscillation can be described by a sinusoid (Fig. 2.1). Each cycle of the sound wave defines the periodicity of change in the amplitude of the sound's vibration around a baseline of ambient amplitude that may be characterized by any number of quantities, including pressure, displacement, velocity, and acceleration. A sound wave's frequency (f) is the number of cycles/second, which is defined in Hertz (Hz); its period (T) is the reciprocal of frequency, or $1/f$, and is the duration of a single cycle, measured in seconds. The distance covered by one full cycle of the sound wave is its wavelength (λ), which is equal to c/f or cT, where c is the speed of sound propagation.

Most natural sounds are not pure sinusoids but complex, nonsinusoidal waveforms. Complex waveforms can be represented by the sum of a weighted series of sinusoids, i.e., the Fourier transform (Bradbury and Vehrencamp 1998). Fourier analysis often reveals that a complex sound is

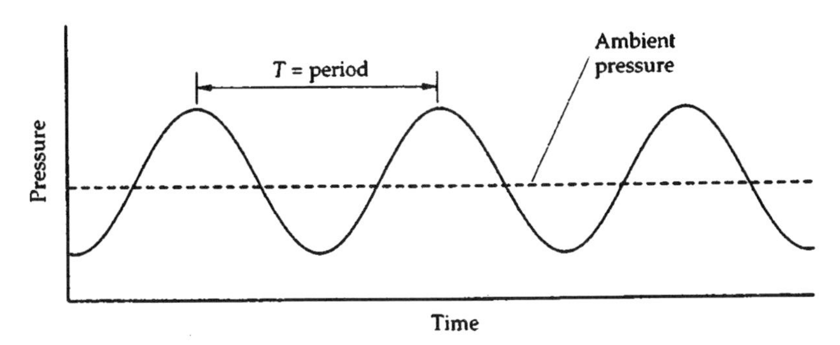

FIGURE 2.1. A sine wave showing the relationship between sound pressure and time in seconds. One cycle of a sine wave defines the period (T) of a sound's vibration around a baseline of ambient pressure. Period is the reciprocal of frequency, or $1/f$, and is the duration of a single cycle measured in seconds. (From Bradbury and Vehrencamp 1998, Fig. 2.1.)

composed of a fundamental frequency (f_0), and a number of harmonics which are integer multiples of the fundamental (i.e., $2f_0$, $3f_0$, $4f_0$ etc.).

2.1. Sound Intensity and Pressure

Two of the quantities most frequently used to describe the magnitude of a sound wave are intensity and pressure. Sound intensity is directional (i.e., a vector quantity) and specifies the average amount of power that is transferred per unit area to an acoustic medium from an acoustic source; the units of measure for intensity are erg/sec/cm^2 or watts/cm^2. Pressure, a scalar quantity, indicates the amount of force per unit area; the unit of measure is 1 Newton per square meter (N/m^2) or 1 Pascal (Pa). One Newton is "the force required to accelerate a mass of 1 kg from a velocity of 0 m per sec to a velocity of 1 m per sec in 1 sec of time" (Speaks 1992, p. 22). One Newton or Pa equals 100,000 dynes; one dyne is "the force required to accelerate a mass of 1 gram (g) from a velocity of 0 cm per sec to a velocity of 1 cm per sec in 1 sec of time" (Speaks 1992, p. 22).

Sound intensity (I) is related to the effective sound pressure p_e by the following equation that applies to a sound wave in a homogeneous and boundless medium (see Kinsler and Frey 1962, p. 162)

$$I = p_e^2/2\rho_0 c \qquad \text{erg/sec/cm}^2 \tag{1}$$

where ρ_0 is the density of the medium, c is the propagation velocity of the sound wave in that medium, and $\rho_0 c$ is also known as the medium's characteristic acoustic impedance ($2\rho_0 c$ refers to spherical waves; see Section 4.1.1). This impedance term varies with each medium and influences the received intensity of an acoustic signal transmitted through that medium. One should use caution in the use of the term impedance because there are a number of different kinds; the reader is referred to Beranek (1988) for their definitions and applications to acoustics. As Bradbury and Vehrencamp (1998) point out, a medium's impedance has important implications for sound communication in water as compared to air. Water's characteristic impedance is almost 5,000 times greater because of its 1,000-fold greater density and the fivefold greater speed of sound in water compared to air (also see Section 2.2). Hence, an animal generating underwater sound will have to produce much higher levels of pressure to attain the same stimulus energy at a receiver's ear. The trade-off here is that it takes much less energy to produce a sound at a given level of pressure in water.

Typically, sound intensity is presented in units of decibels (dB), where intensity in dB is the logarithm to the base 10 ($\log_{10}$) of the ratio of two intensities

$$dB = 10 * \log_{10}(I_1/I_2) \quad (2)$$

where I_1 is the intensity being measured and I_2 is a standard reference intensity. Given Equation 1, the pressure expressed in decibels is

$$\begin{aligned} dB &= 10 * \log_{10}(p_1^2/2\rho_0 c)/(p_2^2/2\rho_0 c) \\ &= 10 * \log_{10}(p_1^2/p_2^2) \\ &= 10 * \log_{10}(p_1/p_2)^2 \\ &= 10\,(2) * \log_{10}(p_1/p_2) \\ &= 20 * \log_{10}(p_1/p_2) \end{aligned} \quad (3)$$

where p_1 is a measured sound pressure relative to p_2, a standard reference pressure. Bioacousticians typically do not measure sound intensity but rather sound-pressure level, or SPL (see below).

The magnitudes of intensity and pressure as expressed in dB are usually shown graphically as logarithmic units in part because of the wide range of sensitivity exhibited by the human auditory system (10^{15} units of power). When using the dB scale, it is essential to denote the reference measure. The standard in-air reference for the decibel (0 dB) is set at the audible threshold for human hearing, which is 0.0002 dyne/cm^2; this corresponds to a reference intensity level (IL) in air of 10^{-12} watts/m^2 or a sound-pressure level (SPL) of 20 μPa (see Table 2.1). Typically, we designate airborne sounds in terms of SPL; an SPL of 30 dB (re 20 μPa) means that a sound is 30 dB above the reference measure of 20 μPa. In water, the standard reference is 1 μPa, and the difference between the in-air and in-water standard reference translates into a difference of 26 dB. Thus, a sound-pressure level of 0 dB (re 20 μPa) in air is equal to 26 dB (re 1 μPa) in water (Table 2.1). As a rule of thumb, remember that a twofold change in pressure is equal to 6 dB (e.g., 10 μPa to 20 μPa) and a tenfold change is equal to 20 dB (e.g., 1 μPa to 10 μPa; see Table 2.1 and Eq. 3). Intensity changes will be 3 dB and 10 dB, respectively, for twofold and tenfold changes in intensity (see Eq. 2). It is always important to indicate if the value being presented is either for

TABLE 2.1. Acoustic Measurements. The standard references are in bold. In air, the standard reference is 20 μPa, whereas in water it is 1 μPa. Thus, the difference between the in-air and in-water standard reference translates into a difference of 26 dB. Modified from Table 2.1 in Richardson et al. (1995).

Pascal (Pa)	dynes/cm^2	dB re 1 μPa	dB re 20 μPa or 0.0002 dynes/cm^2
1,000,000	10^7	240	214
100,000	1,000,000	220	194
10,000	100,000	200	174
1,000	10,000	180	154
100	1,000	160	134
10	100	140	114
1	10	120	94
0.1	1.0	100	74
0.01	0.1	80	54
0.001	0.01	60	34
0.0001	0.001	40	14
20 μ	**0.0002**	**26**	**0**
10 μ	0.0001	20	−6
1 μ	0.00001	0	−26

intensity or pressure. By indicating the accepted standard reference values for intensity or pressure, it will be immediately obvious which measure is being reported for either air or water.

The relationship between intensity and pressure is easily confused. Remember that: (1) as sound intensity increases by some factor, pressure increases only by the square root of the factor; and (2) as pressure increases by some factor, sound intensity increases by the square of that factor (after Speaks 1992, p. 130).

2.2. *Sound Velocity*

Sound velocity varies with the transmission medium and, in general, is higher in liquids than gases but highest in solids. It can be affected by temperature, salinity, and pressure or depth (see Beranek 1988); typically, these variables are designated when reporting sound velocity. Thus, in air, at sea level and 20°C, the speed of sound is 343 m/sec. In this case, velocity, c, is determined as (Kinsler and Frey 1962, pp. 116 and 503)

$$c = \sqrt{\gamma P_0 / \rho_0} \tag{4}$$

where γ is the ratio of specific heats (1.402), P_0 is the standard barometric pressure ($1.013 \times 10^5\,\mathrm{N/m^2}$), and ρ_0 is the density ($1.21\,\mathrm{kg/m^3}$).

At sea level and 32% salinity, the speed of sound in water is 1,518.06 m/sec at 20°C, almost fivefold greater than in air. At first glance, we would expect that sound speed would be less in water than air, given water's higher density. However, the elasticity or resistance to distortion in a fluid has a large contribution to the determination of sound velocity. The theoretical determination of sound velocity in water includes the "isothermal bulk modulus" that measures the compressibility of a liquid (see Kinsler and Frey 1962, p. 117). Sound velocity is then determined as

$$c = \sqrt{\gamma B_T / \rho_0} \tag{5}$$

where B_T is the bulk modulus. For seawater at 13°C, B_T is $2.28 \times 10^9\,\mathrm{N/m^2}$, ρ_0 is $1{,}026\,\mathrm{kg/m^3}$, and γ is 1.01 (Kinsler and Frey 1962, p. 503).

2.3. *Sound-Velocity Profile*

Under natural conditions, sound velocity within a medium is not uniform. As noted above, sound velocity in water changes as a function of temperature, depth, and salinity. Variation in sound velocity as a function of water depth is referred to as the *sound-velocity profile* (SVP). The SVP for an environment is important for predicting sound propagation and potential intensity losses both as a function of depth and range from a sound source (Fig. 2.2; also see later discussions on refraction and the deep sound channel). There are important implications that can be derived from knowing how variations in SVP affect the propagation of different frequencies through an environment. For example, daily and seasonal fluctuations in air temperature and wind conditions can lead to a surface layer of warmer water. As Urick (1983, p. 117) explains: "The surface layer may contain a mixed layer of isothermal water that is formed by the action of wind as it blows across the surface above. Sound tends to be trapped or channeled in this mixed layer. Under prolonged calm and sunny conditions the mixed layer disappears, and is replaced by water in which the temperature decreases with depth." A second layer, "the seasonal thermocline," shows decreasing sound velocity with increasing depth. [Jensen et al. (1994) recognize a distinct "mixed layer" that is a combination of Urick's first two layers with a "surface duct profile" that shows increasing velocity with depth due to the influence of pressure (see Fig. 1.1 in Jensen et al. 1994; also see Section 4.3 in Richardson et al. 1995).] A third layer is the "main thermocline," where the most dramatic decrease in temperature, and hence sound velocity, occurs in the deep ocean; there is little impact of seasonal fluctuations on water temperature in this zone. A "deep isothermal layer" lies below the main thermocline and is a zone where temperature is fairly constant at around 3–4°C. Since the temperature remains fairly stable, sound velocity increases again within the deep isothermal layer with increasing pressure. The position of this deep sound channel will vary dramatically with latitude and season (see Figs. 5.14–5.16 in Urick 1983). The implications of the SVP for acoustic communication will be discussed later mainly in the context of sound propagation in the deep sound channel that is located within the region of minimum sound speed of the SVP (Section 9).

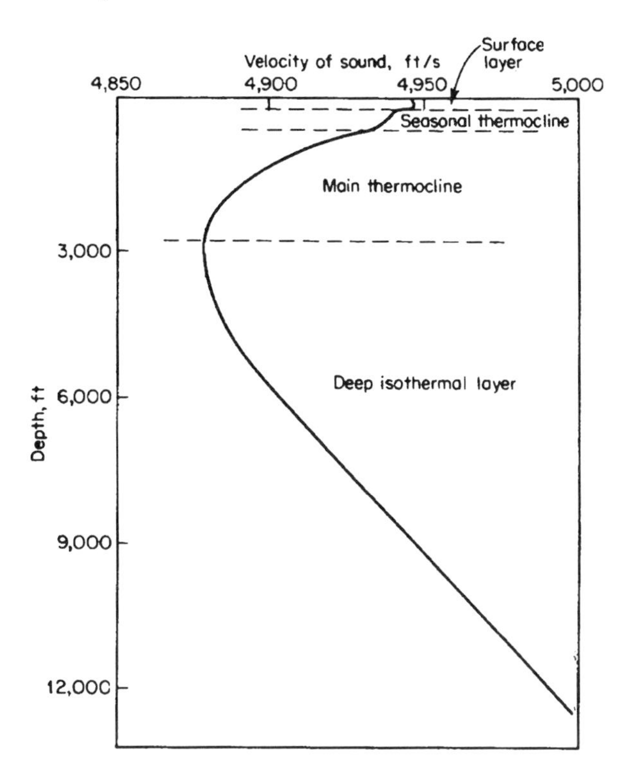

FIGURE 2.2. In the ocean, the sound speed or velocity profile (solid line) changes because of the competing influences of temperature, salinity, and depth (pressure). The dashed lines indicate approximate boundaries between different thermoclines. (From Principles of Underwater Sound by R.J. Urick, Fig. 5.12, © 1983. Reproduced with permission of The McGraw-Hill Companies.)

3. Near Field and Far Field: Monopoles and Dipoles

Two concepts central to all theoretical treatments of sound transmission in an aquatic environment are the near field and the far field. Kalmijn (1988) provides a comprehensive review of this topic, whereas Coombs and Janssen (1988) and Rogers and Cox (1988) provide more concise treatments. For historical purposes, the reader is also referred to the early essays of van Bergeijk (1964), Harris (1964), and Siler (1969) on this subject.

The movements of sound sources in an ideal, boundless medium that lacks any obstacles (a free field) have been modeled as either a pulsating or a vibrating sphere. A pulsating sphere is a monopole source of sound that uniformly expands and contracts, whereas a vibrating sphere is a dipole source that vibrates back and forth along one axis. The sonic swim bladder of some fish has often been compared to a monopole source (Harris 1964; Kalmijn 1988), and recent work on baleen whale sound production models the mechanism as a volume-driven monopole source (Aroyan et al. 2000). The swimming motion of aquatic animals and non–swim bladder sound-producing mechanisms in fish have been compared to dipole and higher-order sources (Harris 1964; Kalmijn 1988; Hawkins 1993).

3.1. Monopole Sound Source

Close to the surface of a pulsating sphere, the motion of the sphere's surface forces the medium to move (that is, accelerate) away from the sphere. The displacement of water molecules adjacent to the sphere's surface establishes the "local hydrodynamic flow" of an "acoustic near field" (Kalmijn 1988, pp. 85–86). All molecules in the near field essentially move in phase with the source itself (Fig. 2.3A(a)). The pulsating sphere also produces a "propagating sound wave" (Kalmijn 1988, p. 88) or "pressure wave" (van Bergeijk 1964, p. 283) that also originates at the source but has a greater range, extending beyond the near field into a far field. In the far field, the mechanical disturbance generated by a pulsating sphere spreads in a periodic, sinusoidal manner as alternating bands of compression (increased density of water molecules) and rarefaction (decreased density of water molecules) (Fig. 2.3A(b)). The use of the terms "near field" and "far field" implies whether the spreading sound wave is dominated by either the non-propagating, hydrodynamic flow or by the propagating sound wave. In theory, both could extend to infinity in an ideal environment. Hence, it is essential to remember that there is no clear boundary between the near and far fields because they are "physically inseparable components of one and the same acoustic field" (see Kalmijn 1988, p. 89).

With increasing distance from the pulsating sphere, acoustic energy is distributed over a greater and greater area so that the amount of energy per unit area dissipates with increasing range. However, the rate of decrease in net fluid displacement, velocity, and acceleration with increasing source distance differs between the nonpropagating, hydrodynamic flow and the propagating sound wave. Thus, the magnitude of these quantities falls off quickly as the square of the radial distance from the source or $1/r^2$ for the hydrodynamic flow, but only as $1/r$ for the propagating wave. Pressure decreases at the rate of $1/r$ for both the local flow and the propagating sound wave.

3.2. Dipole Sound Source

Now consider a vibrating sphere that is modeled as a dipole, a motion that more closely resembles many forms of vibration in water that are of biological origin (Fig. 2.3B). A dipole source is represented by a sphere vibrating mainly along one axis. It is essentially "the equivalent of two equal-strength monopole fields of which the sources are separated only by a short distance and that pulsate 180° out of phase" (Kalmijn 1988, p. 93).

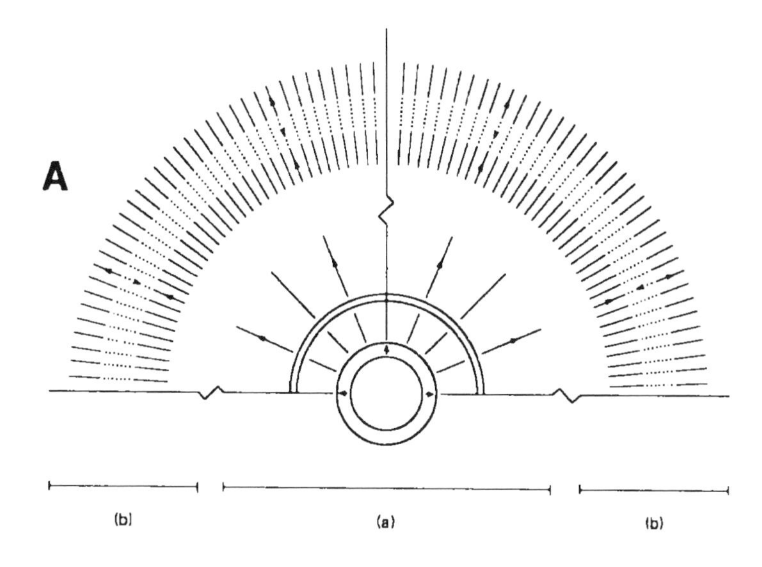

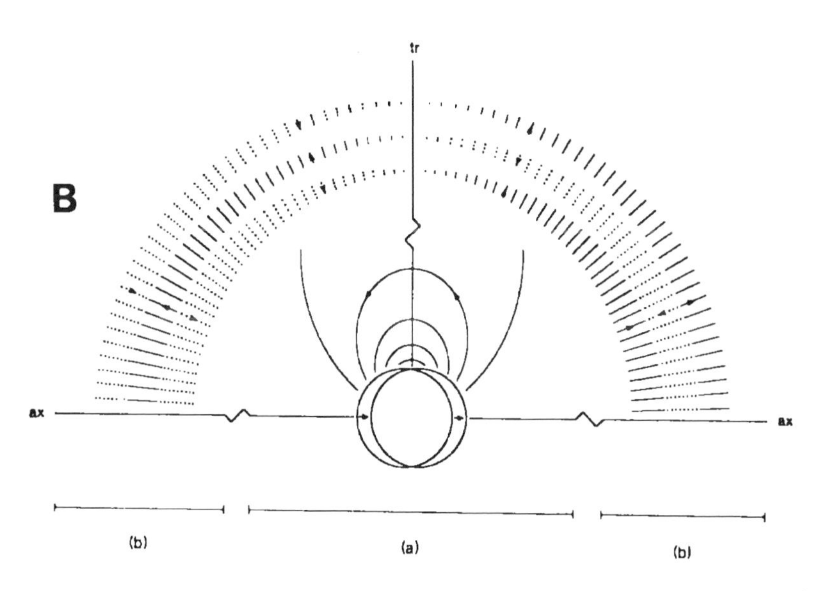

FIGURE 2.3. Theoretical models for the monopole field of a pulsating sphere and the dipole field of a vibrating sphere. (**A**) Shown here for one half of a monopole field are the local flow region in the near field (a) and the propagating sound wave in the far field (b). Arrows show the direction of particle velocities, which are radially symmetrical around the source. Solid and dotted lines represent alternating shells of compression and rarefaction, respectively. (**B**) Shown here for one half of a dipole field are the local flow region in the near and intermediate fields (a) and the propagating sound wave in the far field (b). Arrows along the main axis of vibration (ax) show the direction of vibration. Flow lines directly above the source show the dipole nature of the field; the direction of particle velocities close to the source follows a path from the front of the sphere (defined by the main axis of displacement) back to its rear. In the far field, vibration is maximal along the main axis of vibration (ax) and then gradually decreases toward the transverse axis (tr). Solid and dotted lines represent alternating shells of compression and rarefaction, respectively. (From Kalmijn 1988, Figs. 4.2, 4.4.)

A dipole source can vibrate in any axis and thereby differs from the symmetrical, radial motion of a pulsating sphere, where the magnitude of motion is equal in all directions. The direction of particle displacement in the near field close to the source follows a path from the front of the sphere (defined by the initial direction of displacement) back to its rear (Fig. 2.3B(a)).

Kalmijn (1988) recognizes an additional near-field region of "intermediate flow" for a dipole source whose fluid velocities are like those of the rest of the near field. Also, unlike the monopole source, the amplitude of particle motion in the dipole's far field is asymmetrical and increases from a minimum along an axis perpendicular to the direction of motion to a maximum along the axis of motion (Fig. 2.3B(b)).

Net fluid displacement, velocity, and acceleration attenuate at the rate of $1/r^3$ in the dipole's near-field region dominated by hydrodynamic flow and as $1/r^2$ in the intermediate flow region. Unlike the monopole source, the local flow pressure falls off at the rate of $1/r^2$ in the near field. However, like a monopole, the propagating sound wave's displacement, velocity, acceleration, and pressure all attenuate at a rate that is proportional to $1/r$ with increasing source distance.

The reader is encouraged to consult Kalmijn's (1988) essay for a complete explanation of the physical principles, namely the inertial and compressional forces, that establish the local flow field and the propagating sound wave. In Section 4.1, we will focus on the loss of acoustic energy for the propagating sound wave mainly because it is the component of the sound field commonly measured for the characterization of acoustic communication signals in water and air. The reader is also encouraged to refer to reviews by Coombs and Janssen (1988) and Kalmijn (1989) for a consideration of dipoles in regard to the operation of the mechanosensory lateral-line system.

3.3. Wavelength Dependency of Near- and Far-Field Range

Because the amplitude of particle motion in the local flow region of the near and intermediate fields decreases much faster than that of the propagating sound wave (see above), the latter comes to dominate the spread of acoustic energy as distance from the source increases for either a monopole or a dipole source. The change in strength of the local flow and propagating sound wave with increasing source distance is wavelength-dependent (see Fig. 2 in Coombs and Janssen 1988). For a monopole source, the theoretical prediction for the point at which the local flow and the propagating sound wave contribute equally to the sound wave is at a distance of about $\lambda/2\pi$ from the source; the propagating wave dominates beyond that point. For a dipole source, the local flow region of the near field predominates up to a distance of about $\lambda/2\pi$ from the source, whereas the intermediate region dominates between $\lambda/2\pi$ and λ/π; the propagating sound wave then rules beyond λ/π into the far field. Hence, the transition point at which the propagating sound wave becomes the most influential is at a farther distance from the source for a dipole than for a monopole.

Given these relationships, longer-wavelength (lower-frequency) sounds will have more extensive near and intermediate fields than shorter-wavelength (higher-frequency) ones, and the extent of these fields will vary depending on whether the source is a monopole or a dipole. Because underwater sounds have an almost fivefold greater wavelength than airborne sounds, the local flow region will extend farther and can therefore potentially play a more prominent role in underwater communication. By contrast, the much shorter wavelengths of sound in air, together with the more pronounced drop-off in sound amplitude in the local flow field, mean that the potential influence of the near field is limited to cases where animals are within several body lengths of each other. Thus, the far field tends to dominate the communication systems of most terrestrial species, with insects being an exception to this condition (see Bailey 1991).

4. Transmission Loss

A number of factors influence the process of sound transmission in a non-ideal, complex environment. Several terms are fundamental to this discussion, including transmission loss (TL), source level (SL), and received level (RL) (the sound-velocity profile also influences this process; see Section 2.3). These terms are related in the simplest form as

$$RL = SL - TL \tag{6}$$

For communication between an animal making a sound and one receiving that sound, the received level (RL) is the difference between the level of the sound produced, the source level (SL), and the transmission loss (TL) that occurs between the sender and the receiver (Eq. 6). Transmission loss specifically refers to the change of sound intensity with increasing distance relative to a reference point. In water, transmission loss is often complex and is significantly influenced by the hard structures in the environment (e.g., bottom topography, water surface condition), the characteristics of the water through which most of the sound energy flows, and the depth of the sender and receiver. The reader is referred to Urick (1983) and Jensen et al. (1994) for detailed treatments of this important but often complicated subject.

4.1. Geometric Spreading

For the purposes of this discussion, transmission loss is the sum of the loss due to geometric spreading and attenuation (see below). Two kinds of geometric spreading loss are typically used to describe transmission loss in its simplest form, namely spherical spreading and cylindrical spreading (also see Urick 1983 for descriptions of "no spreading" and "time spreading").

4.1.1. Spherical Spreading

Spherical spreading describes the reduction in intensity of a propagating sound wave in a free field in which there are no boundaries, the medium is homogeneous, and the distance over which the sound is spreading is large compared with the source (see Richardson et al. 1995). Sound spreads uniformly in a spherical manner from a point source that can be easily modeled as the center of a sphere. Because intensity is inversely proportional to area, and the surface area of a sphere is $4\pi r^2$, where r represents the radius of a sphere, intensity falls off by $1/r^2$ with distance from the source; this is known as the "inverse square law."

For acoustic measurements, r translates into the ratio of two distances, namely r_x/r_0, where r_x is the distance of interest that represents the distance from a reference point r_0. Thus, intensity is inversely proportional to $1/(r_x/r_0)^2$. The standard reference distance is 1 meter so that for the purposes of our discussion $1/(r_x/r_0)^2$ reduces to $1/r^2$, where r represents r_x. In decibels, transmission loss for spherical spreading from a point source in a free field is predicted to be proportional to $20 * \log_{10}(r)$. This can be derived from the inverse square law (above) and Equation 2 (Section 2.1) as follows (Note: This theoretical treatment only assumes transmission loss due to spreading and does not account for other losses associated with attenuation as discussed in Section 4.2)

$$I_x = I_0/r^2 \tag{7}$$

where I_x represents the measured intensity at distance x relative to the intensity at our reference point (I_0). The term transmission loss (TL) specifically refers to the ratio of the final intensity (I_x) to the original intensity (I_0); expressed in dB

$$\mathrm{TL} = 10 * \log_{10}(I_x/I_0) \tag{8}$$

If we place Equation 7 into Equation 8, then

$$\begin{aligned}\mathrm{TL} &= 10 * \log_{10}[(I_0/r^2)(1/I_0)] \\ \mathrm{TL} &= 10 * \log_{10}(1/r^2) \\ \mathrm{TL} &= 10 * \log_{10}(r^{-2}) \\ \mathrm{TL} &= -20 * \log_{10}(r) \end{aligned} \tag{9}$$

Often, transmission loss is discussed in the context of sound pressure. Because $I = p^2/2\rho_0 c$ (Eq. 1, Section 2.1), then Equation 7 can be

$$I_x = (p_0^2/2\rho_0 c)(1/r^2) \tag{10}$$

If we place Equation 10 into Equation 9, then

$$
\begin{aligned}
TL &= 10 * \log_{10}[(p_o^2/2\rho_0 c)(1/r^2)][2\rho_0 c/p_o^2] \\
TL &= 10 * \log_{10}(1/r^2) \\
TL &= 10 * \log_{10}(r^{-2}) \\
TL &= -20 * \log_{10}(r)
\end{aligned}
\tag{11}
$$

Hence, although TL formally applies to the loss of sound intensity, a $-20 * \log_{10}(r)$ relationship can be applied to considerations of pressure loss as well.

4.1.2. Cylindrical Spreading

Cylindrical spreading occurs when the transmission medium has upper and lower boundaries such that spreading is confined to an ever-expanding cylinder. In this case, spreading is inversely proportional to the surface of a cylinder of radius (r) and depth (d), or $1/2\pi rd$ (see Jensen et al. 1994, pp. 12–13 and Richardson et al. 1995, pp. 62–63). Therefore, Equation 7 becomes

$$I_x = I_0/r \tag{12}$$

and the derivation of Equation 9 becomes

$$
\begin{aligned}
TL &= 10 * \log_{10}[(I_0/r)(1/I_0)] \\
TL &= 10 * \log_{10}(1/r) \\
TL &= 10 * \log_{10}(r^{-1}) \\
TL &= -10 * \log_{10}(r)
\end{aligned}
\tag{13}
$$

Expressing this in terms of pressure (see Eqs. 10 and 11), then

$$
\begin{aligned}
TL &= 10 * \log_{10}[(p_o^2/2\rho_0 c)(1/r)][2\rho_0 c/p_o^2] \\
TL &= 10 * \log_{10}(1/r) \\
TL &= -10 * \log_{10}(r)
\end{aligned}
\tag{14}
$$

Hence, TL for cylindrical spreading is proportional to $-10 * \log_{10}(r)$ and thus occurs at a slower rate than in cases with spherical spreading. Thus, sound intensity or pressure decreases by 6 dB ($-20 * \log_{10}2$) and by 3 dB ($-10 * \log_{10}2$) per doubling of distance ($r = 2$), respectively, for spherical and cylindrical spreading.

A simple rule in a bounded medium, such as occurs in shallow water habitats bounded by the air–water interface and the substrate, is that spherical spreading occurs out to a range equal to the water depth at the source, after which there is cylindrical spreading (see Fig. 2.3 in Richardson et al. 1995).

4.2. Attenuation

There are a number of factors that have been proposed to influence sound attenuation (Urick 1983; Jensen et al. 1994). Most important here is what is referred to as absorption, or the transformation of acoustic energy into heat energy either in the transmission medium or at its boundaries (see Kinsler and Frey 1962). Boundary losses will be mentioned below in the context of reflection (Section 4.3). Kinsler and Frey (1962, pp. 217–218) recognize three basic types of loss in the transmission medium. (1) Loss due to viscosity arises from the "relative motion occurring between various portions of the medium during compressions and expansions that accompany transmission of a sound wave." (2) Loss due to heat conduction arises from the "tendency for heat to be conducted from regions of condensation where temperature is raised to neighboring regions of rarefaction where the temperature is lowered. In the process of this heat transfer there is a tendency towards pressure equalization, which reduces the amplitude of a wave as it is propagated through the medium." (3) Loss due to the "molecular exchanges of energy" arise from "the finite time required for a portion of the compressional energy of the fluid to be converted into internal energy of molecular vibration, then correspondingly during the expansion cycle some of this energy will be delayed in restoration so as to be returned to the fluid during a time of rarefaction. Such a delay will result in a tendency towards pressure equalization and an attendant reduction in pressure amplitude of the wave."

The attenuation coefficient, sometimes referred to as the absorption coefficient, α, because this factor dominates most of the frequency band of interest, describes how much energy of the incident wave is absorbed by the transmission medium per unit distance (for use in transmission-loss equations, see Kinsler and Frey 1962). Overall, attenuation, expressed in terms of dB per unit distance, increases linearly with frequency. Attenuation coefficients in seawater are very small; for example, they are (see Urick 1983; Spiesberger and Fristrup 1990): 0.0002 dB/km for 10 Hz, 0.0015 dB/km for 100 Hz, 0.025 dB/km for 500 Hz, 0.06 dB/km for 1,000 Hz, and 0.2 dB/km for 4,000 Hz. (For comparative purposes, we note that the attenuation coefficient at 500 Hz in air is 20–30 times greater for typical conditions.) Thus, absorption is not a major factor for many whales and fishes that produce sounds in the low (<100 Hz) and intermediate (100–1,000 Hz) frequency regions.

4.3. Reflection

Reflection occurs when a sound wave encounters a medium with a different impedance. Of particular interest here would be the reflection of sound waves at boundaries such as the surface or bottom. In shallow water, where a sound wave will interact with the surface and bottom multiple times, reflection has an especially important influence on frequency-dependent propagation.

Reflected waves create even more complicated sound fields, depending on the relative texture of the surface that is contacted and the time required for the reflected wave to decay as it moves away from the obstacle's surface. For example, the sound reflected off the surface can combine with sound from the direct path to form an interference pattern representing the cancellation and summation of the two waves. This phenomenon is referred to as the Lloyd mirror effect and can lead to as much as a 6dB increase in sound pressure (complete summation and doubling of pressure) or a total cancellation of sound pressure (see Jensen et al. 1994). Transmission loss associated with the reflection (and absorption) of sound along the bottom is even more complicated than that for the surface because the bottom may vary widely in its composition and layering of different solids.

5. Refraction

In cases where sound moves between media with different specific acoustic impedances, the direction of the incident sound wave will change because of changes in the speed of propagation of the sound wave according to what is known as Snell's law (Weidner and Sells 1965; see discussion in Speaks 1992). This angular change in the wave's direction is known as refraction. Refraction is important when considering the behavior of sound traveling through water in which the sound-velocity profile varies strongly as a function of depth (Section 2.3). Because water is essentially incompressible, sound speed is the dominant variable when considering underwater acoustic refraction, and the sound-velocity profile is a critical measure for predicting refractive effects.

One mechanism for modeling sound propagation is to treat a sound source as emitting rays of sound. Ray paths in the ocean can be categorized into four different types (from Jensen et al. 1994): (a) refracted-refracted rays that propagate only along refracted paths, (b) refracted surface-reflected rays that bounce off the sea surface but refract in the deep water, (c) refracted bottom-reflected rays that bounce off the sea bottom and are refracted in the upper layers of the water column, and (d) surface-reflected bottom-reflected rays that reflect off the surface and the bottom with little influence of refraction. The refracted-refracted rays (Fig. 2.4A) suffer the least loss because they are only affected by attenuation within the water column, whereas the surface-reflected bottom-reflected rays (Fig. 2.4B) suffer the most because they are affected by the losses associated with multiple reflections. As discussed later, each of these conditions can potentially have profound effects on sound communication among cetaceans in the deep ocean and teleost fishes in very shallow water, respectively.

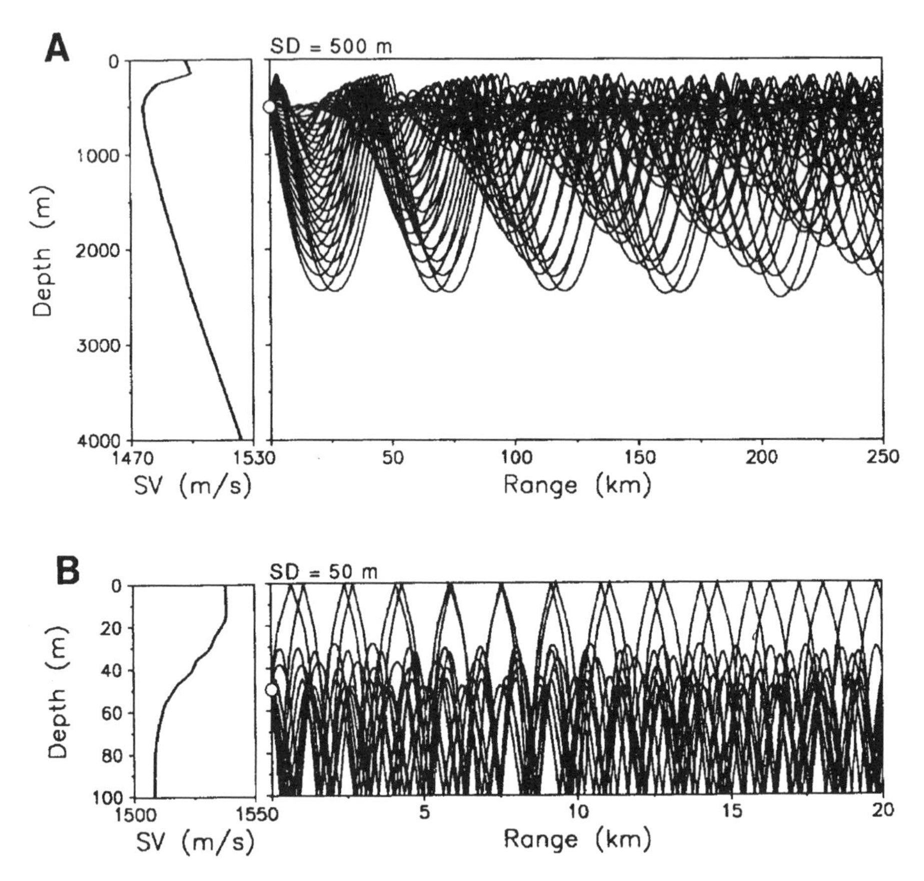

FIGURE 2.4. Ray paths (right column) for sound in deep (**A**) and shallow (**B**) water. To the left of each series of ray paths is the associated sound-velocity (SV) profile (see Fig. 2.2). (**A**) Propagation in the deep sound channel in the Norwegian Sea. The source depth (SD) is at 500 m. Propagation is dominated by refracted-refracted ray paths. (**B**) Propagation in the Mediterranean Sea in summer for a source depth (SD) of 50 m. In this case, propagation is dominated by bottom-reflected ray paths with additional contribution from surface-reflected rays. (From Jensen et al. 1994, Figs. 1.11, 1.14.)

This representation of sound propagation, although simple and effective for illustrating different paths, is insufficient for estimating transmission loss, primarily because propagation depends on range, depth, and frequency, and the interactions of all of the various terms are complicated. Propagation prediction methods include both ray and wave modeling, with modal analysis and the use of parabolic equation methods that are beyond the scope of this chapter (see Jensen et al. 1994; Richardson et al. 1995). However, we will later discuss the implications of these propagation-model predictions and include several key empirical examples specifically related to marine mammal communication.

6. Diffraction

Many of the sounds generated by teleost fishes and baleen whales are of low frequency and therefore of sufficiently long wavelengths that diffraction, or the scattering of sound waves around objects, is unlikely to be much of a factor unless the object encountered is larger than approximately half the sound's wavelength. For example, diffraction should not become significant for a 100-Hz signal ($\lambda \sim 15$ m, which is typical for the fundamental frequency of many teleost fish sounds) unless it contacts some object with a size close to 8 m.

7. Ambient Noise

In deep water, the major natural sources of noise are from waves generated by tidal cycles and wind, seismic disturbances such as earthquakes and volcanism, lightning strikes, rain, oceanic turbulence, and marine mammals (Urick 1983; Bannister 1986; Dyer 1997; Curtis et al. 1999). In shallow water, the major natural sources of noise are from waves impacting the shore, local wind, rain, and biological sounds such as from snapping shrimp and marine mammals. Within the last century, vessel noise, especially from commercial shipping, has increased the ambient noise levels at frequencies below 1,000 Hz (see Payne and Webb 1971; Myrberg 1990; Curtis et al. 1999). Noise from rain is fairly constant across all frequencies (see Fig. 6.15 in Urick 1983), while noise from wind is one of the predominant natural factors influencing low-frequency ambient noise levels (Urick 1983; Bannister 1985; Curtis et al. 1999). The influence of noise on the detection of acoustic signals has attracted much attention, generally within the context of auditory masking (e.g., see discussions in Myrberg 1990 and Fay and Simmons 1999).

8. Received Level, Detection, and Recognition

Effective communication depends on detection and recognition of the acoustic event. Detection requires that the signal's level at the receiver (i.e., received level or RL) be above both the local ambient noise level and the receiver's auditory sensitivity in the signal's frequency band. In simple cases, when the sum of transmission loss (TL) and local ambient noise is greater than the source level (SL), the detection threshold is greater than the sound's received level and no detection occurs (see Eq. 6). Detection alone does not assure effective decoding or recognition of the signal. Signal recognition requires that the sound be more than just detected; it must be properly encoded and then decoded from a suite of other potential signals.

The physical attributes of communication sounds are a balance among multiple, often conflicting, selective pressures. For the terrestrial habitat. Morton (1975) was the first to relate the selective influence of the environment on acoustic characteristics of birds. In the case of underwater acoustic transmission, where the physical environment has, in general, a greater influence on sound transmission than in air, there are cases in which dramatic improvements in detection range are available depending on where the sender and receiver are located and what sounds are produced. Assuming that there is an advantage to greater detection range (larger audience, greater interindividual spacing, greater population range), we should expect to find cases in which animals take advantage of unique underwater acoustic-propagation conditions by optimizing the characteristics of their acoustic signals and their signaling behavior.

Here, we examine two cases of bioacoustic signaling in two habitats with divergent physical acoustic properties. We will first discuss the general characteristics of sound transmission for deep and shallow water marine habitats before turning to specific examples of vocal signaling in those sound channels by cetaceans and teleost fishes.

9. The Deep Sound Channel

When considering sounds in the deep ocean, frequencies less than approximately 100 Hz are considered low-frequency and can be detected at ranges of thousands of miles depending on the source level, sound-speed profiles along the sound path, and local ambient noise conditions. Under certain deep water (>2,000 m) conditions, influenced primarily by a spreading loss that is approximately cylindrical ($-10 * \log_{10}(r)$), most of the sound energy propagates within a certain depth regime known as the deep sound channel (see Urick 1983 and Jensen et al. 1994). The deep sound channel, also known previously as the SOFAR (SOund Fixing And Ranging) channel, is located within the region of minimum sound speed of the sound-velocity profile (see Fig. 2.2 and Section 2.3). Low-frequency sounds traveling in the deep sound channel do not experience reflection off the sea surface or ocean bottom but follow refracted-refracted ray paths, with the result that the sound energy remains within a sound channel (Fig. 2.4A). For low-frequency sounds, this constraint due to refraction in combination with extremely low levels of absorption leads to exceptionally low levels of transmission loss and extremely long ranges of acoustic detection (see Officer 1958, pp. 159–160).

There is some question as to whether whales are physically capable of producing sound at deep sound-channel depths (Aroyan et al. 2000; Thode et al. 2000). In temperate environments, access to the deep sound channel is enhanced along shelf edges and near underwater sea mounts and islands where the channel rises close to the surface, whereas at high latitudes the sound channel is typically within 100–200 m of the surface. Baleen whales are known to routinely dive to depths of 100–200 m and have been recorded feeding near shelf edges at depths of at least 200–500 m (Panigada et al. 1999; Croll et al. 2001).

10. Sound Transmission in Shallow Water

A second example of a sound channel is found in shallow water, where, for the purposes of this paper, shallow water refers to depths of 10–200 m. Coastal species of cetaceans are discussed in this context. In shallow water, unlike the case for the deep sound channel, the water surface and bottom serve as distinct reflective boundaries that interact with sound waves (Fig. 2.4B). Sound transmission in shallow water is strongly influenced by the local conditions of depth, surface roughness (wind-dependent), and the composition of the bottom substrate. In shallow water, the boundary constraints act as a waveguide that effectively filters out water-borne low-frequency sounds. The combined effects of this filtering due to reflection, bottom interaction, and high-frequency absorption result in selective propagation of frequencies in the 50–500-Hz band (see Jensen and Kuperman 1983). Therefore, for whales living in a shallow, coastal environment, selection should favor long-range communication signals in the 50–500-Hz band. There are some intriguing shallow water observations suggesting that low-frequency sounds from fin whales propagate within the substrate or undergo frequency dispersion and time dispersion (D'Spain et al. 1995; Premus and Spiesberger 1997). These propagation phenomena are being exploited by acoustical oceanographers, but whether whales actually take advantage of the information available from these propagation modes remains to be determined.

For teleost fishes, communication occurs in very shallow water (e.g., intertidal and subtidal zones, lakes, and flood plains) with water depths <5 m. In this environment, the same physical constraints apply as in the 10–100-m case, but the effects on propagation do not result in a waveguide. Under these very shallow water conditions, low-frequency propagation within the water column is restricted to distances of several times the water depth. This suggests that limited communication range and not maximum communication range has been the dominant selective feature. However, communication range may be enhanced by sound propagation within the substrate, and this could be especially important for animals that lie quietly near or on the bottom substrate (e.g., see Premus and Spiesberger 1997).

Overall, sound transmission in shallow and very shallow water remains one of the least understood areas of underwater acoustics because many of the details regarding the physical attributes of the important parameters are not well-known and are site-specific and difficult to generalize.

10.1. Surface Temperature

In cases when surface water is warmer than deeper water, sound velocity decreases with increased depth and sound in the upper surface layer will be refracted down into the cooler water. Situations in which cold water nears the surface occur seasonally along shelf breaks and sea mounts as a result of upwellings or other oceanographic changes. Acoustically, such conditions bring the deep sound channel closer to the surface thereby increasing an animal's access to a long range communication path. Biologically, such conditions are often associated with areas of high productivity and concentrations of marine animals (e.g., Croll et al. 1999).

10.2. Depth and Substrate: Cutoff Frequencies

Both the composition of the bottom substrate and water depth influence the frequency below which sound transmission is negligible, referred to as the cutoff frequency. The cutoff frequency, f_0, can be estimated by the following equation (from Rogers and Cox 1988; also see Jensen et al. 1994).

$$f_0 = (c_w/4h)/\sqrt{1-c_w^2/c_s^2} \tag{15}$$

where c_w = speed of sound in water, c_s = speed of sound in the bottom substrate, f_0 is in Hz, and h = water depth. Jensen and Kuperman (1983) provide a more complete discussion of the parameters affecting sound speed for different substrates (also see Jensen et al. 1994 and Premus and Spiesberger 1997). We do, however, want to emphasize the potential influence of these parameters on propagation because the bottom can be a major source of transmission loss. Ocean bottoms are modeled using several geoacoustic parameters, each of which varies with geographical locale. "Clearly, the construction of a detailed geoacoustic model for a particular ocean area is a tremendous task, and the amount of approximate (or inaccurate) information included is the primary limiting factor on the accurate modeling of bottom-interacting sound transmission in the ocean" (Jensen et al. 1994, p. 40).

In general, water depth is considered the dominant factor influencing cutoff frequency up to water depths of about 100m (Jensen and Kuperman 1983). However, at very shallow depths such as those inhabited by many sonic teleost fishes, both depth and substrate likely influence the cutoff frequency. Rogers and Cox (1988) show for a wide range of depths that cutoff frequency is predicted to decrease as one moves from a clayey silt or sandy bottom to a more rigid bottom. Forrest et al. (1993) determined cutoff frequencies for a linearly swept, sinusoidal signal (100 Hz–20 kHz) in a very shallow (5–45 cm depth) freshwater pond with a layered substrate of soft sediment (clay silt and leaf litter) over hard, packed clay. Their results support the prediction that the effect of substrate on cutoff frequency varies with depth, acting more like a rigid or soft bottom with decreasing or increasing depth, respectively. Given a constant depth and no temperature gradient, the wavelength of the cutoff frequency was predicted to lie between two and four times the depth of the water as predicted for soft and rigid bottoms, respectively.

10.3. Geometrical Spreading

It has been proposed that the propagation of the sound wave in shallow water might be best modeled as cylindrical spreading because the air–water interface and the bottom act as boundaries that limit sound radiation (Urick 1983). In this case, pressure is expected to decrease by $10 * \log_{10}(r)$ rather than $20 * \log_{10}(r)$, which is associated with spherical spreading (see Sections 4.1.1 and 4.1.2). Banner (1970) used pressure-sensitive hydrophones to map the sound field of pulsed and continuous noise signals in a very shallow (20–60 cm) bay with a flat, sand-mud bottom. Pressure loss was 8–12 dB per distance doubling over the 80–320-Hz frequency band for a continuous random noise, clearly exceeding the predictions for either spherical or cylindrical spreading losses; transmission loss (TL) in this case was about 25 to $40 * \log_{10}(r)$. By contrast, as predicted for cylindrical spreading, pressure loss was 3 dB per distance doubling for pulsed signals such as the splashes from leaping fish, chewing or grating sounds from feeding fish and crustaceans, and swimming-associated sounds represented by 0.01–1.2 sec pulses of broadband noise. Given the very shallow depths for this study, TL values were likely closer to theoretical predictions because of their higher-frequency content. Other studies of transmission loss associated with the propagation of natural sounds will be discussed in a later section.

11. Baleen Whale Acoustics

As discussed above, low-frequency sound in deep, temperate water is refracted into a sound channel where transmission loss is on the order of $10 * \log_{10}(r)$. Payne and Webb (1971) hypothesized that the loud, infrasonic sounds of certain balaenopterid whales are adapted to take advantage of this physical acoustic niche. They further postulated that, prior to modern shipping, whales might have communicated across ocean basins (see also Patterson and Hamilton 1964; Norris 1966). Here, we will further explore this idea through a simple approach of comparing extant species that inhabit either shallow or deep water habitats. Is there a consistent relationship between the general acoustic characteristics for different whales and the environments in which they principally live? Can this relationship be used to predict the communicative functions of the different types of signals produced?

There are only 11 recognized extant species in the baleen whale (Mysticeti) group (Ridgeway and Harrison 1985). All are known to produce sounds, but representations of vocal repertoires are still incomplete for six species (Thompson et al. 1979; Watkins and Wartzok 1985; Clark 1990; Edds-Walton 1997). Full repertoires are available for five species: southern and northern right whales, bowhead whale, gray whale, and humpback whale (Clark 1982; Ljungblad et al. 1982; Tyack 1983; Clark and Johnson 1984; Dahlheim et al. 1984; Silber 1986; Chabot 1988). Good acoustic representations are available for the blue whale (*Balaenoptera musculus*), fin whale (*Balaenoptera physalus*) (Cummings and Thompson 1971; Watkins et al. 1987; Edds 1988; Thompson et al. 1992; Stafford 1999), and minke whale (*Balaenoptera acutorostrata*) (Winn and Perkins 1976; Mellinger et al. 2000; Edds-Walton 2000). There are a few sample recordings for Bryde's (*Balaenoptera edeni*) (Cummings 1985; Edds et al. 1993), sei (*Balaenoptera borealis*) (Thompson et al. 1979; Knowlton et al. 1991), and pygmy right (*Caperea marginata*) (Dawbin and Cato 1992) whales.

Baleen whale sounds are often very intense, with maximum reported band levels as high as 188 dB re 1 μPa (see Richardson et al. 1995, Table 7.1). Typical source levels range from 155 to 180 dB, but more research is needed to document natural variability in source levels under different contexts (e.g., Thode et al. 2000). Increasing the source level is one direct mechanism for reaching a greater audience. This is especially true for deep-ocean conditions where there are fewer surface and bottom reflections. However, when water depth is poorly matched to a signal's frequency band, the energetic cost of increasing source intensity may outweigh the advantages of increased communication range. Under such circumstances, selection should favor changes in other signal characteristics (e.g., frequency band or duration) to achieve greater communication range.

For purposes of this discussion, we refer to species that prefer coastal and/or shallow water habitats during major portions of their lives as coastal and species that spend major portions of their lives in deep water as pelagic. In general, coastal species are easier to study, and more is known about them, than pelagic species. All five species with essentially complete acoustic repertoires are coastal. This includes the bowhead, northern right, southern right, gray, and humpback whales. The southern right and bowhead whales are used as representative coastal species. The humpback whale, a species more closely related to blue and fin whales than to right whales, is also used as a coastal species while recognizing that it spends a considerable portion of the year migrating across pelagic regions. All balaenopterid species spend a large proportion of their lives in the open ocean and for this reason alone are difficult to study. With few exceptions, the mating and calving areas and behaviors for these species are largely unknown. The most recent data on balaenopterid populations and behaviors come from studies conducted during seasonal episodes of feeding along a coastal zone of upwelling (Croll et al. 1999; Gedamke et al. 1999). We will use blue and fin whales as representative pelagic species.

11.1. Baleen Whale Songs

The term song refers to long, patterned sequences of sounds in which individual sound units are arranged in a hierarchical order (see Payne and McVay 1971). Songs are typically produced by males during the breeding season and are assumed to be a male reproductive display. Two coastal species, humpbacks and bowheads, are known to sing long, patterned combinations of sounds (Payne and McVay 1971; Ljungblad et al. 1982; Würsig and Clark 1993). These songs are composed of complex vocalizations combined into phrases that are repeated to form themes. In humpbacks, a song lasts anywhere from 10 to 45 minutes and is composed of up to 7–8 themes, each consisting of several phrases. The song changes annually, and there is a great deal of intra- and inter-individual variability. In bowheads, the song lasts on the order of a minute. It has as many as two themes, with each one composed from a single phrase. As with humpbacks, bowhead songs change annually, and there is a high degree of variability such that no two songs from the same animal are ever exactly the same.

Blue and fin whales produce long, patterned sequences of sounds composed of repetitions of nearly identical 2–3 note phrases (Watkins et al. 1987; Thompson et al. 1992; Clark and Fristrup 1997; Rivers 1997; Moore et al. 1998; Stafford et al. 1998). There is less inter–annual, intra–individual, and inter–individual variability for blue and fin whales compared to the coastal species, but there are measurable differences in vocal features among populations (Thompson et al. 1992). For fin whales in the northwest Atlantic ocean, Watkins et al. (1987) used the coincidence between the seasonal occurrence of these sounds and breeding to suggest that these are male reproductive displays equivalent to humpback songs. However, the sex of the animals producing these long, patterned sequences needs to be determined, and both species are vocally active during the summer feeding season, which suggests that not all of this vocal activity is related to breeding (Ljungblad et al. 1998; Croll et al. 2001). Figure 2.5 presents spectrographic representations of songs for blue, fin, humpback, and bowhead whales.

11.2. Baleen Whale Calls

For simplicity, we refer to sounds other than songs as calls, with the assumption that the sounds are communicative, produced by males and females, and not restricted to the reproductive context. By far, the majority of information on whale sounds comes from work on coastal species, where

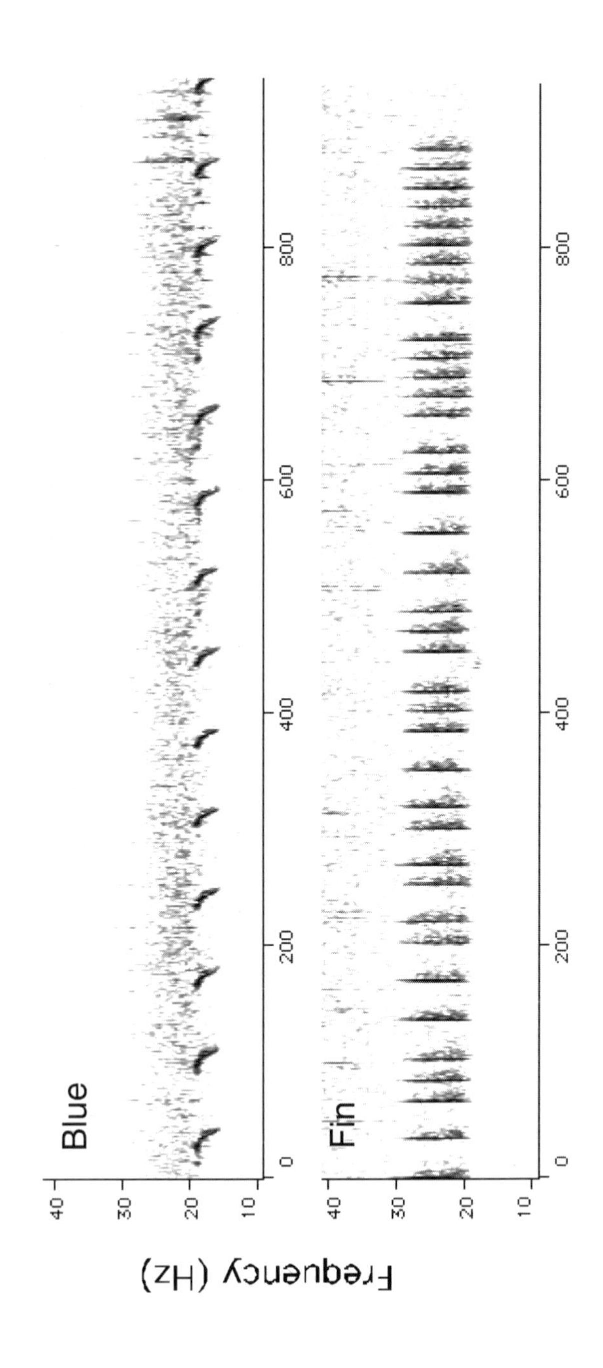
Blue
Fin
Frequency (Hz)
40
30
20
10
0
200
400
600
800

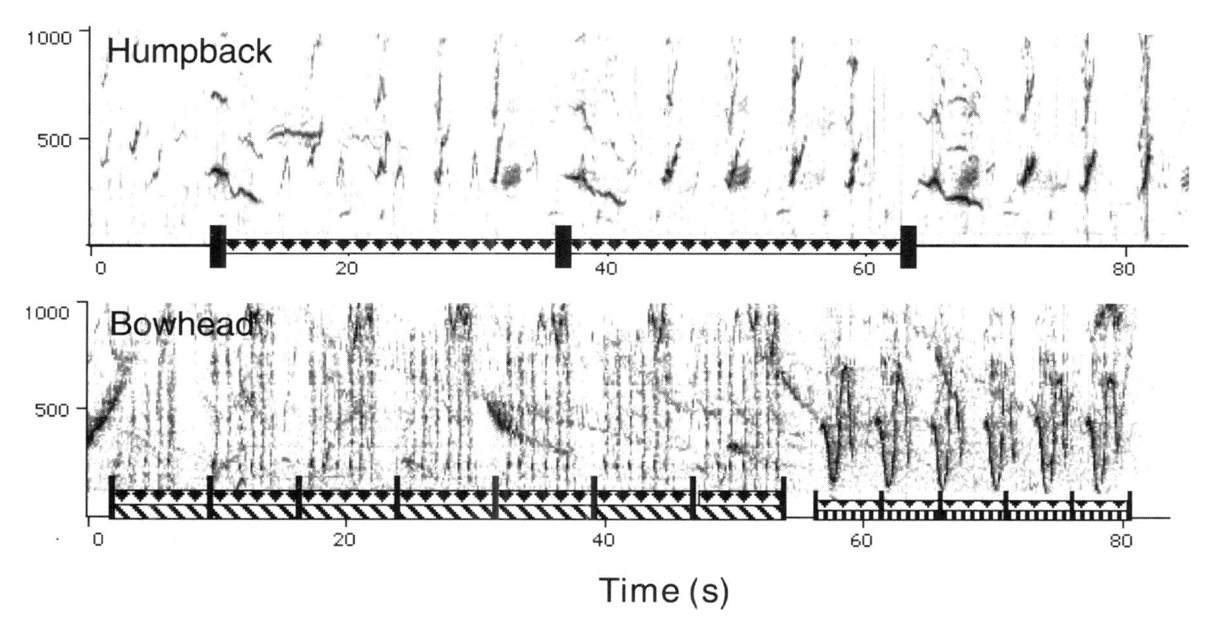

FIGURE 2.5. Spectrographic examples of songs from four different species of baleen whales. The time and frequency scales are the same for the blue and fin whale examples but different for the humpback and bowhead whale examples. Phrases in the humpback and bowhead songs are underlined with a sawtooth pattern, and the two themes in the bowhead song are underlined with striped patterns. Blue and fin whale samples recorded by C. Clark off San Nicolas Island in 1997 and Loreto, Mexico, in 1999, respectively; humpback whale recorded off Kauai in 1998; bowhead whale recorded off Point Barrow, Alaska, in 1988. All recordings made using digital recording systems with flat frequency response in the 10–5,000-Hz band.

researchers have been able to reliably observe and record individual animals over long periods of time (e.g., Clark 1982; Tyack 1983; Dahlheim et al. 1984; Clark et al. 1986; Silber 1986). There have been a few observations of blue and fin whales in surface-active groups with associated recordings of transient sounds (Watkins 1981; Watkins et al. 1987; Edds 1988). Transient signals from coastal species are highly variable in their acoustic structures. In contrast, transient signals from at least three of the pelagic species are much simpler and share common acoustic attributes. Contact calls from right and bowhead whales are simple, frequency-modulated sweeps, whereas feeding calls from humpbacks, which serve to attract distant whales to the aggregation, are long, constant-frequency sirens. At a qualitative comparative level, the most variable transient signals from blue and fin whales are equal to the simplest transient signals from right and humpback whales. Values for pelagic species are either from recordings during late winter or late summer when groups of animals were observed and feeding was the primary activity. Figure 2.6 presents spectrographic representations of calls for blue, fin, humpback, right, and bowhead whales.

There is enough evidence to make comparisons between the sounds of coastal and pelagic mysticetes. Here, we will not rely on statistical analyses (e.g., discriminant function, principle components; see Cortopassi and Bradbury 2000) to reveal acoustic groupings. Instead, we will plot several basic acoustic attributes by species and label them according to the preferred environment in which those sounds are typically produced. We proceed with this mechanism under the assumption that environmental influence on acoustic features at the evolutionary level should be relatively obvious, especially when the differences in the acoustic environments are so drastic. More subtle differences resulting from immediate behavioral influences (e.g., motivation) are expected to be graded (see Clark 1983) and, given the small species sample size, not so easily revealed.

In Figures 2.7 and 2.8, we present several graphical representations of acoustic features for the five representative species. Figure 2.7 emphasizes the importance of inter–note interval, frequency bandwidth, and duration for songs. It shows that song notes for coastal species have greater time-bandwidth products with shorter inter–rate intervals than songs for pelagic species and that, for these two features, the humpback and bowhead whales are remarkably similar while dissimilar from the blue and fin whales. Humpbacks and bowheads are rarely found in the same regions, whereas blue and fin whales are sympatric. In Figure 2.8, species-specific spectral profiles (solid lines) are contrasted with ambient noise spectra (hatched lines) from the deep and shallow water environments. The left-hand column illustrates these spectra for songs, whereas the right-hand column illustrates these spectra for calls. This figure indicates that the overall frequency compositions of the different species' song energies are well-matched to the ambient noise properties of the environment. One immediate advantage of producing song in a frequency band of low ambient noise is a greater range of detection. In this case, the benefit is on the order of at least 6–10 dB, which in deep water represents a fourfold increase in range and 16-fold increase in area. Similar descriptions of animals taking advantage of a low noise window have been observed in both vertebrate and invertebrate groups (Morton 1975; Michelsen 1992; Gerhardt 1994).

The right-hand column in Figure 2.8 shows spectral profiles for the calls of four species (right and bowhead whale call spectra are essentially identical) with the ambient noise for their respective deep water and shallow water environments. This figure illustrates that for coastal species (humpback and right whales) the overall frequency band of call energy is again well-matched to the ambient noise properties of the environment. However, it shows that for blue and fin whale calls, the frequency band of greatest energy occurs in a region of high ambient noise. This discrepancy is somewhat surprising. Perhaps the explanation is related to the specific contexts in which these calls were recorded. For the coastal species, many of the calls in this category are known to be contact calls (e.g., Clark 1983), occurring when animals were countercalling while separated by many miles. For the blue and fin whales, however, the animals presumed to be responsible for the calls were in surface-active groups in which the whales were within close proximity to one another. In this latter context, long-range transmission is not necessarily optimized because there is no advantage to producing low-frequency sounds. In fact, if there is an advantage to restricting the communication range to a local audience, then selection should favor sounds in a frequency band with high ambient noise.

Overall, this relatively simple comparative evaluation using the animal signal characteristics of bandwidth, duration, and intensity, together with ambient noise characteristics for the two habitats, reveals several striking results. First, the sounds from shallow water and deep water species occupy different regions of the frequency band, and these bands are well-matched to the general ambient noise and transmission properties of the two environments. The three coastal species (bowhead, humpback, and right whales) produce most of their energy in the 100–500-Hz band, a region in which propagation is optimum and ambient noise is low (Clark 1983; Jensen and Kuperman 1983). The two pelagic species (blue and fin whales) produce most of their sounds in the 15–30-Hz band, a frequency band for deep water in which transmission loss is low, but there is a trade-off with ambient noise, which is naturally dominating in this low-frequency band. The reverse is not true. That is, the signals from blue and fin whales would not propagate efficiently in the coastal habitat used by bowhead, humpback, or right whales because of the influence of water depth on cutoff frequency. Humpback whale songs are broadband, variable, and repetitive and appear to be adapted for both deep water and coastal habitats. High-frequency portions of humpback song (i.e., >500 Hz) do not propagate efficiently in the pelagic habitat (i.e., the deep sound channel). However, all humpback songs contain a significant amount of energy in the 100–500 Hz band, and some

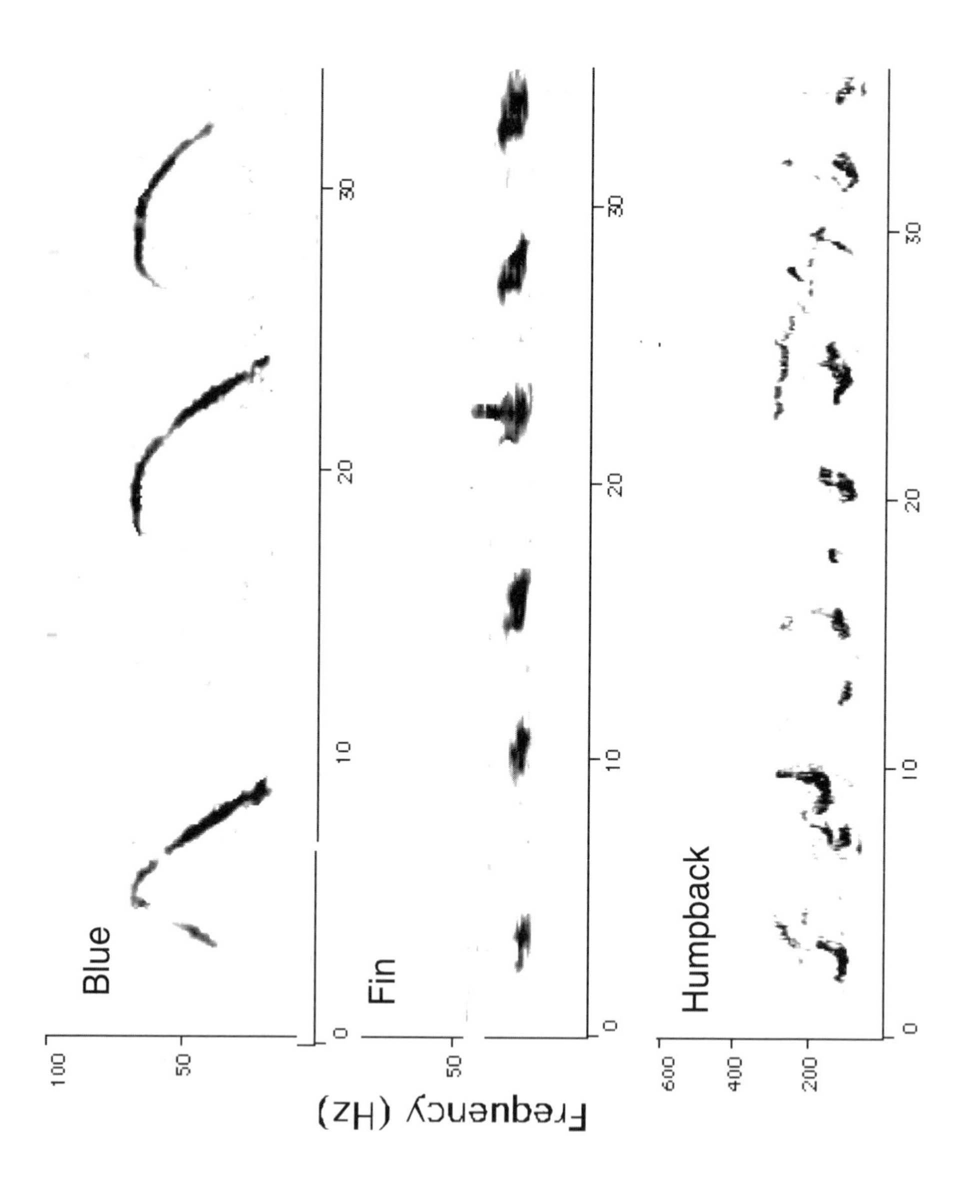
Blue
Fin
Humpback
Frequency (Hz)
100
50
50
600
400
200
0
10
20
30

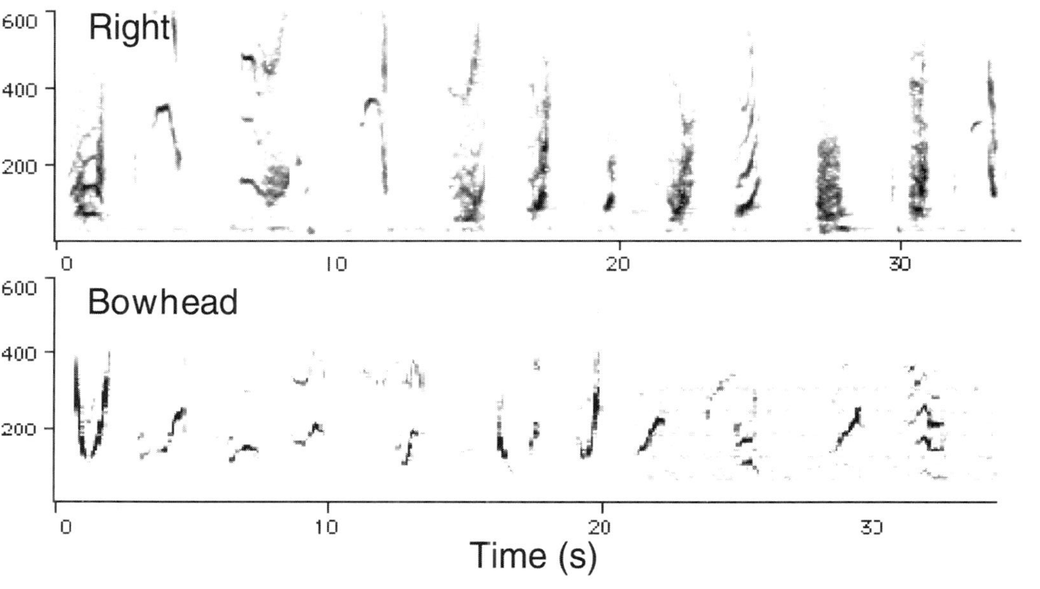

FIGURE 2.6. Spectrographic representations of calls from five different baleen whale species. The time and frequency scales are the same for the blue and fin whale examples but different from that for the humpback, right, and bowhead whales. Blue and fin whale samples recorded off San Nicolas Island in 1997 and Loreto, Mexico, in 1999, respectively; humpback whales recorded off Hawaii in 1995; right whales recorded in Gulfo San Jose, Argentina, in 1977; bowhead whales recorded off Point Barrow, Alaska, in 1988. All recordings made using recording systems with flat frequency response in the 10–5,000-Hz band.

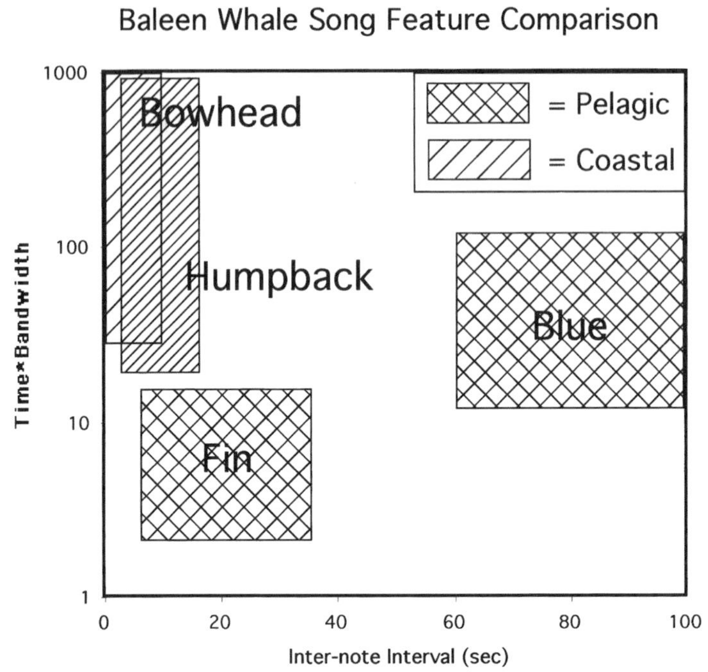

FIGURE 2.7. Schematic diagram illustrating differences in four baleen whale song features, where two of the species, humpback and bowhead, are coastal singers and two others, blue and fin, are deep ocean singers. Inter–note interval is the time in seconds between adjacent song elements, and time-bandwidth product, a dimensionless unit, is the product of element duration and bandwidth.

humpback singers produce sounds in the 20–30-Hz band. It remains to be determined whether these infrasonic singers are older, larger, or more successful breeders than noninfrasonic singers. Another recent observation for humpback singers is that there is as much as a threefold difference in song intensity, a difference that would have profound differences in detection range and therefore audience (Clark unpublished).

12. Acoustic Signaling in Teleost Fishes

12.1. Signal Parameters

Sound production and the underlying sound-producing mechanisms, both neural and mechanical, have been identified widely among distantly related groups of teleost fishes, suggesting that it has been independently evolved

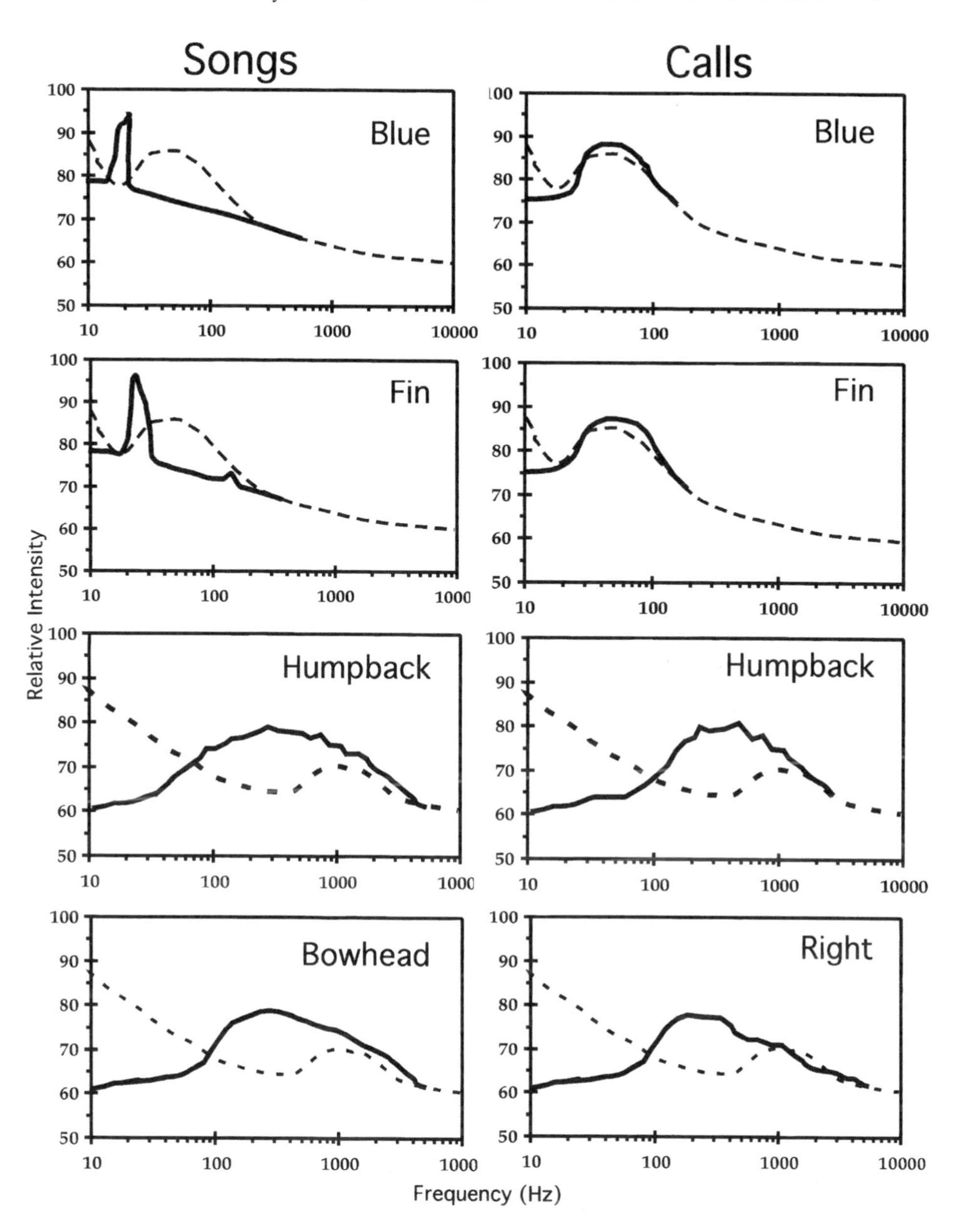

FIGURE 2.8. Spectral energy distributions for songs and calls (solid line) from different baleen whale species together with ambient noise spectra (dashed line) for their respective habitats, namely abyssal temperate ocean for blue and fin whales (Curtis et al. 1999) and coastal temperate ocean for humpback, bowhead, and right whales (Clark 1983).

a number of times (Fish and Mowbray 1970; Myrberg 1981; Bass and Baker 1991; Ladich and Bass 1998). The majority of well-documented teleost sounds are brief (50–200 msec), pulse-like signals that are broadband, with fundamental frequencies typically less than 400 Hz (Winn 1964; Fish and Mowbray 1970; Fine et al. 1977; Myrberg 1981; Ladich 1997). These sounds have been documented in aggressive and/or reproductive contexts. Long-duration signals on the order of seconds to minutes have been more specifically associated with courtship. This includes the boatwhistles of toadfish (Fine et al. 1977), the hums of midshipman (Bass et al. 1999), and the moans of mormyrids (Crawford et al. 1997). Sound intensities near the source are typically in the 120–130 dB re 1 μPa range (e.g., see Crawford et al. 1997; Barimo and Fine 1998; Bass et al. 1999). Most sonic species of teleost fishes are found in shallow water habitats either seasonally or on a perennial basis.

As with the vocalizations of cetaceans and other vertebrates, those of teleost fishes may vary in a number of parameters, including repetition rate, duration, amplitude, and frequency (see references above). To illustrate the nature of some of this variation, we present examples of the vocal signals from one species, the plainfin midshipman fish (*Porichthys notatus*), that has been extensively studied. *P. notatus* is reproductively active at night in water depths up to 5 m at high tide in some locales (Bass and Marchaterre unpublished). Midshipman have two male phenotypes or morphs, type I and type II (Bass 1996). Type I males build nests in the intertidal zone, from which they acoustically court females (Fig. 2.9). Type II males neither build nests nor court females. The vocal repertoire of type II males, like females, is fairly limited; they produce isolated, low-amplitude grunt-like signals (Brantley and Bass 1994). Type I males have the far more dynamic repertoire; they produce long-duration (sometimes > 1 h) "hums" during courtship (top panel, Fig. 2.10). Hums are multiharmonic signals with a fundamental frequency near 100 Hz and several prominent harmonics. Brief-duration "grunts" are produced either individually or in trains of more than 100 during agonistic encounters (bottom panel, Fig. 2.10); their fundamental frequency is also close to 100 Hz. "Growls" are intermediate in duration between single grunts and hums and show frequency modulation with significant energy well below 100 Hz (center panel, Fig. 2.10). Most teleosts generate nonoverlapping sounds (e.g., see Fine et al. 1977). However, an interesting example of the interaction of two conspecific signals is the overlapping hums of neighboring males, which establish acoustic beats with a modulation frequency equal to the difference frequency between the fundamental frequencies of each individual's call (Fig. 2.11).

12.2. Shallow Water Sound Communication

A variety of studies have used pressure-sensitive hydrophones to study transmission loss of naturally occurring vocalizations in shallow water habitats. Fine and Lenhardt (1983) studied transmission loss for pure tones and

FIGURE 2.9. Midshipman fish build nests in the intertidal zone. Shown in this drawing of a site photographed in northern Washington state are several parental, type I males that have been revealed by turning over their rocky shelter. Females are attracted to the nests by the humming mate call of the type I males. Once inside a nest, a female deposits each egg on the roof of the nest (the underside of the rock), and the male then fertilizes the egg; the female leaves the nest soon after depositing her eggs. The type I male remains to guard the eggs (depicted as round circles) and embryos (teardrop shaped) and to sing again to attract more females; the number of fertilized eggs can be in the thousands for any one nest by the end of a breeding season.

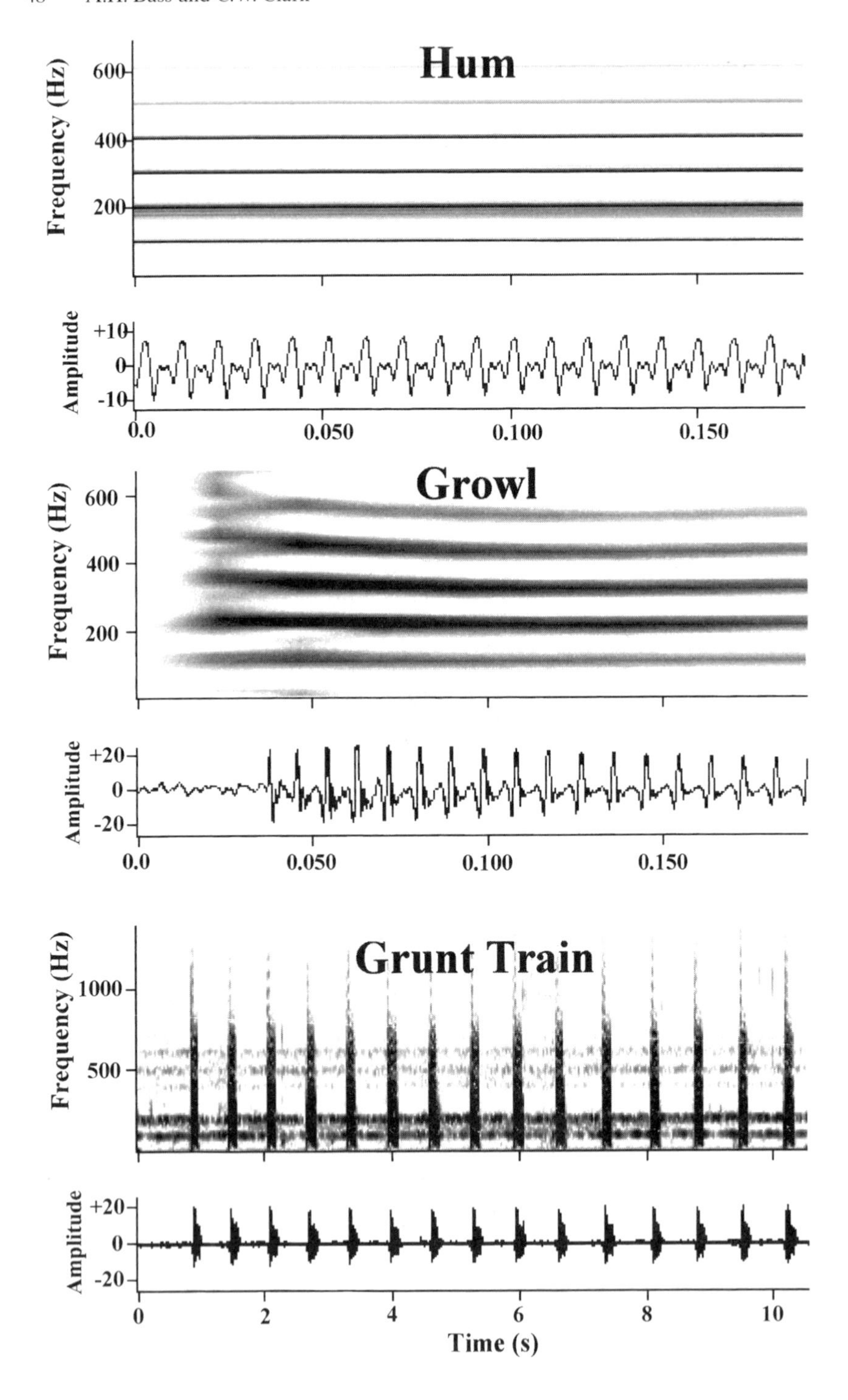
Hum
Frequency (Hz)
600
400
200
Amplitude
+10
0
-10
0.0
0.050
0.100
0.150
Growl
Frequency (Hz)
600
400
200
Amplitude
+20
0
-20
0.0
0.050
0.100
0.150
Grunt Train
Frequency (Hz)
1000
500
Amplitude
+20
0
-20
0
2
4
6
8
10
Time (s)

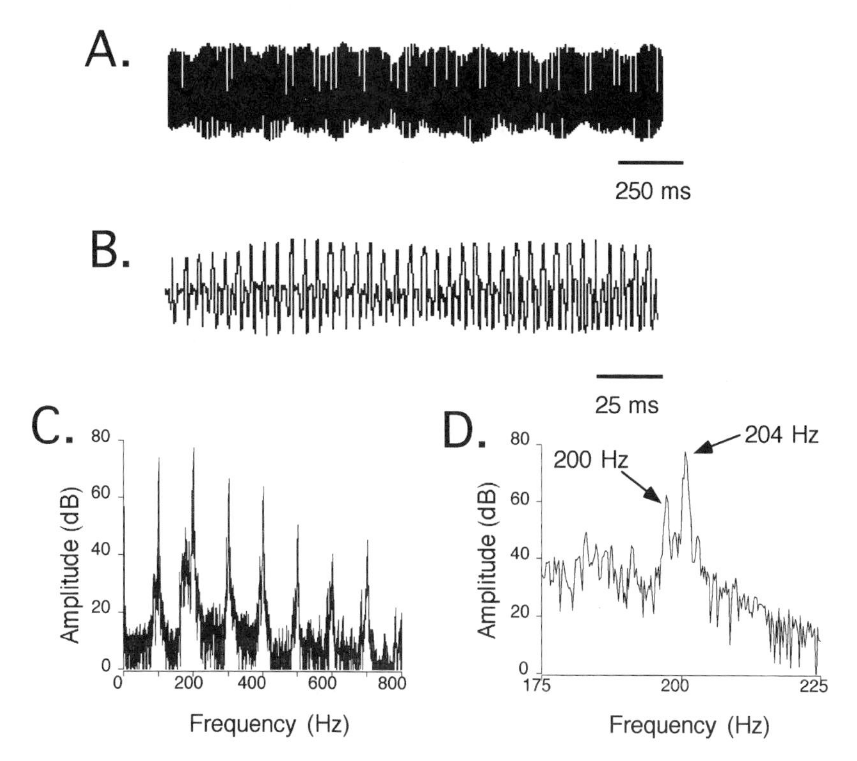

FIGURE 2.11. Shown here are the concurrent hums (acoustic beat) of two neighboring type I male midshipman fish in nests separated by 16 feet 8 inches (nest temperature 16.1°C; Brinnon, WA, June, 1996). (**A, B**) Temporal waveform. (**C**) Power spectrum showing multiple harmonics. (**D**) Expansion of the first peak in (C) shows the two harmonics (200 and 204 Hz) for the fundamental frequencies of the two hums, which were 100 and 102 Hz. Recordings courtesy of M. Marchaterre, Cornell University. (Modified from Bass et al. 1999.)

FIGURE 2.10. Acoustic signals of the plainfin midshipman fish, *Porichthys notatus*. Shown here are spectrograms (top) and oscillograms (bottom) of a representative hum, growl, and grunt train from the nest of the same type I male midshipman (21.5 cm standard length); nest temperature was 16.0°C (Brinnon, WA, June, 1997; recorded between 12:00 and 3:00 h). The amplitude scale is relative (but see Fig. 2.12). The hum shown here is a short segment from one that lasted 1 min 48 sec; the fundamental frequency was 102 Hz, with several prominent higher harmonics. The growl is a segment from one that lasted 4.64 sec with a fundamental frequency of 75.4 Hz. The grunt train shows 15 consecutive grunts from a grunt train that included a total of 22 grunts. The average duration of each grunt is 48.64 msec; the fundamental frequency is 107 Hz. The background hums of neighboring males can be seen in the grunt train record. Recordings courtesy of M. Marchaterre, Cornell University.

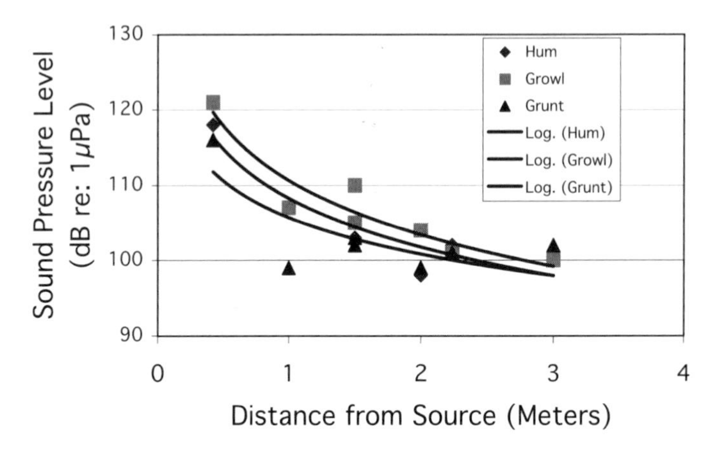

FIGURE 2.12. Transmission-loss profiles for the hum (total duration = 6 min; fundamental frequency = 109 Hz), growl (total duration = 1.93 sec; fundamental frequency = 62.1 Hz), and grunt (total duration = 167 msec; fundamental frequency = 106 Hz) from a single nest containing two type I males (28.0 cm and 18.0 cm standard length). Also shown from top to bottom are curve fits for the growl, hum, and grunt, respectively. Nest temperature was 17.3°C (Brinnon, WA, June, 1999). Recordings courtesy of M. Marchaterre, Cornell University.

prerecorded advertisement calls (boatwhistles) of a marine species, the oyster toadfish (*Opsanus tau*), in very shallow water sites (0.75–1.2 m) with sandy-silt bottoms (water temperatures 24–27°C). All sounds were played through a J-9 transducer. For pure tones, transmission loss was greatest within the first 3 m from the transducer. For boatwhistles, transmission loss was more gradual out to a distance of 5 m from the J-9 transducer, at which point the signals were no longer detectable above background noise levels. The boatwhistle's fundamental frequency was attenuated more quickly than either its second harmonic or the entire call. Hence, higher-frequency components of the call were transmitted over a greater distance, consistent with the predicted influence of water depth on cutoff frequency. Signal attenuation values with distance doubling were about 10 dB more (approximately $30 * \log_{10}(r)$) than that predicted for either spherical or cylindrical spreading and similar to the range reported by Banner (1970) for continuous, random noise over a similar frequency range (see Section 10.3).

Figure 2.12 shows decrements in sound-pressure levels for the calls of midshipman males at increasing distances from their nests. In this case, sound levels were determined using calibrated amplifiers (Shure Brothers Inc.) and a digital tape recorder (TEAC Corp.). Sound-pressure levels fall off at a rate close to that predicted for spherical spreading, namely 6 dB per distance doubling. Compared with the toadfish study summarized above, these sounds were recorded in areas where nests are positioned in greater water depths (~5 m) and have a harder substrate (rocky-gravel), both of which may contribute to attenuation rates that are closer to theoretical predictions. The fundamental frequency of midshipman calls hovers near 100 Hz (Fig. 2.10), which is close to the predicted cutoff frequency for calls in this environment (Bass unpublished).

Crawford et al. (1997) report a transmission loss close to 16 dB per distance doubling (about $50\log_{10}(r)$) for the naturally generated sounds of weakly electric mormyrids (*Pollimyrus isidori*) in 2–3 m water of a freshwater flood plain with a dense clay-like bottom in Mali, West Africa (water temperature approximately 28°C). This value, like that for the calls of toadfish, far exceeded the predicted values for either spherical or cylindrical spreading, in both cases likely resulting from the combined influence of very shallow depths and substrate composition. Crawford et al. (1997) estimated the cutoff frequency as 335 Hz at 2 m for mormyrids in this habitat. As the authors note, the fundamental frequency of many of the mormyrid sounds is close to 300 Hz and so is not expected to propagate very far except when the water depth exceeds 2 m.

Mann and Lobel (1997) studied the propagation of naturally generated courtship sounds of another marine species, the damselfish (*Dascyllus albisella*), at water depths of 7 m; the substrate was a mix of sand, shells, live coral, and coral rubble. As expected, transmission loss for the two highest frequency components (398 and 501 Hz) of this species' pulse-like calls was the most within the first 4 m from the reference point and showed little attenuation at distances up to 11–12 m. The two lowest frequency components (251 and 316 Hz) continued to attenuate beyond 4 m. Signal attenuation approximated theoretical losses that were intermediate between cylindrical and spherical spreading.

A recent study of sound propagation in bullfrogs in a freshwater pond also provides some basic information on underwater sound transmission (Boatright-Horowitz et al. 1999). Sound propagation was studied in three different ponds with depths of 1.25 m (mud bottom covered with detritus), 2.0 m (sand bottom), and 8.0 m (sand and stone bottom). For pure tones, the cutoff frequency was 1.8–2 kHz at a depth of 1.25 m and 1.6–1.8 kHz at depths of either 2 m or 8 m. Transmission loss in the 2-m and 8-m ponds was consistent with theoretical values for loss due to cylindrical spreading, whereas it was much greater than predicted in the 1.25-m pond. As the authors point out, the latter was likely due in part to the "softer" bottom of the 1.25-m pond. Bullfrog advertisement calls underwent losses that approached values predicted for spherical rather than cylindrical spreading.

Together, the variance in the results across the range of studies discussed above is consistent with the predicted influence of depth and bottom-substrate composition on signal transmission.

12.3. Increasing Signal Detection

Wiley (1983) points out a number of traits that can lead to the increased "reliability of detection" of signals, including small repertoires and redundancy. Both of these traits apply to teleosts such as the plainfin midshipman that only generate a few highly stereotyped and redundant call types (Fig. 2.10). The tone-like hum has a prominent fundamental frequency and several harmonics that are unchanging throughout the duration of the call, which can last on the order of minutes to over one hour (Bass et al. 1999). Increased hum duration, together with its temporal and spectral simplicity, should increase detection (see McKibben and Bass 1998 for underwater playbacks that examine the influence of duration on responsiveness). Similarly, many teleosts, including midshipman (Fig. 2.10), generate long, repetitive trains of stereotyped, pulse-like grunts or growls that should also lead to increased detection. These signals, like those of other teleosts, are also fairly broadband at frequencies above any predicted cutoff frequency.

Increased detection could be achieved by increasing the high-frequency content of a signal and thereby shifting the signal's spectrum farther away from the cutoff frequency at any one depth. This can be accomplished by moving to warmer water. The fundamental frequency of individual vocalizations increases with increasing temperature (e.g., Fine 1978; Torricelli et al. 1990; Brantley and Bass 1994; Crawford et al. 1997). The mechanism underlying this trait is an elegant example of how the abiotic environment can directly affect acoustic signaling via its influence on the neural control of sound production. As ambient water temperature increases, there is an increase in the discharge frequency of a vocal pacemaker circuit in the brain that controls the contraction rate of sonic muscles, which in turn establish a sound's fundamental frequency (Bass and Baker 1991). Although temperature affects the calls of other poikilothermic animals, including anuran amphibians, the effect here is mainly on call-repetition rate (see Gerhardt 1983; Zelick et al. 1999).

A decrease in cutoff frequency could be attained by moving to deeper water or by vocalizing at times of the day when water depth is greatest. For example, midshipman fish build nests in the intertidal zone and exhibit a peak in the occurrence of courtship and associated vocal behaviors after sunset (Ibara et al. 1983; Brantley and Bass 1994; Bass et al. 1999) when the high tides are greater in depth. Water depth, however, does not seem to place any limitations on mate calling because individual males will generate hums in 80-liter aquaria (Ibara et al. 1983; Brantley and Bass 1994).

Might sound transmission and detection be influenced by the nest that an individual chooses? For example, the composition of the nest's substrate material itself might affect the cutoff frequency. Intertidal species such as midshipman build nests under shelters with rocky-gravel bottoms (Bass 1996). In comparison with a softer, mud-like substrate, a more rigid substrate should contribute to a decrease in the cutoff frequency (Rogers and Cox 1988). These nests are also often clustered, which should enhance detection of a nearby neighbor's vocal signals (see Fig. 2.9 and Bass 1996).

Barimo and Fine (1998) mapped out the sound field for toadfish (*Opsanus tau*) nesting in terra cotta drainage tiles on a flat, sandy substrate at depths of 1–2 m. The sound field is bilaterally symmetrical around an individual's drainage tile, with sound amplitude at a maximum level directly behind a nest. Observations of acoustic courtship in midshipman showed that females are immediately directed to the nest entrance (a broken cement block on a sandy substrate) by a humming male that is perched at an artificial nest's entrance with his head oriented to the front of the nest (Brantley and Bass 1994). Additional phonotaxis experiments also demonstrated that females directly approach the front of a speaker, where sound pressure is greatest (McKibben and Bass 1998). Together, these studies point to the need for more analyses of sound fields and acoustic-related behaviors in more complex, naturalistic habitats.

13. Complementary Explanations for Low-Frequency Acoustic Signaling in Teleosts

A number of complementary explanations at different levels of analysis may be put forth to explain the existence of low-frequency signaling in an underwater environment (Fig. 2.13). Because of the greater diversity of mechanistic information available for teleost fishes, this portion of our essay focuses on their acoustic signals, thereby providing an example of a pluralistic approach to explaining existing vocal phenotypes. We have adopted this research strategy to explain the existence of vocalizations and mate choice among midshipman fish (Bass 1998; Bass et al. 1999). Here, we focus mainly on mechanistic explanations that identify linkages between the ecological environment and behavioral or structural (e.g., neural, endocrine, genetic) characters. Mechanisms are categorized as either behavioral–ecological, behavioral–structural, or ecological–structural. There are also fitness explanations that identify the consequences of mechanisms in terms of survival or reproductive success and historical explanations that provide adaptive interpretations for patterns observed over an individual's lifetime or geological time (see Bass 1998).

13.1. Behavioral–Ecological Mechanisms

The physical acoustics of signal propagation predicts that teleosts in very shallow water habitats should utilize vocal signals with high-frequency content, and yet the opposite appears to be the case for most species studied so far. Why? One behavioral–ecological explanation is that low-frequency calling behaviors lead to short communication distances in shallow water habitats because of the severe influence of water depth on the range of signal transmission. Short communication distances, in turn, may guard against detection by competitors or predators (Forrest et al. 1993). The short communication distances associated with low-frequency signals might also reduce the negative impact that reflections and reverberations at the

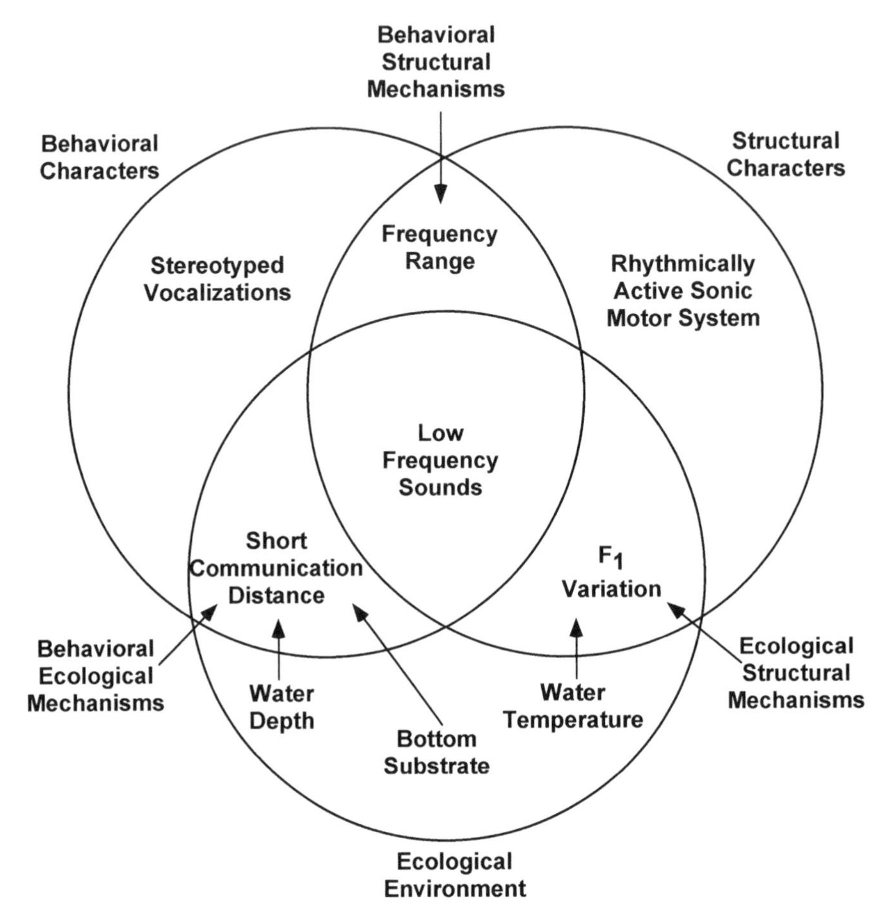

FIGURE 2.13. Complementary explanations for existing phenotypes. Multiple explanations can be provided to account for the existence of phenotypic characters and are best illustrated by a Venn diagram with three overlapping circles (after Bass 1998). The potential range of variation for behavioral and structural characters and the ecological environment is represented by one of the circles. The center represents the existing phenotype, in this case low-frequency calling. Mechanistic explanations are categorized by the overlap between two circles identified here as behavioral–structural, behavioral–ecological, and ecological–structural mechanisms. For example, a behavioral–ecological mechanism that explains low-frequency calling in shallow water is the influence of water depth and the substrate composition on short communication distances that guard against detection by competitors or predators. A structural–behavioral explanation for low-frequency calling is a rhythmically active neuromuscular system (sonic swim bladder) that generates highly stereotyped signals. An ecological–structural explanation for low-frequency calling is found in the direct influence of ambient water temperature on the discharge frequency of the hindbrain vocal pacemaker circuit and, in turn, an acoustic signal's fundamental frequency. This mechanism permits an individual to actively vary the frequency content of its signals.

surface and bottom would have on the temporal properties of these signals (see Bradbury and Vehrencamp 1998).

13.2. Behavioral–Structural Mechanisms

Another explanation for the existence of low-frequency acoustic signaling (behavioral characters) among teleosts is the neuromuscular systems that lead to sound production (structural characters). Teleosts generate sounds using diverse neuromuscular mechanisms, including pectoral fin stridulation, pectoral girdle vibration, and swim bladder vibration (Demski et al. 1973; Ladich and Bass 1998). The most extensively studied among these is the sonic swim bladder, which includes a pair of skeletal, sonic muscles that are either attached to the walls of the swim bladder or to a bony element that vibrates against the bladder (the swim bladder is considered an efficient mechanism for generating sound in a large volume of water and, as mentioned earlier, is generally modeled as a pulsating monopole; see Harris 1964).

The sonic swim bladder muscles of batrachoidids (midshipman and toadfish) are considered the fastest-contracting vertebrate muscles (Skoglund 1961; Rome et al. 1999). They show a number of structural adaptations at both light and electron-microscopic levels that are consistent with high contraction rates that may last as long as 1 h, as in the midshipman (Fawcett and Revel 1961; Ibara et al. 1983; Bass and Marchaterre 1989; Walsh et al. 1995; Bass et al. 1999; Rome et al. 1999). Each muscle contraction generates a single sound pulse so that the contraction rate can be directly translated into a sound's fundamental frequency. An advantage of high contraction rates is the production of a sound with a maximal fundamental frequency that is separated as far as possible from the cutoff frequency at any one depth. However, as others point out (e.g., Bradbury and Vehrencamp 1998), a limitation to such a muscle—based mechanism is an upper limit to the contraction mechanics of striated muscle and, in turn, a sound's fundamental frequency.

The production of highly stereotyped acoustic signals depends on the operation of a rhythmically active vocal neuron network. This vocal control circuitry, again best studied among batrachoidids, extends along the entire axis of the brain and shares a number of traits with the vocal circuitry of other vertebrates (Demski and Gerald 1972; Bass et al. 1994; Bass and Baker 1997; Goodson and Bass 2000a, 2000b; Fig. 2.14). Neurophysiological studies show a pacemaker-motoneuron circuit positioned in the hindbrain that establishes the rhythmic firing frequency of vocal motoneurons that determine the contraction rate of sonic muscles, which, in turn, sets the fundamental frequency of emitted sounds (Fig. 2.14). This hindbrain circuitry provides for the extensive coupling among vocal neurons that leads to the simultaneous contraction of the paired sonic muscles and maximal output (Bass and Baker 1990, 1991; Bass et al. 1994).

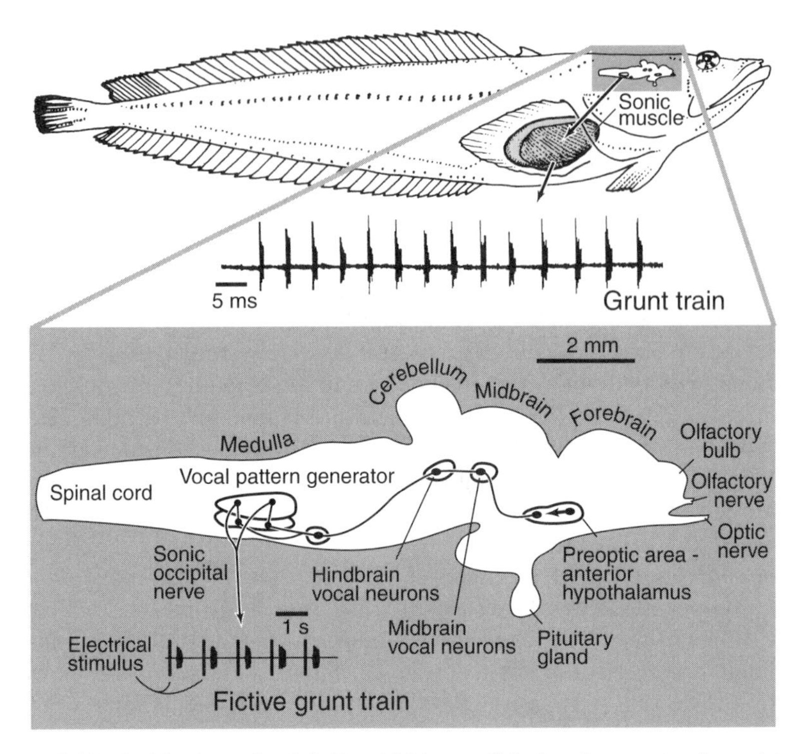

FIGURE 2.14. A side view of a plainfin midshipman fish showing a natural grunt train (also see Fig. 2.12). An expanded view of the brain shows the location of neurophysiologically and anatomically identified brain sites that evoke a rhythmic, patterned output from a hindbrain vocal pattern generator (see Bass and Baker 1990; Bass et al. 1994; Goodson and Bass 2000a). Each half of the SMN projects to the ipsilateral sonic muscle attached to the lateral walls of the swim bladder via a ventral sonic occipital nerve. The output of the hindbrain circuit can directly establish the fundamental frequency and duration of a vocalization (Bass and Baker 1990). Shown here are extracellular recordings from the surface of one occipital nerve. Such recordings, referred to as fictive vocalizations, precisely mimic the temporal structure of natural vocalizations, which in this case are grunts. For this experiment, the anterior hypothalamus was electrically stimulated; each stimulus is followed by a fictive grunt, which is a cluster of closely spaced nerve potentials that represent the synchronous activity of the motoneurons on that side of the brain. Both sides of the brain and hence both nerves and sonic muscles are simultaneously active (see Bass and Baker 1990 and Goodson and Bass 2000a).

The swim bladder is considered to affect mainly the amplitude of a signal, although it may yet contribute to other features of species-specific signals, including harmonic structure (see Demski et al. 1973; Kalmijn 1988; Hawkins 1993). Harmonics may prove to be an important component of signaling in shallow waters to increase call detection via increased energy at upper frequencies that transmit over greater distances (see Fine and Lenhardt 1983).

In summary, the adoption of a sonic swim bladder mechanism favors both the production of low-frequency signals for short-distance communication and stereotyped signals that will enhance detection and recognition by potential mates or nest intruders.

13.3. Ecological–Structural Mechanisms

Kalmijn (1988, 1989) proposes that near-field detection is the original function of the inner ear of fishes, a function that has been maintained among extant species. Thus, although the lateral line functions in the detection of local hydrodynamic flow in the very near field (see Coombs and Janssen 1988), the inner ear functions for detecting the remaining part of the near field. This would explain, in part, the historical origins and current maintenance of the low-frequency range for fish hearing.

The inner ear of teleost fishes includes otolithic end organs that function in hearing and are sensitive to particle motions that are prominent in the near field (Popper and Fay 1993). Although several distantly related teleosts possess structural adaptations that enhance the detection of the propagating sound wave and are considered "hearing specialists," most species lack these specializations and are classified as "hearing generalists" or "nonspecialists." The absence of accessory organs does not eliminate either detection of the propagating sound wave or a hearing sensitivity comparable to that of specialists in the relevant frequency range (see discussion in McKibben and Bass 1999).

Given that the range of particle motion in the near field (an abiotic ecological variable) should increase as a sound's wavelength increases (Section 3.3), low-frequency sounds should extend the detection range of the teleostean inner ear (structural character).

14. Summary Comments

We reviewed some of the basic principles underlying sound transmission in any medium and more specifically addressed the physical attributes of aquatic environments that could influence transmission and hence acoustic communication in underwater environments. We considered the influence of physical acoustics on sound production in two divergent vertebrate groups, cetaceans and teleost fishes, that are found in either deep (cetaceans) or coastal/very shallow (cetaceans and teleost fishes) water habitats.

The available evidence suggests that cetacean vocalizations are well-matched to the general transmission properties of their environments. Thus, a behavioral–ecological explanation for the physical attributes of cetacean vocalizations is that they optimize the range of transmission and detectability at increasing distances from the source by adapting the characteristics of their sounds to the particular environment (see Clark and Ellison in press; Tyack 2000; Tyack and Clark 2000). By contrast, although the physical acoustics of very shallow water favor high-frequency vocal signaling, the majority of teleost fishes produce sounds below 500 Hz in the intermediate- or low-frequency ranges. A number of explanations are suggested to account for this phenotype, including short communication distances to avoid detection by predators and competitors, increased range of detection by the inner ear, and neural-mechanical structures for generating stereotyped signals (Fig. 2.13).

The study of underwater sound communication is indeed in its infancy. Progress has been hindered, in part, by the inherent difficulties of working in aquatic, and especially marine, habitats as well as by the complexity in describing the impact of physical parameters, such as the bottom-substrate composition, on transmission loss. There are now technologies (e.g., hydrophone arrays) for investigating the details of animal movements and distributions (e.g., Clark et al. 1986; McDonald et al. 1995; Stafford et al. 1998; Thode et al. 2000). Such techniques, when merged with empirical studies of underwater sound transmission and applied to the incredible diversity of social systems, encourage greater pursuit of studies to identify the specific influence of abiotic and biotic ecological factors on vocal communication in an environment that accounts for nearly two-thirds of the earth's surface.

Acknowledgments. The authors' work is supported by grants from NIH (DC00092) to AHB and from ONR N00014-94-1-0872, N00014-94-C-6016, N00014-99-1-0244 and the North Slope borough, Barrow, Alaska, to CWC. We thank Margaret Marchaterre for the data in Figures 2.10–2.12; Adam Frankel, M. Marchaterre, Margy Nelson, and Terri Natoli for help with the figures; Terri Natoli, Connie Gordon, and Melissa Fowler for help with the references and data collation; and William Ellison for many inspirational conversations concerning whale sound propagation in the ocean. We also thank Deana Bodnar, Jud Crawford, Richard Fay, Jessica McKibben, Andrea Simmons, and Matthew Weeg for their many thoughtful comments on the manuscript, and especially Sheryl Coombs and Alejandro Purgue for their most generous time spent in discussions and comments.

References

Aroyan JL, McDonald MA, Webb SC, Hildebrand JA, Clark D, Laitman JT, Reidenberg JS (2000) Acoustical models of sound production and propagation. In: Au WWL, Popper AN, Fay RR (eds) Hearing by Whales and Dolphins. New York: Springer-Verlag, pp. 409–471.

Bailey WJ (1991) Acoustic Behaviour of Insects. London: Chapman and Hall.

Banner A (1970) Propagation of sound in a shallow bay. J Acoust Soc Am 49: 373–376.
Bannister RW (1986) Deep sound channel noise from high-latitude winds. J Acoust Soc Am 79:41–48.
Barimo JF, Fine ML (1998) Relationship of swim-bladder shape to the directionality pattern of underwater sound in the oyster toadfish. Can J Zool 76:134–143.
Bass AH (1996) Shaping brain sexuality. Am Sci 84:352–363.
Bass AH (1998) Behavioral and evolutionary neurobiology: A pluralistic approach. Am Zool 38:97–107.
Bass AH, Baker R (1990) Sexual dimorphisms in the vocal control system of a teleost fish: Morphology of physiologically identified cells. J Neurobiol 21: 1155–1168.
Bass AH, Baker R (1991) Evolution of homologous vocal control traits. Brain Behav Evol 38:240–254.
Bass AH, Baker R (1997) Phenotypic specification of hindbrain rhombomeres and the origins of rhythmic circuits in vertebrates. Brain Behav Evol 50:3–16.
Bass AH, Marchaterre MM (1989) Sound-generating (sonic) motor system in a teleost fish (*Porichthys notatus*): Sexual polymorphism in the ultrastructure of myofibrils. J Comp Neurol 286:141–153.
Bass AH, Marchaterre MA, Baker R (1994) Vocal-acoustic pathways in a teleost fish. J Neurosci 14:4025–4039.
Bass AH, Bodnar D, Marchaterre M (1999) Complementary explanations for existing phenotypes in an acoustic communication system. In: Hauser MD, Konishi M (eds) The Design of Animal Communication. Cambridge, MA: MIT Press, pp. 493–514.
Beranek LL (1988) Acoustical Measurements, New York: American Institute of Physics.
Boatright-Horowitz SS, Cheney CA, Simmons AM (1999) Atmospheric and underwater propagation of bullfrog vocalizations. Bioacoustics 9:257–280.
Bradbury JW, Vehrencamp SL (1998) Principles of Animal Communication. Sunderland, MA: Sinauer Associates.
Brantley RK, Bass AH (1994) Alternative male spawning tactics and acoustic signalling in the plainfin midshipman fish, *Porichthys notatus*. Ethology 96:213–232.
Chabot D (1988) A quantitative technique to compare and classify humpback whale (*Megaptera novaeangliae*) sounds. Ethology 77:89–102.
Clark CW (1982) The acoustic repertoire of the Southern right whaler: A quantitative analysis. Anim Behav 30:1060–1071.
Clark CW (1983) Acoustic communication and behavior of the southern Right whale (*Eubalaena australis*). In: Payne RS (ed) Communication and Behavior of Whales. Boulder, CO: Westview Press, pp. 163–198.
Clark CW (1990) Acoustic behavior of Mysticete whales. In: Thomas J, Kastelein RA (eds) Sensory Abilities of Cetaceans. New York: Plenum Press, pp. 580–583.
Clark CW, Fristrup K (1997) Whales '95: A combined visual and acoustic survey of blue and fin whales off southern California. Rep Int Whaling Comm 47: 583–600.
Clark CW, Johnson JH (1984) The sounds of the bowhead whale, *Balaena mysticetus*, during the spring migrations of 1979 and 1980. Can J Zool 62:1436–1441.
Clark CW, Ellison WT (in press) Potential use of low-frequency sounds by baleen whales for probing the environment: Evidence from models and empirical

measurements. In: Thomas JA, Kastelein RA (eds) Advances in the Study of Echolocation in Bats and Dolphins. New York: Plenum Press, pp. ••–••.

Clark CW, Ellison WT, Beeman K (1986) Acoustic tracking and distribution of migrating bowhead whales, *Balaena mysticetus*, off Point Barrow, Alaska in the spring of 1984. Rep Int Whaling Comm 36:502.

Coombs S, Janssen J (1988) Water flow detection by the mechanosensory lateral line. In: Stebbins WC, Berkley MA (eds) Comparative Perception, Vol 2. New York: John Wiley and Sons, pp. 89–123.

Cortopassi KA, Bradbury JW (2000) The comparison of harmonically rich sounds using spectrographic cross-correlation and principal coordinates analysis. Bioacoustics. 11:89–127.

Crawford JD, Jacob P, Bénech V (1997) Sound production and reproductive ecology of strongly acoustic fish in Africa: *Pollimyrus isidori*, Mormyridae. Behaviour 134:9–10.

Croll D, Benson S, Marinovic B, Demer D, Hewitt R, Chavez F (1999) From wind to whales: Foraging ecology of rorquals in the California current. 13th Bienn Conf Biol Mar Mamm Conf Abstr, p. 41.

Croll DA, Clark CW, Calambokidis J, Ellison WT, Tershy BR (2001) Effect of anthropogenic low frequency noise on the foraging ecology of *Balaenoptera* whales. Anim Conserv. 4:13–27.

Cummings WC (1985) Bryde's whale—*Balaenoptera edeni*. In: Ridgeway SH, Harrison R (eds) Handbook of Marine Mammals, Vol 3: The Sirenians and Baleen Whales. London: Academic Press, pp. 137–154.

Cummings WC, Thompson PO (1971) Underwater sounds from the blue whale, *Balaenoptera musculus*. J Acoust Soc Am 50:1193–1198.

Curtis KR, Howe BM, Mercer JA (1999) Low-frequency ambient sounds in the North Pacific: Long time series observations. J Acoust Soc Am 106:3189–3200.

Dahlheim ME, Fisher HD, Schempp JD (1984) Sound production by the gray whale and ambient noise levels in Laguna San Ignacio, Baja California Sur, Mexico. In: Jones ML, Swartz SL, Letherwood S (eds) The Gray Whale *Eschrichtius robustus*. New York: Academic Press, pp. 511–541.

Dawbin WH, Cato DH (1992) Sounds of pygmy right whale (*Caperea marginata*), Mar Mamm Sci 8:213–219.

Demski LS, Gerald JW (1972) Sound production evoked by electrical stimulation of the brain in toadfish (*Opsanus beta*). Anim Behav 20:507–513.

Demski LS, Gerald JW, Popper AN (1973) Central and peripheral mechanisms of teleost sound production. Am Zool 13:1141–1167.

D'Spain GL, Kuperman WA, Clark CW, Mellinger DK (1995) Simultaneous source ranging and bottom geoacoustic inversion using shallow water, broadband dispersion of Fin Whale calls. J Acoust Soc Am 97: 3353.

Dyer I (1997) Ocean ambient noise. In: Crocker MJ (ed) Encyclopedia of Acoustics. New York: John Wiley and Sons, pp. 549–557.

Edds PL (1988) Characteristics of finback, *Balaenoptera physalus*, vocalizations in the St. Lawrence Estuary. Bioacoustics 1:131–149.

Edds PL, Odell DK, Tershy BR (1993) Vocalizations of a captive juvenile and free-ranging adult-calf pairs of Bryde's whales, *Balaenoptera edeni*. Mar Mamm Sci 9:269–284.

Edds-Walton PL (1997) Acoustic communication signals of mysticete whales. Bioacoustics 8:47–60.

Edds-Walton PL (2000) Vocalizations of minke whales (*Balaenoptera acutorostrata*) in the St. Lawrence Estuary. Bioacoustics 11:31–50.
Fawcett DW, Revel JP (1961) The sarcoplasmic reticulum of a fast-acting fish muscle. J Biophys Biochem Cytol 10:89–102.
Fay RR, Simmons AM (1999) The sense of hearing in fishes and amphibians. In: Fay RR, Popper AN (eds) Comparative Hearing: Fishes and Amphibians. New York: Springer-Verlag, pp. 269–318.
Fine ML (1978) Seasonal and geographical variation of the mating call of the oyster toadfish *Opsanus tau* L. Oecologia 36:45–57.
Fine ML, Lenhardt ML (1983) Shallow-water propagation of the toadfish mating call. Comp Biochem Physiol 76A:225–231.
Fine ML, Winn HE, Olla BL (1977) Communication in fishes. In: Sebeok TA (ed) How Animals Communicate. Bloomington: Indiana University Press, pp. 472–518.
Fish MP, Mowbray WH (1970) Sounds of Western North Atlantic Fishes. Baltimore and London: The Johns Hopkins Press.
Forrest TG, Miller GL, Zagar JR (1993) Sound propagation in shallow water: Implications for acoustic communication by aquatic animals. Bioacoustics 4:259–270.
Gedamke J, Costa D, Dustan A (1999) Minke whale vocal behavior and remote acoustic monitoring. 13th Bienn Conf Biol Mar Mamm Abstr, p. 65.
Gerhardt HC (1983) Communication and the environment. In: Halliday TR, Slater PJB (eds) Animal Behaviour, Vol 2: Communication. New York: WH Freeman, pp. 82–113.
Gerhardt HC (1994) The evolution of vocalization in frogs and toads. Annu Rev Ecol Syst 25:293–324.
Goodson JL, Bass AH (2000a) Forebrain peptides modulate sexually polymorphic vocal circuitry. Nature 403:769–772.
Goodson JL, Bass AH (2000b) Vasotocin innervation and modulation of vocal-acoustic circuitry in the teleost *Porichthys notatus*. J Comp Neurol 422:363–379.
Harris GG (1964) Considerations on the physics of sound production by fishes. In: Tavolga WN (ed) Marine Bio-Acoustics. New York: The Macmillan Company, pp. 233–247.
Hawkins AD (1993) Underwater sound and fish behaviour. In: Pitcher TJ (ed) Behaviour of Teleost Fishes, 2nd ed. New York: Chapman and Hall, pp. 129–169.
Ibara RM, Penny LT, Ebeling AW, van Dykhuizen G, Cailliet G (1983) The mating call of the plainfin midshipman fish, *Porichthys notatus*. In: Noakes DLG, Lundquist DG, Helfman GS, Ward JA (eds) Predators and Prey in Fishes. The Hague: Dr. W. Junk Publishers, pp. 205–212.
Jensen FB, Kuperman WA (1983) Optimum frequency of propagation in shallow water environments. J Acoust Soc Am 73:813–819.
Jensen FB, Kuperman WA, Porter MB, Schmidt H (1994) Computational Ocean Acoustics. New York: American Institute of Physics.
Kalmijn AJ (1988) Hydrodynamic and acoustic field detection. In: Atema J, Fay RR, Popper AN, Tavolga WN (eds) Sensory Biology of Aquatic Animals. New York: Springer-Verlag, pp. 83–130.
Kalmijn AJ (1989) Functional evolution of lateral line and inner ear sensory systems. In: Coombs S, Görner P, Münz H (eds) The Mechanosensory Lateral Line. New York: Springer-Verlag, pp. 187–215.
Kinsler LE, Frey AR (1962) Fundamentals of Acoustics, 2nd ed. New York: John Wiley and Sons.

Knowlton AR, Clark CW, Karus SD (1991) Sounds recorded in the presence of Sei whales *Balaenoptera borealis*. 9th Bienn Conf Biol Mar Mamm Abstr, p. 40.

Ladich F (1997) Agonistic behaviour and significance of sounds in vocalizing fish. Mar Fresh Behav Physiol 29:87–108.

Ladich F, Bass AH (1998) Sonic/vocal motor pathways in catfishes: Comparisons with other teleosts. Brain Behav Evol 51:315–330.

Ljungblad DK, Thompson PO, Moore SE (1982) Underwater sounds recorded from migrating bowhead whales, *Balaena mysticetus*, in 1979. J Acoust Soc Am 71:477–482.

Ljungblad DK, Clark CW, Shimada H (1998) A comparison of sounds attributed to pygmy blue whales (*Balaenoptera musculus musculus*) recorded south of the Madagascar Plateau and those attributed to 'True" blue whales (*Balaenoptera musculus*) recorded off Antarctica. Rep Int Whale Comm 48:439–442.

Mann DA, Lobel PS (1997) Propagation of damselfish (Pomacentridae) courtship sounds. J Acoust Soc Am 101:3783–3791.

McCracken KG, Sheldon FH (1997) Avian vocalizations and phylogenetic signal. Proc Narl Acad Sci U S A 94:3833–3836.

McDonald MA, Hildebrand JA, Webb SC (1995) Blue and fin whales observed on a seafloor array in the northeast Pacific. J Acoust Soc Am 98:712–721.

McKibben JR, Bass AH (1998) Behavioral assessment of acoustic parameters relevant to signal recognition and preference in a vocal fish. J Acoust Soc Am 104:3520–3533.

McKibben JR, Bass AH (1999) Peripheral encoding of behaviorally relevant acoustic signals in a vocal fish: Single tones. J Comp Physiol A 184:563–576.

Mellinger DK, Carson CD, Clark CW (2000) Characteristics of minke whale (*Balaenoptera acutorostrata*) pulse trains recorded near Puerto Rico. Mar Mamm Sci 16:739–756.

Michelsen A (1978) Sound reception in different environments. In: Ali MA (ed) Sensory Ecology. New York: Plenum Publishing Corporation, pp. 345–373.

Michelsen A (1992) Hearing and sound communication in small animals: Evolutionary adaptations to the laws of physics. In: Webster DB, Fay RR, Popper AN (eds) The Evolutionary Biology of Hearing. New York: Springer-Verlag, pp. 59–77.

Moore SE, Stafford KM, Dahlhei ME, Fox CG, Braham HW, Polovina J, Bain D (1998) Seasonal variation in reception of Fin Whale Calls at five geographic areas in the North Pacific. Mar Mamm Sci l4:617–627.

Morton ES (1975) On the occurrence and significance of motivational-structural rules in some bird and mammal sounds. Am Nat 111:855–869.

Morton ES (1982) Grading, discreteness, redundancy and motivational-structural rules. In: Kroodsma DE, Miller EH (eds) Evolution and Ecology of Acoustic Communication in Birds. New York: Academic Press, pp. 183–212.

Myrberg AA Jr (1981) Sound communication and interception in fishes. In: Tavolga WN, Popper AN, Fay RR (eds) Hearing and Sound Communication in Fishes. New York: Springer-Verlag, pp. 395–425.

Myrberg AA Jr (1990) The effects of man-made noise on the behavior of marine animals. Environ Int 16:575–586.

Norris KS (1966) Some observations on the migration and orientation of marine mammals. In: Storm RM (ed) Twenty-Seventh Annual Biology Colloquium. Corvallis: Oregon State University Press, pp. 101–125.

Officer CB (1958) Introduction to the Theory of Sound Transmission. New York, Toronto, and London: McGraw–Hill.

Panigada SM, Zanardelli M, Canese S, and Jahoda M (1999) How deep can baleen whales dive? Mar Ecol Prog Ser 187:309–311.

Patterson B, Hamilton GR (1964) Repetitive 20 cycle per second biological hydro-acoustic signals at Bermuda. In: Tavolga WN (ed) Marine Bio-Acoustics. Oxford: Pergamon Press, pp. 125–145.

Payne R, Webb D (1971) Orientation by means of long range acoustic signalling in baleen whales. Ann NY Acad Sci 188:110–141.

Payne RS, McVay S (1971) Songs of humpback whales. Science 173:587–597.

Popper AN, Fay RR (1993) Sound detection and processing by fish: Critical review and major research questions. Brain Behav Evol 41:14–38.

Premus V, Spiesberger JL (1997) Can acoustic multipath explain finback (*B. physalus*) 20-Hz doublets in shallow water? J Acoust Soc Am 101:1127–1138.

Richardson WJ, Greene CR Jr, Malme CI, Thomson DH (eds) (1995) Marine Mammals and Noise. New York: Academic Press.

Ridgeway SH, Harrison SR (eds) (1985) Handbook of Marine Mammals, Vol 3: The Sirenians and Baleen Whales. London: Academic Press.

Rivers JA (1997) Blue whale, *Balaenoptera musculus*, vocalizations from the waters off central California. Mar Mamm Sci 13:186–195.

Rogers PH, Cox M (1988) Underwater sound as a biological stimulus. In: Atema J, Fay RR, Popper AN, Tavolga WN (eds) Sensory Biology of Aquatic Animals. New York: Springer-Verlag, pp. 130–149.

Rome LC, Cook C, Syme DA, Connaughton MA, Ashley-Ross M, Klimov A, Tikunov B, Goldman YE (1999) Trading force for speed: Why superfast cross-bridge kinetics leads to superlow forces. Proc Natl Acad Sci U S A 96:5826–5831.

Silber GK (1986) The relationship of social vocalizations to surface behavior and aggression in the Hawaiian humpback whale (*Megaptera nouaeangliae*). Can J Zool 64:2075–2080.

Siler W (1969) Near- and farfields in a marine environment. J Acoust Soc Am 46:483–484.

Skoglund CR (1961) Functional analysis of swim-bladder muscles engaged in sound production of the toadfish. J Biophys Biochem Cytol 10:187–200.

Speaks CE (1992) Introduction to Sound: Acoustics for the Hearing and Speech Sciences. San Diego: Singular Publishing Group Inc.

Spiesberger JL, Fristrup KM (1990) Passive localization of calling animals and sensing of their acoustic environment using acoustic tomography. Am Nat 135:107–153.

Stafford KM (1999) Low-frequency whale sounds recorded on hydrophones moored in the eastern tropical Pacific. J Acoust Soc Am 106:3687–3698.

Stafford KM, Fox CG, Clark DS (1998) Long-range acoustic detection and localization of blue whale calls in the northeast Pacific Ocean. J Acoust Soc Am 104:3616–3625.

Thode AM, D'Spain GL, Kuperman WA (2000) Matched-field processing, geo-acoustic inversion, and source signature recovery of blue whale vocalizations. Acoust Soc Am 107:1286–1300.

Thompson PO, Findley LT, Vidal O (1992) 20-Hz pulses and other vocalizations of fin whales, *Balaenoptera physalus*, in the Gulf of California, Mexico. J Acoust Soc Am 92:3051–3057.

Thompson TJ, Winn HE, Perkins PJ (1979) Mysticete sounds. In: Winn HE, Olla BL (eds) Cetaceans. New York: Plenum Press, pp. 403–431.

Torricelli P, Lugli M, Pavan G (1990) Analysis of sounds produced by male *Padogobius martensi*, and factors affecting their structural properties. Bioacoustics 2:261–275.

Tyack P (1983) Differential response of humpback whales (*Megaptera novaeangliae*) to playback of song or social sounds. Behav Ecol Sociobiol 13:49–55.

Tyack P (2000) Functional aspects of cetacean communication. In: Mann JR, Conner P, Tyack PL, Whitehead H (eds) Cetacean Societies: Field Studies of Whales and Dolphins. Chicago: University of Chicago Press, pp. 270–307.

Tyack P, Clark CW (2000) Communication and acoustical behavior in dolphins and whales. In: Au WWL, Popper AN, Fay RR (eds) Hearing by Whales and Dolphins. New York: Springer-Verlag, pp. 156–224.

Urick RJ (1983) Principles of Underwater Sound. New York: McGraw–Hill.

van Bergeijk WA (1964) Directional and nondirectional hearing in fish. In: Tavolga WN (ed) Marine Bio-Acoustics. New York: MacMillan, pp. 281–299.

Walsh PJ, Mommsen TP, Bass AH (1995) Biochemical and molecular aspects of singing in batrachoidid fishes. In: Hochachka PW, Mommsen TP (eds) Biochemistry and Molecular Biology of Fishes, Vol 4. New York: Elsevier, pp. 279–289.

Waser PM, Brown CH (1986) Habitat acoustics and primate communication. Am J Primatol 10:135–154.

Watkins WA (1981) Activities and underwater sounds of fin whales. Sci Rep Whales Res Inst 33:83–117.

Watkins WA, Wartzok D (1985) Sensory biophysics of marine mammals. Mar Mamm Sci 1:219–260.

Watkins WA, Tyack P, Moore KE (1987) The 20-Hz signals of finback whales (*Balaenoptera physalus*). J Acoust Soc Am 82:1901–1912.

Weidner RT, Sells RL (1965) Elementary Classical Physics, Vol 2. Boston: Allyn and Bacon.

Wiley RH (1983) The evolution of communication: Information and manipulation. In: Halliday TR, Slater PJB (eds) Animal Behaviour. New York: WH Freeman, pp. 156–189.

Wiley RH, Richards DG (1982) Adaptations for acoustic communication in birds: Sound transmission and signal detection. In: Kroodsma D, Miller EH, Luellet H (eds) Acoustic Communication in Birds. New York: Academic Press, pp. 131–181.

Winn HE (1964) The biological significance of fish sounds. In: Tavolga WN (ed) Marine Bio-Acoustics. New York: MacMillan, pp. 213–231.

Winn HE, Perkins PJ (1976) Distribution and sounds of the minke whale, with a review of mysticete sounds. Cetology 19:1–12.

Würsig B, Clark CW (1993) Behavior. In: Burns J, Montague J, Cowles CJ (eds) The Bowhead Whale. Lawrence: Allen Press, pp. 157–193.

Yost WA (2000) Fundamentals of Hearing. New York: Academic Press.

Zelick R, Mann DA, Popper AN (1999) Acoustic communication in fishes and frogs. In: Fay RR, Popper AN (eds) Comparative Hearing: Fish and Amphibians. New York: Springer-Verlag, pp. 363–411.

3 Unpacking "Honesty": Vertebrate Vocal Production and the Evolution of Acoustic Signals

W. TECUMSEH FITCH and MARC D. HAUSER

1. Introduction

When autumn arrives in northern Europe, female red deer (*Cervus elaphus* L.) begin to congregate. The mating season has begun. They are soon joined by males, who have spent the previous ten months in preparation, feeding and sparring. Some of the males herd females into groups, or "harems," which they vigorously defend against other males. A prominent component of this defense is roaring, a powerful, low-pitched groaning sound made only by males and primarily by harem holders. Why do males produce these sounds, and what effect do they have on listeners? Early observers (Darling 1937) suggested that the roars intimidated rivals and repelled intruders without the need for a dangerous fight. However, selection should favor opponents who are not so easily intimidated and base their behavior solely on a balanced assessment of their chances of winning a fight and inheriting the herd (Maynard Smith and Price 1973). In a classic paper, Clutton-Brock and Albon (1979) showed that roaring provides a source of information relevant to this decision: roaring rates of individual males are highly correlated with their fighting ability and thus provide an accurate indication of the males' ability to repel intruders. They also demonstrated, in a series of playback experiments, that rival males attend to this information, responding preferentially to high roar rates and ignoring the roars of young, small males. Red deer roaring has since become a classic example of "truth in advertising" in an animal vocalization.

"Honest signaling" in animal communication refers to signals that provide accurate information to perceivers either about the quality or properties of the signaler itself (e.g., advertisement calls) or about something in the environment (e.g., alarm calls). The degree to which animal signals are honest in this sense has been a perennially provocative problem and has generated significant theoretical advances, along with some empirical work, in the last few decades. Early workers in ethology were primarily interested in the historical evolutionary origins of particular communicative displays (e.g., via "ritualization" of intention movements; Cullen 1966) and devoted little attention to questions of honest or deceitful information transfer (Hinde 1981). To the extent that information is transferred, it was often implicitly assumed to be in both parties' interests that the information be accurate. However, the red deer example illustrates the weakness of this assumption. What prevents every harem holder from roaring at very high rates and thus repelling all comers without regard to his fighting ability? Such a deceitful mutant would be spared the costs of fighting, and his genes should spread rapidly through the population (Dawkins and Krebs 1978). This in turn would generate strong selection for "skeptical" males who ignored roaring altogether (Clutton-Brock and Albon 1979; Hinde 1981; Krebs and Dawkins 1984). Thus, theory predicts that selection resulting from receiver skepticism would eliminate the potential benefits of dishonest signaling. The only communication systems that would be stable in the long run (an "evolutionarily stable strategy," or ESS) are those in which honesty is ensured by some mechanism. This search for such honest ESSs in animal communication has been a major preoccupation of the field for more than a decade.

The best-known mechanism by which honesty in communication could be ensured was proposed by Zahavi (1975). Although initially repudiated by many researchers (e.g., Maynard Smith 1976), Zahavi's "handicap principle" received support from a mathematical model developed by Grafen (1990) and has now become widely cited as a possible source of honesty in communication. The handicap principle proposes that only heritable signals that bear a high cost (the "handicap"), thereby reducing their bearer's fitness, can be stably honest. Of course, all signals bear some cost (Bradbury and Vehrencamp 1998), so this statement alone has little explanatory value. In Grafen's (1990) analysis, a "handicap" signal of male quality must be more expensive for low-quality males to perform than for high-quality males. Although there has been growing agreement that Zahavian handicaps may play a role in the evolution of honest signaling, a more recent paper by Siller (1998) exposes important flaws in the Grafen (1990) model and again casts some doubt on the basic logic of the handicap principle. In particular, Siller (1998) demonstrates that there is no guarantee of a single ESS under Grafen's model conditions and opens up the possibility of multiple coexisting ESSs. More generally, if there are mechanisms that allow honest communication without a handicap, these less expensive alternatives should be favored over handicap signals, which can impose an arbitrarily high cost on their creators.

Again, the roaring of red deer illustrates the point. Harem-holding males rarely feed during the rutting period and lose up to 20% of their body weight. Toward the end of this exhausting period, male fighting ability drops rapidly, with harem holders tending to tolerate rival males and lose fights with males they had previously beaten. This decline occurs at different times for different individuals but in each case is associated with a drastic reduction in roaring rates (Clutton-Brock and Albon 1979). This suggests that roaring rate is an honest signal of fighting ability because both are dependent on bodily condition and stamina. Clutton-Brock and Albon (1979) suggested that this results from the fact that the same thoracic musculature is used for fighting and roaring but commented that "current knowledge of cervid physiology is inadequate" to evaluate that hypothesis. Unfortunately, this comment is still mostly true twenty years later, although recent work has revealed some interesting adaptations for sound production in red deer (Fitch and Reby 2001; see below). Resolution of such questions demands an understanding of the mechanisms used to produce sound and their physiological and anatomical relationship to other systems. Considerations such as these have led to an increased interest in production mechanisms in recent years and to a growing consensus that our understanding of the evolution of acoustic signaling systems will remain incomplete until the physics and physiology of signal production are better understood (Hauser 1996; Krebs and Davies 1997; Bradbury and Vehrencamp 1998).

In this chapter, we will address the production of acoustic signals from a dynamic evolutionary perspective, paying close attention to the role of physical and phylogenetic constraints on the evolution of acoustic signals and the mechanisms that produce them. This choice of perspective is somewhat atypical and perhaps requires justification. First, why focus on signal-production mechanisms? For researchers in the behavioral ecology tradition, who typically seek ultimate evolutionary explanations for a given pattern of behavior, the proximate mechanisms responsible for a behavior have often appeared irrelevant. However, researchers in evolutionary bioacoustics can expect at least two benefits from a basic knowledge of sound-production mechanisms. First, our ability to conduct research in bioacoustics depends crucially on our ability to analyze animal sounds precisely and in some cases to synthesize them, both of which hinge critically on a solid understanding of the acoustic mechanisms that generate them. Second, an understanding of mechanism offers crucial insights into the adaptive landscape of a communication system: what sounds are easy or impossible to produce? What are the costs and benefits of a given sound type in terms of energetics, predator detection, environmental transmission, and receiver characteristics? What are the possibilities, for a given species, to cheaply produce honest signals or to cheaply mimic honest signals? The data we will review below indicate that the laws of physics, and the structure of sound-producing organs, place strong constraints on the possible evolution of acoustic signals, determining what information is available for perceivers to exploit initially, what deceptive or skeptical mutations can subsequently arise, and the costs and benefits of producing and responding to a given signal. Far from being a wide open field, the adaptive landscape for vertebrate acoustics seems to be characterized by a circumscribed range of biologically relevant and potentially honest signals and an even narrower range of potentially deceptive mechanisms (both morphological and behavioral). Thus, we argue, an understanding of signal-production mechanisms and physical acoustics can provide crucial insights into the evolution of acoustic communication.

Second, why adopt an evolutionary perspective? Researchers interested in the mechanisms underlying vocal production and/or auditory perception might argue that the ultimate evolutionary forces structuring a communication system offer little insight into the proximate morphological and neuronal mechanisms that underlie acoustic behavior. One reason that an evolutionary viewpoint is valuable is that many aspects of animal morphology and behavior may appear nonoptimal from an engineering perspective but can be understood as an optimal solution to a problem, given a certain phylogenetic starting point and well-defined developmental, physical, or mechanistic constraints. No animal has wheels, despite the fact that wheels would be adaptive for large grassland herbivores. The absence of wheels results from developmental and physiological constraints operating over evolutionary time. More prosaically, small body size will forever prevent many species from exploiting infrasonic communication, which might otherwise be optimal in terms of attracting mates but not predators. An evolutionary perspective also encourages an exploration of interspecific similarities and differences, providing a comparative framework to address many mechanistically based questions that have already been answered by experiments of nature. Could an animal with a larynx half the size of its body still breathe and eat? The fruit bat *Hypsignathus monstrosus* shows us that the answer is yes (see Section 2.2.2). What is the relationship between perceptual and production systems in the evolution of acoustic communication? Ryan's (Ryan 1985, 1988; and Ryan and Kime, Chapter 5) and Gerhardt's (1982, 1991) elegant comparative work on anuran bioacoustics has increased our understanding of the coevolution of production and perception mechanisms in frogs as well as anuran evolution more generally. Finally, certain adaptive problems are so persistent and pervasive that they have spawned numerous, independently evolved solutions, and we need to identify these powerful evolutionary forces if we are to understand the broad patterns of diversity seen in animal behavior and morphology. Tracheal elongation in birds, which has independently evolved at least eight times, provides a possible example that will be discussed below.

1.1. Evolutionary Constraints

The notion of restrictions or constraints on evolution has been with us since Darwin (1859, Chapter 13). Although most researchers agree on their importance, detailed analysis and quantification of the role of evolutionary constraints has proved an elusive goal (see Maynard Smith et al. 1985; Carroll 1997), and many different classifications of constraints have been proposed. In this chapter, we will distinguish functional constraints from phylogenetic constraints. Functional constraints result from physical, anatomical, and physiological factors that limit the range of possible forms and functions; for example, physical constraints on evolution stem from properties or laws of the physical world that do not depend on properties of living organisms. For example, hydrodynamic forces have a profound influence on the shape of aquatic organisms. The functions relating pressure drag and friction drag to velocity lead to an ideal length/width ratio of 4.7 for a streamlined body (Hildebrand 1974). Although this ratio is equally valid for a swimming fish and an inert object dragged through the water, it is clearly no accident that streamlined body forms of this aspect ratio have evolved repeatedly in fast-swimming organisms (fish, penguins, dolphins, and ichthyosaurs; Carroll 1997). The study of such physical constraints is an important component of the discipline of functional morphology.

Phylogenetic constraints, on the other hand, result from developmental and historical factors. They stem from the gradualistic principles that underlie the generation of variation in evolution: recombination and mutation can only explore that small portion of adaptive space that is adjacent to a particular species' current position. The fact that insects have six legs is not due to any physical constraint (as the existence of quadrupeds, spiders, crabs, and millipedes clearly demonstrates) but to the strong canalization of the developmental pathways that generate adult insect forms. Although small mutations can have drastic phenotypic effects (e.g., mutations of the *Antennipedia* gene lead to flies with legs in place of their antennae), the chance of such macromutational changes leading to increased fitness is vanishingly small. Another example of a phylogenetic constraint is the genetic code itself: there is no reason in principle for the nucleotide-to-amino acid code to be shared by all life on this planet but, practically speaking, any mutant with a deviant code would be eliminated very early in development. Such constraints are in some sense arbitrary results of the particular evolutionary history of a species but nonetheless create extremely powerful limitations on the viable genotypes available by mutational changes from a given parental lineage.

A major goal of this chapter is to explore the role of physical, physiological, and cognitive constraints in shaping vertebrate communication systems over time. Because many relevant constraints are shared by all vertebrates (and physical constraints are shared by all organisms), sampling a wide range of species may allow us to uncover and understand such constraints, and outline their influence on the shape of adaptive space, with some accuracy. Thus, we will cast a broad net, considering the communication systems of most terrestrial vertebrates at least briefly, although focusing on tetrapods, and birds and mammals in particular (see Ryan and Kime, Chapter 5, for a fuller discussion of anuran communication, Bass and Clark, Chapter 2, for discussion of underwater organisms, and Tyack and Miller, in press, for more on marine mammals). In Section 2, we will outline a set of relatively well-understood constraints that follow from the physics of sound, relating body size to signal frequencies. Using the comparative method, we show how these constraints have led to the evolution of cheap, honest signals in birds and mammals. In Section 3, we turn to perceptual and cognitive mechanisms in communication, showing how perceptual constraints can leave perceivers vulnerable to "sensory exploitation" and how learning mechanisms can allow perceivers to short-circuit such manipulation. We conclude that an understanding of constraints in signal evolution, together with a dynamic view of evolution, can provide deeper insights into animal communication systems. We hope to show that, when integrated with the comparative evolutionary perspective advocated above, the study of physical and phylogenetic constraints can provide rich insights into the evolution of acoustic communication systems.

1.2. Distinctions and Definitions

There are many important distinctions to be made in discussions of the evolution of honest signals. Here, we will focus on three: arbitrary versus direct signals, lies of omission versus commission, and external- versus self-reference. These distinctions do not represent hard and fast categories that exist in nature but rather lay out a continuum on which any given signal can be located. A signal might be intermediate or extreme along any of these continua. Thus, these distinctions must be seen as aids to communication and understanding and not claims about dichotomies in the world.

1.2.1. Arbitrary and Direct Signals

One continuum along which we can analyze acoustic signals concerns the degree to which the mapping between signal and referent is arbitrary. In human language, word meaning is generally not tied in any direct way to the acoustic structure of that sound. Thus, except for a small number of onomatopoetic or sound-symbolic words, one cannot guess the meaning of a word in a foreign language based only on its sound. It is often assumed that the information in animal calls is similarly arbitrary. There is, however, significant evidence that at least some of the information in animal calls is "direct," with a nonarbitrary mapping between sound and meaning. The most obvious example is the association between low frequency and large size. Because only large objects can produce and radiate low-frequency signals effectively, there is a natural mapping between size and frequency, particularly large body size and low-frequency calls (e.g., elephants and infrasound). We will argue that physical facts such as the size/frequency relationship have played a pervasive role in the evolution of communication systems and that a solid understanding of production mechanisms is necessary to uncover and understand such relationships. In contrast, the difference between the vervet monkey's leopard and snake alarm calls probably has little or nothing to do with the sounds themselves (Cheney and Seyfarth 1990). Rather, these calls are more like human words, providing an example of an arbitrary sound/meaning pairing that has been conventionalized over evolutionary time.

1.2.2. External and Internal Reference

Because "reference" in humans is typically construed to involve consciousness and intentionality, researchers wishing to remain agnostic about these issues in animals often adopt the term "functionally referential" when discussing this issue (Marler et al. 1992; Hauser 1996). We will use the word "refer" here only in a limited sense of causal correlation between an object or event and a corresponding animal signal. Thus, a bee's waggle dance "refers" to flowers in the sense that it enables other bees to locate and feed from the flowers despite a presumed lack of self-awareness or high-order intentionality on the part of the dancing bee. Most traditional work in animal communication has concentrated on signals that indicate the signaler's species, individual characteristics or motivational state, and we will consider many such internally referential signals in the first part of this chapter. Recently, there has been much interest in signals that seem to refer to the world outside the signaler (Seyfarth et al. 1980a, 1980b; Gyger et al. 1987; Hauser and Marler 1993a, 1993b). The classic example is alarm calling in vervet monkeys, where different calls produced by signalers viewing eagle, snake, or leopard predators elicit appropriate escape reactions from listeners (Struhsaker 1967, see Section 3.2). These calls are externally referential in that, regardless of the state of the caller, listeners respond to them as if they indicated something about the world outside of the caller (playback experiments that simulate a predator encounter provide further evidence that the calls alone are sufficient to elicit appropriate responses from listeners). We will consider such signals in some detail in the second half of this chapter.

1.2.3. Lies of Omission and Commission

In their recent textbook on animal communication, Bradbury and Vehrencamp (1998) define deceit as "the provision of inaccurate information by a sender to a receiver." However, it is often possible to deceive by failing to provide information (e.g., failing to give food calls upon discovering food) (Hauser 1992b; Hauser and Marler 1993a). We thus distinguish between lies of commission (the provision of inaccurate or misleading information) and lies of omission (withholding expected information) (Hauser 1997). There are good empirical data concerning both types of "deceit" in animal communication (see Section 3). It may be quite difficult to locate and identify individuals who withhold signals, which may make the evolution of lies of omission easier. The evolution of receiver skepticism is not as common as might be expected. In particular, receivers who are able to learn about signalers through repeated interactions should quickly be able to learn to identify signalers who "cry wolf" because calls typically carry information about signaler identity along with any external referentiality. Our review of the literature will suggest that the development of such facultative skepticism is surprisingly uncommon, with clear evidence only in a few primate species. This suggests either that the cost of being deceived may be very low or that perceptual or cognitive constraints may limit the ability of perceivers to become effective skeptics.

2. Physical and Anatomical Constraints on Signal Production: The Physics of Honesty

This section is focused on the mechanisms involved in vertebrate vocal production and is divided into three parts. First, we briefly review the basic acoustics of vocal production, showing how a few easily understood physical principles combined with anatomy and physiology can have profound consequences on the sounds a particular animal is able to produce. Then, we examine the morphology of vertebrate sound-production systems in greater detail, surveying the vast and mostly unexplored diversity of vocal tract anatomy in terrestrial vertebrates. Finally, we attempt to make some sense of this diversity, describing how evolutionary constraints can act in some cases to enforce honest communication in the absence of any specific selection for honesty. We also describe how novel morphological or physiological mechanisms can allow the evasion of certain anatomical and physiological constraints, suggesting that many morphological oddities are best understood as constraint-evasion mechanisms. For example, the "key innovation" of the syrinx has opened the door to vocal adaptations in birds that are inaccessible to other tetrapods due to their reliance on the larynx as a sound-producing source.

2.1. Vertebrate Vocal Production: Anatomy and Acoustics

Our intent in this section is to survey vocal production in tetrapods. With 9,000 species of birds, 6,000 reptiles, 4,500 amphibians, and 4,000 mammals, we could not hope to be exhaustive. Due to space limitations and the significant differences between sound production and transmission in air and water, we will have little to say about fish or cetaceans (see Bass and Clark, Chapter 2). Furthermore, available research on terrestrial vertebrate vocal production is unevenly distributed: the best-researched groups are oscine birds (Nowicki and Marler 1988; Gaunt and Nowicki 1998), anurans (see Ryan and Kime, Chapter 5; see also Schneider 1988), and, among mammals, echolocating bats (Suthers and Fattu 1973; Suthers 1988) and humans (Fant 1960; Lieberman and Blumstein 1988; Titze 1994). Much less is known about reptile vocal production (see Gans and Maderson 1973), and little or nothing is known about vocal production in most nonpasserine birds and most mammalian orders. Significant unresolved questions remain about production in virtually all vertebrate groups other than for humans. Even in our own species, vocalizations other than speech and singing are little studied. However, because research in speech science also has provided the basic concepts and analysis tools for the rest of bioacoustics, we follow tradition (e.g., Greenewalt 1968; Lieberman 1968; Hartley and Suthers 1988) in using data from human vocal production to ground our discussion of call production in other vertebrates. Where appropriate, we point out the many significant differences between human speech and animal calls and call attention to the importance of developing new methodological and theoretical tools to cope with these differences.

2.1.1. Generating Power: The Lungs

Voice production is the conversion of air flow into acoustic energy (that is, longitudinal pressure waves in the audible frequency range). Typically, this air flow emanates from the lungs. Tetrapod lungs are filled during normal respiration by various means, including diaphragmatic contraction in mammals, buccal pumping in some reptiles and amphibians, intercostal contraction in birds, and even a piston-like retraction of the liver in crocodilians (Liem 1985). Due to the lung's elastic recoil resulting from alveolar elasticity and surface tension, optionally combined with muscular compression from intercostal or abdominal muscles, this air can be pressurized, resulting in a flow outward through the glottis. It is this air flow that typically provides the energy for vocalization, either directly or indirectly by filling air sacs. The diversity of systems for moving air in and out of the lungs is of less relevance in understanding vocal diversity than diversity in the vertebrate voice source. Broad comparative treatments of diversity and function in the vertebrate respiratory system can be found in Gans (1970), Liem (1985), or Perry (1989), or Lasiewski (1972) for birds.

Although most tetrapod vocalizations (e.g., human speech) are apparently generated upon expiration, inspiration also plays an important role in vocalization in some anurans, mammals, and birds. In anuran advertisement calls, for example, air typically flows outward from the lungs into a distensible submandibular air sac, which can inflate in some cases to the size of the animal itself (Dudley and Rand 1991). It is this outward flow that fuels vocal cord vibration and vocalization. The air captured in the sac is then returned, via deflation, to the lungs, where it can then fuel another vocalization (Gans 1973). This conservation of air serves at least two functions, which are discussed in more detail below. First, it enables anurans to vocalize at higher rates and for longer than would otherwise be possible (due to the inefficiency of lung inflation in this group; Rand and Dudley 1993). The relevance of call duration to both energy expenditure and to female choice has been documented in anurans (Klump and Gerhardt 1987; Welch et al. 1998; see also Ryan and Kime, Chapter 5) and provides a good example of a nonarbitrary signal parameter. Second, the inflated air sac may serve as an impedance-matching device, more efficiently transferring acoustic energy to the environment (Watkins et al. 1970). A similar mechanism may operate in nonhuman primates with distensible air sacs (Gautier 1971).

Nonhuman primates such as chimpanzees (*Pan troglodytes* and *P. paniscus*) produce vocalizations, such as low hoots and pant hoots, that have both inspiratory and expiratory components (Marler 1969; Marler and Tenaza 1977; de Waal 1988), as do human infant cries (Truby and Lind 1965; Wolff 1969). In birds, which have an extremely complex and efficient respiratory system, the respiratory dynamics underlying vocalization appear to be equally complex. Many songbirds appear to produce shallow, rapid respiratory cycles called "mini-breaths," which allow extended periods of unbroken song and suggest a level of respiratory/vocal coordination far superior to that seen in other tetrapods (Calder 1970; Hartley and Suthers 1989).

2.1.2. The Voice Source

In terrestrial vertebrates, vocalizations are initially generated by a structure that converts air flow from the lungs (or air sacs) into acoustic energy. This structure is known as the acoustic source, or voice source, and its anatomical location varies among tetrapods (Fig. 3.1). In amphibians, reptiles, and

FIGURE 3.1. Representative tetrapods showing (in gray) the different anatomical sound sources in each group (schematic). Left: Anuran amphibians use a larynx with vocal folds to produce sound and often vocalize into an inflatable vocal sac. Middle: Birds (a passerine is shown here) have an evolutionarily novel structure, the syrinx, which is located at the base of the trachea. Right: Mammals (a rodent is shown) use a larynx and vocal folds as the sound source.

mammals, the source is typically the larynx. In birds, an evolutionarily novel structure called the syrinx serves as the voice source. In both cases, the source contains mobile elastic structures that act as mechanical vibrators and can reduce or stop the passage of air through the source by constricting its lumen. In the larynx, these vibrators are the vocal folds, sometimes called vocal cords. In the syrinx, the identity of the vibrators was long thought to be the medial tympaniform membranes (Miskimen 1951; Greenewalt 1968; Gaunt and Gaunt 1985). However, more recent direct visualization via endoscopy suggests that the vibratory structures are the syringeal labia in passerines (Goller and Larsen 1997b) and the lateral tympaniform membranes in pigeons (Goller and Larsen 1997a). Although the medial tympaniform membranes may play some acoustic role, their complete ablation does not prevent nearly normal vocalization (Larsen and Goller 1999). In both the larynx and syrinx, energy created by the passage of air through the constriction between the vibrators acts to set them into motion. When the vibrators collide (or approach close enough to modulate air flow), they generate acoustic energy. The main acoustic difference between the larynx and syrinx is their location: the larynx is located at the top of the trachea, whereas the syrinx is located at its base. Although birds also possess a larynx, there is little evidence that the avian larynx is used as a sound-producing source (see White 1968 for a possible exception).

There is a long history of scientific attempts to understand the functioning of the human voice source, starting with Johannes Müller's pioneering work with excised larynges (Müller 1848). The further work of van den Berg (1958, 1968) and Titze and colleagues (Titze 1994) has deepened this understanding, and we now have a detailed and accurate model of vocal production at the voice source, which is called the "myoelastic–aerodynamic theory." The vocal folds act as mechanical vibrators, with their own elasticity and mass, that are coupled to the aerodynamic flow from the lungs to generate self-oscillation. Before oscillation can begin, the vocal folds must be placed in an appropriate "phonatory" position, closer to one another than during normal breathing. Once this position is attained, air flow from the lungs can set up sustained oscillations. The critical factor is that energy must be pumped into the system in phase with the vocal fold oscillation: there must be a greater pressure pushing the folds apart as they are opening than while they are closing. In human speech, this is made possible by a change in the vertical geometry of the folds, with a divergent glottis during closing leading to less force than the convergent glottis during opening. Although it is frequently stated that the Bernoulli force is adequate to maintain oscillation, this alone is not adequate because it is equally strong during opening and closing (Titze 1976, 1980). Similar considerations must apply in all tetrapod sound sources; Titze (1994) is recommended as an excellent introduction to the topic.

The quantitative details of laryngeal vibration are still an area of active investigation, even for humans. Investigations of nonhuman larynges usually stem from the difficulty of obtaining or working with human cadaveric larynges rather than from an interest in comparative physiology (Brown and Cannitto 1995 and Mergell et al. 1999 are exceptions). Nonetheless, over the years, the larynges of a wide variety of mammalian species have been investigated, and all available data are consistent with the hypothesis that the vibratory mechanics of the larynx are fundamentally similar among all mammals, including humans. Nonhuman mammals whose larynges have been experimentally studied include baboons, sheep, dogs, horses, cattle, and Syke's monkeys (Slavitt et al. 1990; Hirano 1991; Brown and Cannitto 1995; Bless et al. unpublished). In all cases, the vocal folds act as vibrators, and the myoelastic–aerodynamic theory applies. Although current evidence suggests that similar considerations apply to the avian syrinx, it is only very recently that *in situ* vibrations of the syrinx have been directly observed (Larsen and Goller 1999), and the basic mechanisms underlying avian sound production are still the subject of active investigation. Because the vocal folds are solid masses of tissue, whereas the syringeal membranes are relatively thin and light, it is quite likely that the details of their vibratory patterns will differ in some ways. Nonetheless, current data for all tetrapods are consistent with the idea that the voice source involves movement of a set of vibrators (vocal folds or syringeal membranes) that modulate air flow and thus generate acoustic energy.

Often the oscillation of the vibrators is periodic, with their opening and closing occurring regularly. The time it takes for one open/close cycle is the period, and the rate at which these cycles occur is the fundamental frequency (abbreviated f_0). A fundamental insight of the myoelastic–aerodynamic theory is the realization that this rate of opening and closing is determined passively by the setting of muscle tensions, effective mass of the vibrators, and lung pressure. It is unnecessary for any muscles in the vocal folds to twitch, or the motor neurons to fire, at the fundamental frequency. Indeed, for the sounds of many vertebrates, this would be impossible because the fundamental frequencies are much higher than the maximum rate of muscular tetany or even neural firing. Only relatively low vibration rates can typically be generated by rapid muscle twitching in tetrapods (e.g., 25-Hz purring in cats; Remmers and Gautier 1972). The neurally passive feature of the larynx or syrinx is best demonstrated by the fact that a larynx or syrinx can be removed from the body and deprived of all nervous input and still be induced to produce sound by blowing air through the approximated vocal folds or syringeal membranes (Rüppell 1933; Schmidt 1965; van den Berg 1968; see also Yamaguchi and Kelley, Chapter 6). All that is necessary is to place the vibrators in an appropriate state of tension and approximation, with the proper air flow, and the system will vibrate with a motion and at rates closely approximating those seen *in vivo*. The passive frequency control of the voice source of terrestrial vertebrates contrasts sharply with frequency control of the swim bladder production system found in most fish, where each pulse of acoustic energy is produced by a muscular contraction (Demski et al. 1973; Bass 1989; Bass and Baker 1997). Such a system puts a clear physiological constraint on the highest producible fundamental frequency (although some of the fastest-twitching muscles in the animal kingdom are found in swim bladder muscles; Tavolga 1964). Only a few teleosts produce sound via expulsion of gas from the swim bladder in a manner analogous with vocalization in tetrapods (e.g., minnows through the pneumatic duct, loaches through the anus; Demski et al. 1973).

The oscillations of the source are never perfectly periodic. Even a nearly constant fundamental frequency has small perturbations around the mean frequency, which are called jitter by speech scientists (Lieberman 1961; Baken 1987), and most vocalizations involve large, nonrandom changes in fundamental frequency over time. Nonetheless, to a good approximation, much of normal phonation can be idealized as periodic. In human speech, such quasiperiodic phonation accounts for the vast majority of voiced sounds and can be understood using standard linear systems theory and Fourier analysis. However, in the past decade it has become increasingly clear that certain types of broadband vocal phenomena are the result of deterministic chaos in the (nonlinear) dynamics of the vocal source (see Herzel 1996 for an introduction). In pathological cases in adult human speech (Herzel and Wendler 1991), in human infant crying (Mende, Herzel, and Wermke 1990), and in a variety of animal sounds (Wilden et al. 1998), nonlinearities in the vocal-production mechanism can play an important role in structuring the acoustic morphology of calls. In these cases, quasiperiodic phonation is replaced by one or more of a variety of irregular or aperiodic phenomena, including period doubling, biphonation (the presence of two independent frequencies), and deterministic chaos (Fig. 3.2). Although research into nonlinear phonation is still in its infancy, it appears likely that such vocalizations play an important role in the communication systems of many species (see also Tyack and Miller, in press).

The most common example of irregularities in the voice source is provided by calls such as those shown in Figure 3.2, which shows calls from a normal rhesus macaque. Although the last call is "noisy" in the sense that it lacks clear periodicity, it has too much spectral structure to be caused simply by turbulent noise generated by vocal tract constrictions. Instead, such calls appear to be generated by irregular opening and closing of the glottis due to strong nonlinearities in the dynamics of the vocal folds; they are examples of deterministic chaos in glottal dynamics. This hypothesis has been confirmed in the human voice by direct observation of the vocal folds with high-speed video endoscopy (Tigges et al. 1997) and is supported in the case of mammalian vocalization by abundant evidence summarized in Wilden et al. (1998). Other types of phenomena, such as period-doubling bifurcations shown in the second call of Figure 3.2, lend further credence to this hypothesis. Despite the ubiquity of such calls, they have received little attention in the bioacoustics literature, perhaps because the appro-

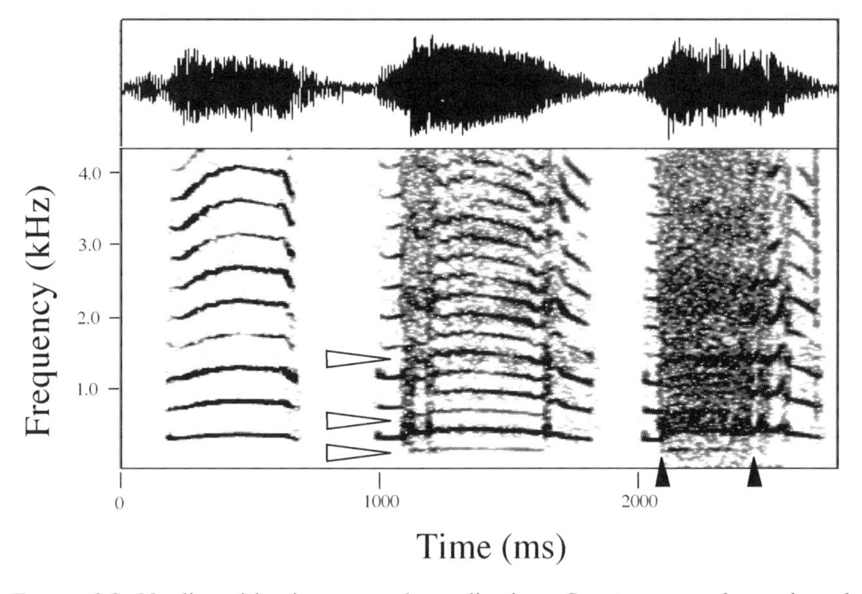

FIGURE 3.2. Nonlinearities in mammal vocalizations. Spectrogram of a series of three consecutive calls by a normal adult rhesus macaque female. The individual was approached by a dominant male during this call series. The first call is a prototypical "coo" call, and the subsequent calls show intrusions of nonlinear phenomena. Call 2: after normal onset, subharmonics (indicated with light arrows) intrude. Call 3: after normal onset, deterministic chaos appears (indicated with heavy arrows).

priate conceptual framework to understand their production is quite new and has until recently been confined mostly to physics journals. Fortunately, however, one of the ideal tools to analyze such calls is the familiar narrow-band spectrographic representation. Consequently, the next few years will bring a much more detailed understanding of both the acoustic production of nonlinearities in animal vocalizations and their behavioral and evolutionary significance.

In addition to the larynx or syrinx, there are other possible sources of acoustic energy available in all terrestrial vertebrates. Given adequately high flow, a narrow constriction anywhere along the path from lungs to lips or nostrils can produce turbulent noise (as in human whispers or "s" sounds, or snake hisses), thus providing a set of other possible sources of broad-band noise. Such a turbulent source can operate alone or simultaneously with the laryngeal or syringeal source. For example, the English sound "f" is produced by a turbulent noise source alone, generated at a constriction between the teeth and lips. In contrast, the "v" sound is created by phonating simultaneously with "f" and thus is a dual-source sound. Both non-laryngeal sources and dual-source sounds are common in human speech, forming a significant portion of the consonantal repertoire. Much less is known about the use of turbulence in animal communication, although hissing is obviously a widespread type of vocalization among tetrapods. Examples include llamas, cats, viverrids, and sloths among mammals, many snakes, turtles, and crocodilians among reptiles, and oxpeckers (genus *Buphagus*), vultures, geese, swans, and ostriches among birds.

To give a concrete example of the value of mechanistic understanding in addressing evolutionary questions, we will briefly consider the role of fluctuating asymmetry in mate choice. Fluctuating asymmetry (FA), individual deviations from physical symmetry that are hypothesized to provide an indication of developmental stability (see Møller and Thornhill 1998 for a recent review, and Houle 1998 for a critique), has been shown to play an important role in mate choice in an impressive variety of species. FA is an intrinsically unfakeable cue: the simple fact of bilateral symmetry means that most animals have paired structures, and small differences in the developmental environment of these structures can potentially have perceptible effects on the adult that can be used to evaluate FA and provide an indication of developmental stability. The use of FA in mate choice has been documented in a great diversity of species, but in all of these examples visual cues were used to evaluate FA. However, because asymmetries in the vocal folds exist (Hirano et al. 1989) and can have a perceptible effect on the vocal signals (Isshiki et al. 1977; Steinecke and Herzel 1995; Tigges et al. 1997), it is plausible that mistuned vocal folds could provide an acoustic indicator of FA. Similarly, the different lengths of bronchi in oilbirds give acoustic cues to vocal tract asymmetries in the species (Suthers 1994). Such acoustic indicators of FA could theoretically play a role in mate choice in addition to, or instead of, the well-known visual indicators. Such a supposition could be tested via tests of animals that vary naturally in FA, with calls synthesized with vocal tract models possessing varying degrees of asymmetry, or via experimental manipulation of vocal fold asymmetry (e.g., via unilateral vocal fold injections).

2.1.3. The Vocal Tract

The acoustic energy generated at the source must pass through the remainder of the respiratory tract before it can emanate out into the environment. In birds, this portion of the respiratory system is called the suprasyringeal vocal tract, whereas in other terrestrial vertebrates it is the supralaryngeal vocal tract. Although the entire vocal-production system, including lungs, source, and supralaryngeal respiratory passages, is sometimes called the vocal tract, it is convenient when discussing tetrapod vocal acoustics to restrict use of the term "vocal tract" to the suprasyringeal or supralaryngeal air passages and their associated articulators, using the term "vocal-production system" to refer to the entire system. As a broad generalization, there is much more diversity in vocal tract morphology than in the voice source: a large anatomical literature records a huge variety of air sacs, diverticula, elongated snouts or trachea, or other resonating structures among vertebrates (see next section). However, most of these papers are old and many are in German, and these morphological features have received little attention from modern bioacousticians. Furthermore, the functional importance of these features has received almost no study despite the fact that such variation may play a significant role in shaping the communication systems of different species.

The column of air contained in the vocal tract, like any column of air, has elasticity and mass and thus will vibrate preferentially at certain frequencies, called normal modes or "resonances." As the sound energy generated by the source passes through this air column, it may set one or more of these modes into vibration. The presence of the vocal tract will thus enhance the transmission of these frequencies while damping or attenuating others; it acts as a spectral filter on the source signal. In speech science, these filtering frequencies are called "formants," from the Latin *formare*, meaning "to shape," because they sculpt the vocal signal on its way from the source out to the environment. This term is preferable to the term "vocal tract resonances" both due to its brevity and because it highlights the independence of source and filter, which is indicated by most available work on the subject (see Section 2.1.4). Thus, the most basic acoustic model of the vocal-production system has two components: the sound-generating source (e.g., syrinx, larynx) and the filter (the air column contained by the vocal tract). The function of the filter varies among species. Human speech uses changes in formant frequencies to code meaning directly: formants and their movements are the most important acoustic cue in speech, which typically provides external reference. In both humans and other species, formants can also be internally referential, providing cues to identity, body size, age, or sex (Rendall 1996; Fitch 1997; Fitch and Giedd 1999; Riede and Fitch 1999). A different use of the vocal tract filter is to suppress certain frequencies, typically to enhance the salience of some particular source components (e.g., the second harmonic). This is the case in some birds (Nowicki 1987) and bats (Hartley and Suthers 1988). In all cases, it is extremely important to recognize that formants are an acoustic entity independent from the source (the fundamental frequency and its harmonics) in terms of production, acoustic analysis, and perception. Formants can vary independently of the source, and formants have little or no influence on pitch perception (which is determined by source characteristics), at least in humans (Lieberman and Blumstein 1988; Titze 1994).

All terrestrial vertebrates possess a vocal tract that can be predicted from basic physics to have a substantial acoustic effect on production of many call types. In both birds and mammals, the evidence for formant frequencies is abundant based on even a cursory examination of spectrograms. Despite this, there has been little attention to formants, or research on the anatomy, physiology, or acoustics of the nonhuman vocal tract, compared with research on the larynx or syrinx. In humans, the vocal tract plays a far more critical role in speech than does the larynx, and thus we have a detailed understanding of the anatomy and physiology of the human vocal tract and accurate quantitative models of its acoustics at rest and in motion. Thus, compared to our knowledge of the human vocal tract, research on animal vocal tract acoustics and dynamics is in its infancy, and information on comparative anatomy of the vocal tract is scattered throughout works focusing on digestion or respiration. Even the most basic questions have been addressed for only a few species. However, despite a long pause in publications since the late 1960s (Greenewalt 1968; Lieberman 1968; Lieberman et al. 1969), there appears to be a growing realization of the importance of the vocal tract in sound production in birds and mammals, especially in the last decade (Suthers and Fattu 1973; Andrew 1976; Nowicki 1987; Hartley and Suthers 1988; Suthers and Hector 1988; Suthers et al. 1988; Owren and Bernacki 1988, 1998; Owren 1990; Hausberger et al. 1991; Hauser 1992; Hauser et al. 1993; Westneat et al. 1993; Hauser and Schön-Ybarra 1994; Fitch 1994, 1997, 1999, 2000b, 2000c; Fitch and Hauser 1995; Rendall 1996; Owren et al. 1997; Riede and Fitch 1999; Fitch and Reby 2001).

For anurans, the role of supralaryngeal filtering is more difficult to assess, at least in part because of the pervasive use of the term "dominant frequency." In anuran bioacoustics, dominant frequency refers to the highest-amplitude frequency in the spectrum of a call, without regard to whether this is the fundamental frequency, one of its harmonics, a noise- or impulse-excited formant, or a carrier frequency with amplitude-modulation side-bands. Although this term is convenient for acoustic analyses, it obscures the important differences among such acoustic features, both in terms of understanding vocal production and possibly in perception as well. For example, many anuran vocalizations possess features that superficially resemble formants, with a high-amplitude peak at one of the higher harmonics of a series. However, the data of Rand and Dudley (1993) suggest that, at least for the four species they examined, this peak does not represent a formant frequency because the location of the highest-amplitude spectral peak did not change in a helium/oxygen atmosphere (see Section 2.1.4). Such spectral peaks could be caused by low-frequency amplitude modulation (e.g., by the arytenoids) of a higher carrier frequency (e.g., from the vocal cords), as suggested by Schneider (1988) and Ryan (1985). Alternatively, they could result from an interaction between a generalized descending spectral envelope (i.e., the –6 dB/octave amplitude drop-off characteristic of most vocal sources) and impedance characteristics of the frog's body (where low frequencies are radiated poorly due to small body sizes; see Section 2.2.5 below). Our point here is that the abundant and excellent work in anuran bioacoustics could be more easily integrated into the rest of bioacoustics (including work on humans, other mammals, and birds) if explicit production-related terminology were adopted (e.g., separating formants from fundamental frequency or harmonics) rather than relying on the catch-all acoustic term "dominant frequency." Increased precision of acoustic description will enable researchers interested in the anatomy and physiology of sound production to more easily pin down the mechanisms relevant in perception and communication, and thus in the evolution of mate choice or speciation, in this important group of vertebrates.

Another term that is confusing because it is used ambiguously is "pitch." Pitch is defined as "that attribute of auditory sensation in terms of which sounds may be ordered on a musical scale." Pitch is a subjective quality, defined in human terms, that cannot be measured directly. It is, strictly speaking, inappropriate to use the term in animal bioacoustics. However, the term is convenient and its usage widespread, making it unlikely to vanish from the technical literature. Thus, it is critical that bioacousticians use the term consistently and precisely. For most periodic sounds, perceived pitch corresponds to the physical variable fundamental frequency (or its inverse, waveform period). Exceptions include periodic sounds that lack energy at the fundamental frequency, so-called "missing fundamental" stimuli. A sound with energy only at 200, 300, and 400 Hz will often have a perceived pitch corresponding to a sine wave at 100 Hz, despite the lack of physical energy at this frequency, due to perceptual processes that "restore" the missing fundamental frequency. Although such phenomena may be relevant in calls produced by birds or bats, where the fundamental frequency is suppressed (Hartley and Suthers 1988; Nowicki and Marler 1988), in general there is a close correspondence between "pitch" and fundamental frequency. Thus, the use of "pitch" to refer to other acoustic parameters, such as voice timbre or vocal tract resonances (e.g., Hausberger et al. 1991), is to be discouraged.

2.1.4. Independence of Source and Filter

There is a superficial similarity between vertebrate vocal-production systems and wind instruments such as the clarinet or trumpet, where the reed or lips play the role of the source and the column of air contained by the body of the instrument is analogous to the vocal tract. However, there is an important difference in the physics of wind instruments and the vocal tract. In wind instruments, the vibrating frequency of the source is largely determined by the resonant frequencies of the instrument's "vocal tract"; that is, by the air column contained by the body. The instrumentalist manipulates the pitch of the instrument by changing the length of this air column and thus the characteristic frequencies of the column's vibratory modes. In this case, it is appropriate to call the modes "resonances" because the source vibrates in resonance with (at the same frequency as) the air column.

In contrast, the vibratory frequencies of the source and filter appear to be independent in vertebrates, an independence that makes vocal acoustics fundamentally different from the acoustics of wind instruments. There is little evidence for anything but weak coupling either in the human voice or in other vertebrates that have been studied thus far. Thus, to a first approximation, the frequencies produced by the vocal source (typically a fundamental and its harmonics) are independent of the filtering frequencies of the vocal tract (Miller 1934; Sutherland and McChesney 1965; Hersch 1966; Pye 1967; Greenewalt 1968; Capranica and Moffat 1983; Gaunt et al. 1987; Nowicki 1987; Hartley and Suthers 1988; Rand and Dudley 1993; Westneat et al. 1993; Brittan-Powell et al. 1997).

The best evidence for source/tract independence comes from experiments with animals vocalizing in light gases. Typically, researchers have used heliox, a mixture of helium and oxygen with nearly double the speed of sound in air. Because formants are dependent on the transit time of sound waves up and down the vocal tract, raising the speed of sound shortens transit time and thus nearly doubles formant frequencies. In a coupled system such as a wind instrument, doubling the air column resonances also doubles the fundamental frequency at which the source vibrates. However, in the human voice (e.g., Beil 1962) and in those animals tested (birds: Hersch 1966; Gaunt et al. 1987; Nowicki 1987; Brittan-Powell et al. 1997; anurans: Capranica and Moffat 1983; Rand and Dudley 1993; bats: Pye 1967; Hartley and Suthers 1988), the fundamental frequency does not shift appreciably in heliox. Where formant frequencies are present (mammals and birds), they shift upward. In humans, this leads to the peculiar "Donald Duck" quality of helium speech, with a normal, low fundamental frequency and high formants (Beil 1962). In the case of birds and bats, the formant shift often "unmasks" harmonics that are present in the source signal but are normally filtered out as the signal passes through the vocal tract (Pye 1967; Nowicki 1987; Hartley and Suthers 1988; Nowicki and Marler 1988). The only case of which we are aware where the perceptual relevance of heliox-shifted vocalizations has been examined is the work by Strote and Nowicki (1996), who found in a two-speaker choice experiment that song sparrows respond slightly more strongly to normal calls than to helium-shifted calls. There appears to be no consistent, significant effect of helium on vocalizations in the anuran species tested to date (Rand and Dudley 1993).

Despite the consistency of these heliox data in these species, there is little information relevant to source/tract independence for the vast majority of tetrapod species. Less direct analyses suggest independence simply because the relatively short vocal tract of nonavian tetrapods would result in formant frequencies that are high relative to the fundamental frequency in most mammals and anurans. Thus, in macaque and baboon grunts (Andrew 1976; Rendall 1996; Owren et al. 1997) and dog growls (Riede and Fitch 1999), the fundamental frequency falls far below that of the lowest formant. Although Bauer (1987) found a correlation between fundamental frequency and mouth opening in an adult male chimpanzee, there was no indication that the change in fundamental frequency was causally related to changes in formants due to mouth opening, and other time-synched analyses of formant changes with mouth opening indicate that no causal connection is likely (Hauser et al. 1993; see Fitch and Hauser 1995).

Thus, there is a significant body of data indicating independence of source and filter in many vertebrate species. Nonetheless, independence is best considered a working hypothesis at present, given our limited knowledge of animal vocal production. Although source/filter theory is well-tested and well-accepted in speech acoustics, independence of source and filter is only a first-order approximation even in speech, and some interactions between the two do occur (Bickley and Stevens 1986; Mergell and Herzel 1997). Early *in vitro* work in birds provided some evidence of strong source/tract coupling. For example, Rüppel (1933) found that the vibratory frequency of an excised crane syrinx was dependent on the length of the vocal tract attached suprasyringeally. Furthermore, a more recent paper looking at *in vivo* production showed no heliox effect on intact birds but a profound effect on a budgerigar with a denervated syrinx (Brittan-Powell et al. 1997). These data suggest that source and filter may be passively coupled in this species but that the bird normally overrides this coupling via active control of the syrinx. This hypothesis would explain both the *in vitro* results of Rüppel (1933) and the lack of evidence for source/tract coupling in the vast majority of more recent studies. Another possible type of source/tract coupling, suggested by Hartley and Suthers (1988) for an echolocating bat species, is that energy propagating back from the trachea could provide positive feedback to support high-amplitude phonation, which is critical to receiving a sufficiently strong echo from their echolocation cries. Despite its plausibility, this hypothesis remains speculative at present. Finally, there are many bird and mammal species for which the fundamental frequency is close to the predicted formant frequencies based on vocal tract length, suggesting the possibility for source/tract coupling, but whose production has not been experimentally examined.

In conclusion, the accumulated data for terrestrial vertebrates, direct and indirect, suggest that independence of source and filter should be assumed as the working hypothesis of researchers in vertebrate bioacoustics as it is in human speech. Specific data (e.g., derived from vocalizations in heliox) would have to be adduced before rejecting this hypothesis and positing source/tract coupling. This is worth stressing because many physicists and bioacousticians, particularly in the older literature, adopt wind instruments, and therefore coupling between source and filter, as their default model of acoustic production. All current data suggest that the wind instrument analogy is dangerously misleading as a model of vocal production. Moreover, independence of source and filter has an important practical consequence for researchers interested in studying call perception: using well-developed techniques from speech science such as linear prediction (LPC) or cepstral modeling, it is possible to pull a signal apart into source and filter components and independently modify one specific parameter of interest (Markel and Gray 1976; see Owren and Bernacki 1998 for a bioacoustically oriented review). Such analysis/synthesis techniques provide an extremely powerful way to isolate the relevant acoustic parameters in bioacoustic communication systems; these are only starting to be explored (Owren and Bernacki 1988, 1998; Fitch and Kelley 2000).

2.1.5. Vertebrate Vocal Production: Summary

The vocal-production apparatus of terrestrial vertebrates is made up of two components. The source—typically the syrinx in birds and the larynx in mammals, reptiles, and amphibians—creates the sound. When this source-generated signal is periodic, its fundamental frequency (and in some cases the spacing between harmonics) determines the perceived pitch of the signal. The term "pitch" should be used with care and only to refer to source-related acoustic parameters (e.g., the fundamental frequency). This source-generated energy then passes through and is filtered by the vocal tract, which includes the mouth, nasal cavities, and pharynx in all vertebrates, and additionally the trachea in birds. The filtering frequencies of the vocal tract are called formants. There is no colloquial term to refer to their perceptual correlate of formant frequencies, which is nonetheless perceptually salient to humans and to all vertebrates that have been tested. Broadly speaking, formants are one correlate of the percept of "timbre" in animal sounds; in no case are formants correlates of pitch. The frequencies of the source signal appear, in normal situations, to be completely independent of formant frequencies. This is in contrast to the situation in wind instruments, where source and filter are strongly coupled.

2.2. Morphological Diversity in the Vertebrate Vocal-Production System

Although all of the vocal-production systems considered here work in roughly the same fashion and are governed by the same physical principles described above, there is an impressive diversity of form in the vertebrate vocal tract. Below, we will provide a selective overview of this morphological diversity and of some of the hypotheses that have been put forward to account for this variation.

2.2.1. Dual Sources: The Two-Voice Phenomenon in Birds

An important distinction between the syrinx and the larynx is that the typical syrinx contains two independent sets of vibrating membranes, one in each bronchus, which are also under independent nervous control (via left and right branches of the twelfth cranial nerve, the hypoglossal). Based on this anatomy, Greenewalt (1968) proposed the "two-voice" theory, which holds that the two sides of the syrinx in many birds are independent, allowing two independent fundamental frequencies to be produced by one bird. Greenewalt based his theory on observation of spectrograms, but more direct evidence was provided by Nottebohm (1971), who sectioned the right or left hypoglossal nerves in several songbird species. He found that disabling the right nerve had little effect on canary or chaffinch song, whereas sectioning the left nerve produced dramatic effects, with most syllables disappearing entirely from the song. Thus, these birds are lateralized for song production, with one side being dominant. Such asymmetries have also been discovered in other species, but in some species, such as zebra finches, the asymmetry is reversed (Williams et al. 1992). Conclusive evidence for the two-voice theory was provided by Suthers' (1990) elegant experiments with mimic thrushes, which produce elaborate songs including imitations of other species' songs. By implanting pressure and flow sensors in living birds, Suthers was able to definitively observe the two voice sources creating independent portions of the awake, singing birds' final song. Interestingly, mimic thrushes (at least the catbirds and thrashers that Suthers studied) appear to utilize both sides of the syrinx relatively equally, although each side appears to habitually produce certain syllables and not others.

Although full use of a two-voice system appears limited to birds by virtue of the anatomy of the syrinx, it should be noted that the two vocal folds of the mammalian larynx can also vibrate independently in certain cases. Normally, the vocal folds collide with every vibratory cycle, which forces them into the same frequency and phase. However, during breathy voice, or in pathological cases of unilateral laryngeal paralysis, the vocal folds do not collide and have been shown to be capable of vibrating at two independent fundamental frequencies (Tigges et al. 1997). However, mammals appear to lack the fully independent anatomy and nervous control that would allow each vocal fold to generate rapidly varying and independent pitches as in many birds. Thus, true two-voice phonation appears to be limited to the class Aves.

2.2.2. Hypertrophy of the Voice Source

In many mammalian species, the male larynx is enlarged relative to that of female conspecifics. Male-specific enlargement of the larynx is probably common, but we know of no systematic review of this topic. The best-known example is provided by our own species. At puberty, the cartilages of the human male larynx increase rapidly in size (to about 150% of female laryngeal dimensions), and the length of the vibrating portion of the vocal folds increases even more, to nearly twice the female size (Hollien 1960; Titze 1994). This change in vocal fold length leads to a precipitous drop in f_0 at puberty that is one component of the pubertal voice change in males (f_0 about 50% of prepubertal values, Hollien et al. 1994). That this laryngeal enlargement is triggered by androsteroid hormones such as testosterone has been known for centuries, leading to the widespread practice in medieval times of castration of boy singers to produce adult males with high-pitched voices ("castrati"). However, the details of the hormonal mechanisms mediating the pubertal voice shift have not been worked out in any species and are currently receiving intense scrutiny (Yamaguchi and Kelley, Chapter 6).

Other examples of laryngeal hypertrophy are even more extreme. In howler monkeys (genus *Alouatta*), for example, the male hyoid apparatus and larynx are vastly enlarged, with the swollen, hollow hyoid occupying the entire space within the mandible (Schön-Ybarra 1986, 1988). The hyoid contains a laryngeal air sac similar to that seen in many other primates, which may function as some sort of resonator (see below). It is possible that the mandible functions to limit further growth of the hyoid; it also is enlarged relative to other primates. Howler monkeys produce loud, low-pitched roars, which appear to function in intergroup spacing; these calls may have provided the selective force underlying laryngeal hypertrophy in this genus. Finally, the most pronounced laryngeal hypertrophy in the animal kingdom is seen in African epomophorine fruit bats, especially the hammerhead bat (*Hypsignathus monstrosus*), in which the male larynx fills the entire thoracic cavity (more than half of their body, Fig. 3.3). As with the howler, the hammerhead larynx seems to have grown until it reached a bony anatomical limit, namely the rib cage, pushing the heart, trachea, and lungs down into the abdomen. The size of the larynx of female

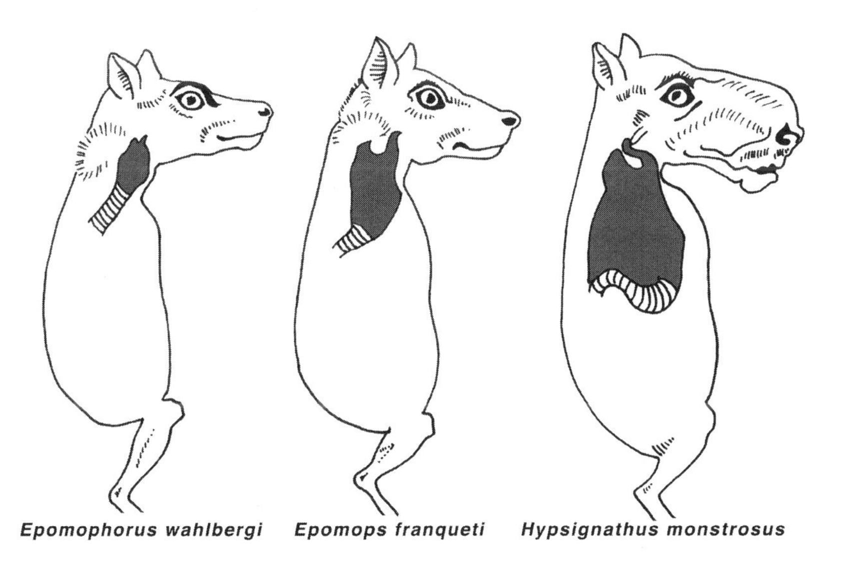

FIGURE 3.3. Enlargement of the larynx in the epomophorine bat *Hypsignathus monstrosus* compared to two related bat species, *Epomops franqueti* (which shows some laryngeal enlargement) and *Epomophorus wahlbergi*. The figure shows the outline of the larynx (black) and the upper portion of the trachea in each species.

hammerheads is 1/3 its size in males. For the hammerhead bat, the selective pressures underlying laryngeal hypertrophy have been quantified by the field work of Bradbury (1977), who studied mate selection in this African species. In trees along riverbanks, male hammerheads form "leks" (areas where males aggregate to attract mates), from which they emit an extremely high-amplitude advertisement call. Females fly up and down the riverine corridor and finally choose a male with which to mate. Males provide no parental care or other resources, suggesting that female choice might rely primarily on the vocal display. Bradbury found that five males in his population of 85 males accounted for 79% of the matings observed. These data suggest that sexual selection on the vocal-production apparatus of this species, and perhaps other epomophorine species, may be extremely intense.

These examples indicate that the mammalian larynx is not tightly constrained by body size. Although there is considerable interspecific variability in the size of the anuran larynx and avian syrinx, we are not aware of any examples of hypertrophy of the voice source as extreme as those seen in mammals. Some groups of birds that are known for having loud or low-pitched voices also have unusually large syringes (e.g., currassows and their allies; Amadon 1969; Delacour and Amadon 1973). Many studies have failed to find a correlation between body size and "dominant frequency" in anurans, but it is unclear to what extent dominant frequency depends on larynx size (see Section 2.1.3). Thus, current data are adequate only to suggest a lack of constraint on source size, suggesting that even mild selection could disturb any primitive correlation between voice pitch and body size (*contra* Morton 1977).

2.2.3. Diversity in Vocal Fold Morphology

Anuran vocal folds have a wide variety of cross-sectional shapes: they can be T- or L-shaped, or rounded—more like the vocal folds in mammals (see Schneider 1988 for examples). Anurans also often have additions to the vocal folds, which modify their oscillatory characteristics. The best-studied example is in the Túngara frog (*Physaleamus pustulosus*). This species has two components in its advertisement call, the "whine" and "chuck." The high-pitched, frequency-modulated whine results from the oscillations of the vocal folds alone. The lower-frequency chuck is hypothesized to result from vibrations of two fibrous masses, coupled to the vocal folds, which are introduced into the air stream late in the advertisement call (Drewry et al. 1982; see also Ryan and Kime, Chapter 5).

Another example of an anatomical modification of the voice source are the vocal membranes found on the vocal folds of many mammalian species. Vocal membranes, sometimes called "vocal lips" or "sharp-edged vocal folds," are thin, upward extensions of the glottal margin of the vocal folds. They vary in thickness and in the details of their histology. They are common in microchiropteran bats, where they appear to subserve the production of ultrasonic echolocation pulses (Griffin 1958). Vocal membranes are also common in primates, where they have been hypothesized to allow individuals to generate calls with very high fundamental frequencies (Schön-Ybarra 1995) or two simultaneous frequencies ("biphonation": Brown and Cannito 1995), and perhaps to create instabilities in the fundamental frequency (Lieberman 1968; Schön-Ybarra 1995; but see Hauser and Fowler 1991). The hypothesis of higher-frequency calling follows directly from the fact that, if the lightweight vocal membranes are free to vibrate independently of the much larger vocal folds, they will do so with a much higher frequency. Similarly, Brown and Cannito's "biphonation" hypothesis follows from the possibility that the vocal folds could vibrate simultaneously with the membranes, leading to two independent frequencies in the vocal output. This effect was inferred from electroglottographs of the vocalizations of the Syke's monkey (*Cercopithecus albogularis*).

A recent modeling study sought to understand the acoustic role of the vocal membranes by simulating their effects in a nonlinear dynamical computer model (Mergell et al. 1999). Mergell and colleagues modeled the membranes as fixed upward extensions of the upper mass in a well-studied two-mass model of the vocal folds (Ishizaka and Flanagan 1972); thus, the membrane was not simply treated as an independent oscillator but as an integral portion of the vocal fold as a whole. This relatively minor geometrical change had significant effects on the dynamics of vocalization. In particular, the addition of vocal membranes enabled the model to support louder and higher-pitched vocalizations. Because echolocating bats need to produce extremely loud and high-pitched calls in order to provide a detailed, long-range "picture" of their surroundings, the functional utility of this enhancement is clear. Mergell and colleagues also found that the addition of membranes increased the possibility of source–tract coupling in the model, resulting in an increased possibility of nonlinear effects and irregularities in vocalizations of species with vocal membranes, as predicted by Lieberman (1968) and Schön-Ybarra (1995). More empirical work is needed to further test these predictions, including simple anatomical measurements of vocal membranes in different species, *in vivo* observations of vocal membranes during vocalization, and investigations of the social consequences of sounds acoustically manipulated to possess, or to lack, the characteristics caused by vocal membranes.

A second, and less common, modification of the mammalian vocal folds is essentially the opposite of vocal membranes: the addition of thick, fleshy pads to the vocal folds rather than thin membranes. Such "vocal pads" are seen in lions and other cats of the genus *Panthera* (Hast 1989; Harrison 1995). No detailed physiological or observational data are available on their function. However, it seems quite likely that, due to their large mass, they play a critical role in the production of the low-pitched roars made by all of these "roaring cat" species (Hast 1989). Finally, there are significant differences in the histological fine structure of the vocal folds of different mammalian species, differences that will have significance for the f_0 range and possibly other aspects of vocal fold dynamics (Hirano 1991).

2.2.4. Modifications of the Vocal Tract

Moving now from the voice source to the vocal tract, there are numerous morphological adaptations of the vocal tract in tetrapods. Many (or most) of these can be interpreted as vocal tract elongation. These include proboscises and descended larynges in mammals (deer and humans), but perhaps the most widespread example is tracheal elongation in birds (Fig. 3.4).

At least 60 species of birds have an elongated trachea that forms loops or coils inside the thorax, coiled between the ventral skin and breast musculature, or invaginating the sternum or clavicle (Niemeier 1979; Fitch 1999). Tracheal elongation (TE) is common among currassows and allies (Cracidae), cranes (Gruidae), and swans (Anatidae) and is also found scattered among many other groups. The diversity of morphology, combined with this scattered phylogenetic distribution, suggests that TE has evolved independently a number of times. Because TE has been known for many

FIGURE 3.4. Tracheal elongation in birds. Species shown (left to right) are crested guinea fowl, *Guttera edouardii*; European spoonbill, *Platalea leucorodia*; trumpeter swan, *Cygnus buccinator*; and trumpet manucode, *Manucodia keraudrenii* (upper right).

years (the first published record, for the European crane *Grus grus*, is from Emperor Friederich II in 1250 A.D.; Niemeier 1979), a wide variety of hypotheses have been offered to explain its function (reviewed in Fitch 1999). Briefly, these can be classified into physiological hypotheses and acoustic hypotheses. "Physiological" hypotheses hold that TE serves some nonacoustic function and include the idea that TE is a respiratory adaptation to retain CO_2, to increase nonvascular respiratory area for cooling, or to retain water, to give a few examples. Because only one sex exhibits TE in many species, all of the physiological hypotheses run into immediate problems explaining the absence of a presumed physiological adaptation in one sex (typically males) and not the other. Even in nondimorphic species, each of the various physiological hypotheses has problems explaining the presence of TE in the wide variety of species and habitats in which it is seen. For example, Schmidt-Nielsen's (1972) hypothesis that TE represents a respiratory adaptation to long, high-altitude migratory flights, while applicable to trumpeter swans and many cranes, cannot explain TE in sedentary rainforest birds such as cracids and manucodes. Although TE may serve some physiological function in some species, it is unlikely to play the same role in all (or many) of them. Considerations similar to these, combined with the problem of explaining sexual dimorphism, have led most workers to dismiss physiological functions as a general explanation of TE. Acoustic hypotheses suggest that the function of TE has to do with modifying vocal output; recall that because the vocal source in birds is the syrinx, at the base of the trachea, the trachea is an integral part of the avian vocal tract. Thus, elongation of the trachea is also vocal tract elongation and may allow an individual to deceptively mimic the vocalizations of a larger bird that lacks this feature. This acoustic hypothesis, which explicitly links production mechanisms to honest signaling, will be taken up below (Section 2.3.6); see Fitch (1999) for a more complete review.

A second relatively common type of vocal tract modification is widespread among geese and ducks. Many of these species show bony enlargements of the syrinx or trachea, which are often confined to males and probably play some as yet undetermined acoustic role. A good review of the anatomy and its relevance in taxonomy is given by Johnsgard (1961, 1971), but like so many of the morphological phenomena reviewed here, there has been no further work elucidating their role in sound production or the evolution of anatid communication systems. Similarly, there are a number of possible vocal tract modifications of unknown significance among reptiles. These include an elongated trachea and bronchi in some tortoises (Siebenrock 1899; Crumly 1984), tracheal diverticula in snakes (Young 1992), the narial excresence of gharials (Martin and Bellairs 1977), and the unusual hollow bony crests of many lambeosaurine dinosaurs (Weishampel 1981).

In mammals, the supralaryngeal vocal tract spans from the larynx to the lips or nostrils. Consequently, the vocal tract can be elongated in three ways:

by elongating the nose, protruding the lips, or by lowering the larynx in the throat. Recent cineradiographic observations of vocalizing mammals reveal that in dogs, cotton-top tamarins, pigs, goats, and deer the larynx is lowered during vocalization (Fitch 2000a; Fitch and Reby 2001). During resting breathing in mammals, the larynx is inserted into the nasopharynx. The epiglottis engages with the velum, forming a tight seal, which separates the nasal/tracheal passageway for air and the oral/esophageal passageway for food. This allows many mammals to breathe and swallow liquids at the same time and is probably particularly important for young mammals, allowing them to suckle (orally) and breathe (nasally) simultaneously. For many years, it was thought that an intranarial larynx was the only normal position in nonhuman mammals, but the cineradiographic data indicate that in fact a lowering of the larynx during calling is typical for these mammals. This descent may subserve the production of high-amplitude calls because the nasal passageways absorb sound much more than the oral cavity, decreasing the amplitude of nasal calls by about 15 dB relative to oral calls (Fitch 2000a).

Another form of laryngeal descent is permanent descent of the larynx. Rather than being dynamically disengaged from the nasopharynx during calling and then reinserted, the larynx is permanently lowered. This is the situation in humans. Although human babies start life with an intranarial larynx and can suckle and breathe simultaneously like other neonatal mammals, the larynx begins to descend caudally starting around 3 months of age. This gives humans an unusually long pharynx, which is believed to subserve the production of a wider range of vocal tract shapes, and thus vowel formant frequencies, than attainable by other mammals (Lieberman et al. 1969; Lieberman 1984; Crelin 1987). Recent data show that there is in fact a second "descent of the larynx" in humans at puberty, limited to males, in which the larynx descends another several centimeters relative to females (Fitch and Giedd 1999). Because adult males are not superior in speech abilities to females (indeed, available evidence suggests the opposite: Koenigsknecht and Friedman 1976; Kimura 1983; Henton 1992), this observation suggests that the function of the elongated vocal tract in humans may not be limited to its effects on the vowel space.

Although the descended larynx has long been thought to be a uniquely human feature, a similar descent of the larynx is seen in several species of Eurasian deer, including fallow (*Dama dama*) and red (*Cervus elaphus*) deer (Fitch and Reby 2001). In parallel with adult male humans, the larynx in adult males of these deer species is enlarged relative to juveniles or females and rests much lower in the throat. During roar vocalizations produced during the rutting period, powerfully developed sternothyroid muscles pull the larynx even lower, until it reaches the sternum in large males. As predicted by acoustic theory, time-synched audiovisual analysis shows that formant frequencies drop as the larynx descends, due to the elongation of the vocal tract.

A final possible example of vocal tract elongation is provided by the proboscises seen in a wide variety of mammalian species, including elephants, elephant seals, elephant shrews, tapirs, male proboscis monkeys, and some extinct but formerly common mammal groups, such as oreodonts. Such elongations of the nasal cavity would inevitably lower the formant frequencies of vocalizations emitted through the nose (although, as mentioned earlier, nasal vocalizations are likely to be considerably quieter than oral vocalizations). Although there is little evidence suggesting that the primary function of proboscises in most species is acoustic, such a hypothesis may be reasonable in sexually dimorphic species such as proboscis monkeys, *Nasalis larvatus*, in which only the male has an elongated nose.

2.2.5. Air Sacs

The final type of morphological modification of the vocal tract that we will consider is vocal air sacs, which exist in a bewildering diversity among tetrapods. We distinguish "vocal" sacs, which at least may have some acoustic function and are typically attached to the larynx or vocal tract, from the respiratory air sacs found in all birds and some reptiles (Lasiewski 1972), which are not likely to serve any acoustic function. There are many types of vocal air sacs, for which we will offer a preliminary classification into oral, nasal, laryngeal, tracheal, and "other," depending on the location of the air sac opening. However, even within the laryngeal sacs (the most common type), there is great diversity of form, including soft-walled versus hard-walled sacs, paired, midline, or asymmetric sacs, and a variety of possible opening locations relative to the glottis (sub-, supra-, or para-glottal). We will review four plausible hypotheses that have been advanced for the acoustic and/or respiratory function of air sacs and give possible examples of each type (for more detail, see Negus 1949; Schneider 1964).

Although our survey of air sacs is organized by possible function for conceptual clarity, we do not mean to imply that air sacs serve a single function in any single species and certainly not across species. For example, in frogs, the air sacs appear to serve both air-recycling and impedance-matching functions. Another good example of multipurpose air sacs is provided by the walrus *Odobenus rosmarus*. Adult male walruses have large pharyngeal air sacs that are outgrowths of the pharyngeal wall, opening just dorsal to the larynx (Sleptsov 1940; Fay 1960). These sacs appear to subserve production of the peculiar "bell" sound made by males during sexual behavior (Schevill et al. 1966), although the mechanism for this is unknown. However, the pouches can also be inflated as "life preservers," allowing the walrus to stay afloat during naps at sea. Fay (1960) reports that he observed walruses sleeping at sea at least eight times, and the pharyngeal air sacs were invariably inflated, holding the shoulders out of the water. Finally, Sleptsov (1940) reported finding the sacs of two hunted walruses filled with food (crustaceans and molluscs) and suggested a third function for the sacs as food-storage devices (similar to the cheek pouches of Old World monkeys, which play no acoustic role; Schön-Ybarra 1995). However, Fay (1960) found this last suggestion unlikely, suggesting that Sleptsov's specimens had regurgitated stomach contents into the sacs in their death throes. In any case, the "bell production" and "life preserver" hypotheses are both well-supported, indicating at least two functions for walrus pharyngeal sacs and suggesting that air sacs may serve multiple functions in other species as well.

The most commonly cited possible function for vocal air sacs is that they play a role in impedance matching from the vocal tract to the atmosphere (see below for more details). Such a role has been suggested for most anurans as well as for the inflatable (soft-walled) air sacs of cercopithecid monkeys such as guenons (*Cercopithecus* spp., Gautier 1971) and siamangs (*Hylobates syndactylus*, Napier and Napier 1985). Impedance matching has also been suggested as a function of the inflatable esophagus of male bitterns (*Botaurus lentiginosus*), which during the breeding season produce a loud, low-pitched booming sound (Chapin 1922). Similar observations have been made concerning the role of the swim bladder in sound-producing teleost fishes (Demski et al. 1973) as well as other birds that inflate the crop or a gular air sac during vocalization (e.g., doves and pigeons, grouse, ostriches, bustards, and other species; Ziswiler and Farner 1972). There are many other species with air sacs attaching to the vocal tract that have not been studied but where a similar impedance-matching function seems plausible, for example in the "drumming" of emus (*Dromaius novaehollandiae*) (McLelland 1989), in some baleen whales (Quayle 1991), or in the pharyngeal sacs of the walrus described above. In several species, an impedance-matching function has been experimentally verified simply by puncturing the air sacs and observing that normal-sounding vocalization continues but at a much reduced amplitude (Gautier 1971; Gans 1973).

Another hypothesized role of air sacs is to allow air recycling, where the same volume of air is used repeatedly to excite the voice source. Air expelled through the lungs passes through the larynx into the elastic sac, which then deflates, returning the air to the lungs. Such a role is clear for anurans, where the recycling of air probably allows much higher rates of vocalization than would be possible given the relatively inefficient mechanisms available to anurans to inflate the lungs; this mechanism may also allow some conservation of mechanical energy (Dudley and Rand 1991). Although most anurans appear to vocalize upon expiration, members of the relatively primitive genera *Discoglossus* and *Bombina* vocalize upon inspiration (Schneider 1988). An air-recycling function also seems very likely for the large laryngeal air sacs seen in Mysticete (baleen) whales (Hosokawa 1950; Quayle 1991), which can vocalize for long periods under water without releasing air. Although no experimental data are available, Mysticete air sacs are heavily invested with muscle, which would aid in returning the expired air to the lungs.

A related possibility, the "accessory lung" hypothesis, is proposed here for the laryngeal air sacs of the great apes. Chimpanzees, orangutans, and gorillas all have voluminous air sacs (6 liters in orangutans, Schön-Ybarra 1995) that can be inflated with air from the lungs. The air sacs connect to the larynx via a long, thin-walled channel that opens directly above the vocal membranes and vocal folds. The air sacs extend into the subdermal space in the pectoral region and are overlaid by the sheet-like platysma muscle. Thus, an ape could inflate the air sacs via lung pressure and then forcibly deflate them by tensing the platysma and other pectoral muscles (or by pounding the chest, as in *Gorilla*). This anatomy suggests that great ape air sacs may act as "accessory lungs," providing an additional source of expiratory air flow and thus of energy into the source. This hypothesis seems more plausible than that offered by Negus (1949), who suggested that ape air sacs act as storage sites for oxygen during vigorous activity. Because the sacs are inflated with exhaled air that has already been in the lungs, and thus will be low in oxygen and high in CO_2, such an air reserve would be of dubious respiratory value (Fitch and Hauser 1995). Air sacs are also found in some pinnipeds, where a gas-storage function would be of clear value during diving (Sleptsov 1940), but Fay (1960) doubted this possibility because the additional oxygen stored even in large sacs would be trivial relative to dissolved blood O_2 in a diving pinniped.

A final class of laryngeal air sacs, found in many nonhuman primate species, are subhyoid air sacs. This type of thin-walled sac opens into the glottis and extends into an enlarged hollow bulla in the hyoid bone. Such a hard-walled laryngeal sac is typical of cercopithecids (Old World monkeys) and is developed to the extreme in New World howler monkeys (*Alouatta* spp.). Because these sacs are surrounded by bone, they would be of little value in radiating sound out to the environment and no value as an accessory lung. We speculate that they could act as Helmholtz resonators and that the small plug of air that vibrates in and out of the narrow neck of the sac would support vocalization at the Helmholtz resonance frequency. If true, this would constitute a form of source tract coupling. Although the opening of these sacs directly at the glottis is consistent with this hypothesis, there are currently no empirical data (e.g., using light gases) available to further evaluate this hypothesis.

2.2.6. Morphological Diversity: Summary

As this brief review makes clear, there is considerable variability in the anatomy of the tetrapod vocal-production system. Unfortunately, little of this impressive morphological diversity has received enough concentrated empirical attention for any firm conclusions to be reached about its proximate, much less ultimate, functions. This is particularly true regarding the significant morphological diversity in the vocal tract. Compared with the relatively conservative tetrapod larynx, there is a bewildering diversity of vocal tract morphologies, but the functional significance of this diversity is only beginning to be explored. Advances in digital signal analysis, techniques for the visualization of the vocal tract in action, and an increasing interest in the role of proximate mechanisms in evolution suggest that progress in understanding this morphological diversity, and correlating it with social behavior and evolutionary history, will be rapid in the coming years.

2.3. Physical and Phylogenetic Constraints on Vocal Production

In this section, we attempt to explicate some of the diversity documented above by integrating the acoustic and anatomical data into a more comprehensive functional and evolutionary framework. In particular, we argue that much of the anatomical diversity seen in tetrapod vocal tracts can be understood from the point of view of ubiquitous selective pressures operating within a framework of physical and phylogenetic constraints together with evasions of those constraints via "key innovations."

Because this is a selective synthesis, there are two potentially relevant topics that we will not cover: (1) adaptations of calls to the transmission characteristics of the environment (Morton 1975; Wiley and Richards 1982; Brown and Gomez 1992; Bradbury and Vehrencamp 1998; see also Ryan and Kime, Chapter 5; and Bass and Clark, Chapter 2) and (2) adaptations of alarm-call morphology that make localization difficult. This last topic was initiated by Marler's (1955) classic observation that the "seep" alarm calls of passerine birds are difficult to spatially localize and has more recently been reviewed by Catchpole and Slater (1995) and Hauser (1996).

2.3.1. Syringeal Diversity and Multiuse Constraints on Laryngeal Function

The primary function of the tetrapod larynx, both functionally and in terms of its history, is as a valve controlling access to and protecting the respiratory tree. Full of sensitive mucosa, the larynx will quickly close and exclude any foreign bodies that near it. In mammals, the larynx also can engage into the nasopharyngeal opening, forming a sealed respiratory passage from the nostrils to the lungs. Nevertheless, during swallowing of large, solid food items, and at all times in humans, food must pass over the opening of the glottis during swallowing before entering the digestive tract. This situation, as noted by Darwin, means that the "gatekeeper" role of the larynx is ever present. Its role as a sound-producing organ must always coexist with this gatekeeping role.

In contrast, the avian syrinx appears to serve only one function: sound production. In birds, the larynx is devoted to the gatekeeping role exclusively, whereas the syrinx is free to create sound. We hypothesize that this freedom is at least partly responsible for the considerable variability of the syrinx as an organ (Wunderlich 1886; Warner 1972a, 1972b), which can have from zero to nine pairs of muscles and is variously located tracheally, bronchially, or tracheobronchially. Raikow (1986) observed a correlation between syringeal complexity and the number of species in various taxa of passerine birds and suggested that morphological changes in syringeal form might facilitate reproductive isolation and thus speciation. In contrast, the mammalian larynx is always made up of the same basic cartilages and muscles, and although the shapes and sizes of these may vary somewhat, the larynx is overall quite a conservative organ. In anurans, the situation appears to be intermediate: there is considerable variability in laryngeal structure, although still minor compared with that seen in the syrinx.

The larynx of mammals and reptiles is under what can be described as a "multiuse constraint": the same structure serves multiple functions with incompatible design requirements, and its function is thus an unhappy compromise between these functions. We suggest that one of the virtues of the avian syrinx was as an evolutionary "key innovation" (Liem 1973) that allowed birds to escape from this constraint. The evolution of a specialized sound-producing organ allowed birds to evade the conservative restrictions imposed on laryngeal anatomy and nervous control by its critical role in swallowing. This constraint was presumably in effect in the ancestors of birds; the closest extant group, the crocodilians, possess a surprisingly mammal-like larynx, including a nonhomologous "epiglottis" and soft palate that allow them to form a sealed nostril-to-lung respiratory pathway, and some crocodilians use the larynx in vocalization. Although little is known about the evolutionary origins of the syrinx, we argue that its freedom from the role of gatekeeper to the trachea has been significant in the evolution of the impressive morphological diversity of the syrinx relative to the anuran or mammalian larynx. In turn, it seems plausible that morphological diversity is tied to repertoire diversity and perhaps has implications for the rapid diversification and speciation of the passerine birds, which have the most complex syrinx (but see Raikow 1986). In contrast to the syrinx, the avian larynx shows almost no functionally significant variation throughout the entire class (McLelland 1989), consonant with its primary and unchanging role as protector of the airway.

2.3.2. Physical Constraints and the Communication of Body Size

A fundamental fact differentiating the physics of sound from the physics of light is that sound waves are about the same size as organisms. For example, an average human female's speaking voice has a fundamental frequency around 220 Hz, with a wavelength of 1.6 m, on the order of her height. In contrast, a spring peeper's (*Hyla crucifer*) 3-kHz call has a wavelength about four times its 3-cm body length. This simple fact has enormous consequences for the production and propagation of sound at an immediate mechanistic (proximate) level and therefore, we will argue, at the ultimate evolutionary level as well. In particular, interactions between sound waves and the vocal-production system place significant constraints on what sounds can be effectively generated or transmitted, thereby rendering a large class of signals that might be theoretically possible and biologically advantageous impossible to produce in practice. On the other hand, interactions between sound and body can in other cases provide information "by default" without any need to invoke biological advantage or selection at all. This is particularly true for information about body size. Finally, because it is the dimensions of the vocal-production system that are acoustically relevant, and not overall body dimensions, it is sometimes possible for organisms to evade physical constraints by changing dimensions of vocal structures independent of body size. Over the course of evolution, nature has been ingenious in finding ways to pack more vocal tract into less body. Thus, the communication of body size provides an ideal arena within which to explore the interactions of physical constraints with ubiquitous selective forces in the evolution of communication.

Body size is a critical parameter in virtually all aspects of biology. An animal's body mass has important implications for its physiology (Schmidt-Nielsen 1984), ecology (Peters 1983), fecundity (Smith-Gill and Berven 1980), and life history (Calder 1984). At the behavioral level, body size plays a role in aggressive interactions and/or mating success (Parker 1974; Clutton-Brock et al. 1977; Clutton-Brock and Albon 1979; Modig 1996; Schuett 1997) Thus, the accurate perception of body size is predicted to be adaptive for a wide variety of organisms for a number of different reasons and should constitute a ubiquitous selective force in the evolution of communication systems. There is also a ubiquitous physical limitation on signals in that the size of various components of the sound-production apparatus has an important effect on the acoustic output (Fant 1960; Lieberman 1984), with larger components producing lower frequencies. Because the size of these production components may in many cases be related to the overall weight or length of the animal, there is good reason to expect that some aspects of the acoustic signal may provide cues to the size of the vocalizer. In particular, we can predict a negative correlation between body size and any of a variety of measures of call frequency (Morton 1977). Such acoustic cues to body size would be internally referential (providing information about the vocalizer itself) and direct or nonarbitrary (because the link between large size and long wavelengths is a fact of physics).

Because of the importance of body size in animal behavior, we expect that there will often be strong selection on perceivers to make use of available acoustic cues to body size. For the same reason, however, once perceivers are using a particular cue, we expect selection on senders to manipulate this cue to their own advantage (Dawkins and Krebs 1978). There may be situations (such as when retreating from a lost aggressive contest or luring in a timid mate) in which it would be beneficial for a sender to seem smaller than it is. In general, however, we expect this manipulation to be in the direction of size exaggeration; regardless of whether the receiver in question is a competitor or a potential mate, it will typically benefit the sender to seem larger than it is. Thus, we will focus on constraints that might prevent the production of low frequencies or morphological innovations that might allow it.

2.3.3. Body Size and Acoustic Impedance Constraints

The most fundamental limitation on the generation and propagation of low-frequency sounds comes from impedance-matching requirements. Although a small body may produce low-frequency oscillations, its ability to convert energy from these oscillations to acoustic energy in the environment is limited by the relationship of oscillator size to the wavelength of the generated sound. In general, wavelengths longer than twice the length of the vibrator will be very ineffectively transmitted to the environment, and lower frequencies will suffer even worse attenuation (Beranek 1954). A good example is provided by a tuning fork, which is nearly inaudible when vibrating freely in air but is quite loud when placed on a large surface (a tabletop or the sounding board of a musical instrument). Mechanical vibrations set up on a large surface couple to the air much more effectively than those isolated to the moving tines of the fork itself. The difficulty in radiating low-frequency sounds, called an impedance mismatch, provides a physical constraint on the production of low-frequency sounds by small animals.

The most frequent evolutionary solution to this problem appears to be the use of various types of air sacs, which are interposed between the vibrating structures (e.g., the vocal folds or the air in the vocal tract) and the environment. A detailed description was given earlier, in Section 2.2. Air sacs are ubiquitous in anurans, and a role as impedance-matching systems appears to be undisputed (see, e.g., Ryan 1985; Bradbury and Vehrencamp 1998). By increasing the size of the vibrating structure, anuran air sacs allow their bearers to more effectively radiate lower frequencies to the environment than would otherwise be possible given their small body size. However, some anurans lack sacs, which may be related to underwater vocalization where there is no impedance mismatch (Hayes and Krempels 1986). A similar example is provided by some nonhuman primates in which puncturing and subsequent deflation of laryngeal air sacs results in an attenuation of the radiated low-frequency sound but no change in pitch (Gautier 1971).

An impedance-matching function was also proposed for those cases of avian tracheal elongation in which the trachea invaginates the sternum (e.g., cranes, trumpeter swans) by Gaunt and colleagues (Gaunt and Wells 1973; Gaunt et al. 1987). Gaunt and his colleagues reasoned that the entire sternum of such birds could be like the sounding board of a stringed instrument, with the coiled trachea serving a function analogous to the bridge. The main problem with this hypothesis is a different impedance mismatch, that between the vibrations in the tracheal air column and the walls of the trachea. In a stringed instrument, mechanical vibrations in the strings are efficiently conveyed to the instrument body and sounding board, where they are then converted to acoustic energy. In contrast, the vibrations in the vocal tract start out as acoustic pressure waves and suffer a large impedance mismatch that prevents these acoustic vibrations from being converted to mechanical vibrations in the bony tracheal walls and sternum. In fact, Gaunt et al. (1987) report that virtually all acoustic energy radiates from the mouth in cranes, not from the chest; see Fitch (1999) for further discussion.

Although impedance-matching systems such as air sacs have been shown experimentally to be effective, and appear to have evolved independently multiple times, the extension of the low-frequency range for a given size air sac is limited. The fully inflated air sac of a 3-cm spring peeper substantially increases the efficiency with which its 3-kHz call is radiated to the environment, but it would have no effect on a 300-Hz call with a wavelength greater than 1 m. Thus, impedance-matching air sacs ameliorate the situation without actually evading the physical constraint relating low frequencies to large bodies. We still expect this constraint to play a significant role over the large range of body sizes seen in terrestrial vertebrates.

2.3.4. Lung Volume and Acoustic Cues to Size

The lungs (along with air sacs in birds) occupy most of the thorax in mammals, reptiles, and birds. Thus, it is unsurprising that the size of the lungs is closely related to body size (Scammon 1927; Krogman 1941; Hinds and Calder 1971). If an acoustic variable directly depended on lung volume, it would also be correlated with body size. The most obvious example is the maximum length of a single call, where one would expect longer calls to indicate larger callers. However, because a quiet call requires less air flow than a loud one, the relevant acoustic parameter might be more complex (e.g., the integral of call amplitude over an entire call). Such details aside, it is reasonable to hypothesize that the production of long, loud calls might be restricted to large individuals and thus provide a cue to body size. We currently lack data relevant to this prediction. The nearest example comes from the classic study on red deer vocalization by Clutton-Brock and Albon (1979) discussed in the introduction to this chapter.

As mentioned earlier (Section 2.2.5), a possible function of the elastic air sacs found in many primate species, including most apes, might be as "accessory lungs" (Fitch and Hauser 1995), either prolonging vocalizations or increasing the intensity of calls relative to those produced solely by lung deflation. There has been no experimental test of this hypothesis to date, although MacLarnon and Hewitt (1999) found that those primates with air sacs do seem to have longer maximum call durations than those without air sacs. Interestingly, humans have the longest "calls" of all primates (our inordinately long single-expiration spoken sentences) but lack air sacs, unlike all of our nearest relatives, the great apes. MacLarnon and Hewitt (1999) suggest that this is due to an increase in breathing control in our species. If this hypothesis is correct, it suggests that other species in which call length plays an important selective role might also be expected to evolve enhanced breath control.

The possible link between call length and body size, or body condition, provides a nice example of an unexplored source of cheap, honest cues in vertebrate acoustic communication. If only animals in good physical condition have larger, healthier lungs and can thus sustain longer calls, or longer bouts of calling, we expect selection for discriminating perceivers who attend to this unfakeable cue. For example, females might compare the length of calls from two competing males in order to choose between them, or males might avoid picking fights with rivals who can call longer than they can.

2.3.5. Source-Related Cues to Body Size

The most frequently cited acoustic parameter that could provide a cue to body size is mean and/or lowest fundamental frequency (Darwin 1871; Morton 1977). In nonavian tetrapods, the lowest producible fundamental frequency of phonation (f_{0min}) is determined by the length of the vocal folds: the longer the folds, the lower is f_{0min} (Titze 1994). Mass plays a role only if it is unequally distributed over the fold, as in *P. pustulosus* discussed above. If the length of the vocal folds is related to the vocalizer's body size, f_{0min} will thus provide an honest cue to body size (Morton 1977; Hauser 1993). This indeed appears to be the case in some species, including some toads and frogs (Martin 1972; Davies and Halliday 1978; Ryan 1988). However, such a relationship between body size and vocal fold size does not seem to be typical in other vertebrates. For instance, there is no correlation between f_0 and body size in adult humans (Lass and Brown 1978; Cohen et al. 1980; Künzel 1989; van Dommellen 1993), red deer (McComb 1991), and amphibian species (Sullivan 1984; Asquith and Altig 1990). This lack of correlation in adult humans may be particularly surprising given the widespread assumption that a "deep" or low-pitched voice indicates large body size.

The lack of correlation between f_0 and size seems less surprising when the anatomy of the vocal folds is considered. The folds are housed within the flexible cartilaginous larynx, which itself floats at the top of a trachea and is unconstrained in size by neighboring bony structures (the hyoid bone, although ossified, grows as a unit with the larynx, Schneider et al. 1967; Schön 1971). Thus, the larynx and vocal folds can grow independently of the rest of the head or body, as indeed occurs in human males at puberty (Negus 1949; Goldstein 1980), where androgen receptors in the laryngeal cartilages respond to increased circulating testosterone with a profound growth spurt (Tuohimaa et al. 1981; Beckford et al. 1985). The result is a typical f_0 for adult males that is about half that of adult females, despite an average difference in body weights of only 20% (Hollien 1960). As mentioned in Section 2.2.2, hypertrophy of the male larynx, out of all proportion to body size, is carried to an absurd extreme in animals such as the howler monkey (*Allouatta seniculus*, Schön 1971) and the hammerhead bat (*Hypsignathus monstrosus*), which clearly illustrate that larynx size, within broad limits, is unconstrained by body size. Although much less is known about the relationship between body size and syrinx size in birds, it seems likely that similar considerations apply. The syrinx, like the larynx, is free from any skeletal constraints on its size and would be expected to respond freely to selection for low voices. For example, both cranes and currassows are groups with unusually large syringes (Delacour and Amadon 1973; Johnsgard 1983; Fitch 1999), and both groups are typified by low-pitched, loud voices. In contrast, other groups, such as Falconiformes (e.g., hawks, eagles), have unusually high-pitched voices for their size. These observations suggest that the syrinx is not under any strong size constraints and can respond to selection by either increasing or decreasing size.

When such developmental flexibility is present, there is clearly no *a priori* reason to expect vocal fold size (and thus f_0) to be well-correlated with body size (Fitch 1994; Fitch and Hauser 1995). Of course, between disparate enough taxa some degree of correlation is inevitable simply due to the very large differences in overall avian body sizes; the syrinx of an ostrich or emu could contain the entire body of a hummingbird. Thus, various researchers have found correlations between body size and some measure of vocal frequency across different avian or mammalian taxa (birds: Ryan and Brenowitz 1985; mammals: August and Anderson 1987; Hauser 1993). Similarly, in species with large size differences between infants and adults, we may expect some differences in pitch between young and old animals, as indeed appears to be the case in humans, where the f_0 of infant cries averages around 500 Hz and adult speech between 100 and 200 Hz (Titze 1994). However, the relevant information for many species in many communicative situations is not the size of young or of members of other species but of conspecific adults. In this domain, and despite the common claim that voice pitch provides an accurate cue to body size (e.g., Morton 1977), the data reviewed above suggest that the voice source (larynx or syrinx) is ill-suited to provide dependable cues to body size in adult terrestrial vertebrates.

2.3.6. Vocal Tract Length and Acoustic Cues to Body Size

A different potential acoustic cue to body size comes from vocal tract length and formant frequencies. If the cross-sectional area function of the vocal tract is constant, the primary determinant of formant frequencies is the length of the vocal tract (Fant 1960; Lieberman and Blumstein 1988; Fitch 1997). In particular, a lengthening of the vocal tract tube will lead to a decrease in the average spacing between successive formants, or "formant dispersion" (Fitch 1997; Riede and Fitch 1999). Thus, if vocal tract length is correlated with body size, there will be an inverse correlation between formant dispersion and body size, and formants will provide honest cues to body size. Such formant cues are completely independent of voice fundamental frequency or perceived pitch.

Formant dispersion is simply the average spacing between successive formants and provides one simple metric for estimating vocal tract length. However, no single number can accurately capture all of the information in a complete list of formant frequencies and bandwidths, and in some cases other statistics that rely only on higher formants, or on the most reliably excited formants, may be preferable. It may appear that the first formant would provide an equally good estimate of vocal tract length. There are two reasons why this is not the case. The first concerns the boundary (end) conditions of an air column contained in a simple tube, which have a drastic effect on the lowest formant but no effect on formant spacing. For example, a 17.5-cm tube that is open at both ends has formant frequencies at 1,000, 2,000, and 3,000 Hz, and so on, while the same tube with one end closed has formants at 500, 1,500, and 2,500 Hz, and so on. The spacing is 1 kHz in both cases, but f_1 varies between 500 and 1,000 Hz. Although the human vocal tract during speaking is often idealized as being closed at the glottal end, this approximation is only strictly correct for a portion of the glottal cycle and may never be true in certain phonatory modes (e.g., the glottis may never close during breathy phonation). The use of formant dispersion avoids the need for any assumptions about glottal state and phonatory mode and is thus preferable to f_1 as a measure of vocal tract length. A second reason that f_1 provides a poor correlate of vocal tract length is the increased role of the yielding walls of the vocal tract at low frequencies. In much the same way as described for the anuran vocal air sac, the soft parts of the vocal tract begin to absorb significant energy from the acoustic signal at lower frequencies. This effect of the vocal tract walls at low frequencies will place a lower limit on f_1, irrespective of total vocal tract length (Fujimura and Lindqvist 1970). This effect will be most pronounced for long vocal tracts, such as in large mammals, or in animals with vocal sacs.

Is there any reason to expect vocal tract length (which determines formant spacing) to be more closely tied to body size than vocal fold length (which determines fundamental frequency)? For mammals, the answer is clearly yes. The mammalian vocal tract is made up of the pharyngeal, oral, and nasal cavities, which are firmly bounded by the bones of the skull, and skull size is closely tied to overall body size (Morita and Ohtsuki 1973; Dechow 1983; Alcantara et al. 1991; Fitch 2000c). Because the facial region of the vertebrate skull is involved in so many other life-critical functions (it houses the sense organs, provides the passageway for water and air, must capture and process food, plays an important role in grooming in many species, and other functions), vocal tract length should be much less free to vary independently of body size than larynx size. According to this hypothesis, mammalian vocal tract length is highly constrained by multiuse factors. Although some evolutionary modifications of facial structure may occur for the purposes of modifying vocal-production acoustics (e.g., the elongated nose in proboscis monkeys or some pinnipeds, or the descended human larynx), in most species the structure of the facial skeleton is largely determined by the more basic needs of prey capture and food processing. We would expect this multiuse constraint to place stringent limits on the ways in which vocal tract structure and function can change in nonavian tetrapods. This may help explain why the vocal tract, like the skull itself, has a rather conservative evolutionary history. Thus, we can expect vocal tract length and the attendant acoustic cue of formant dispersion to provide a correspondingly more robust cue to body size in mammals.

This hypothesis is supported by data from several mammalian species. Fitch (1997) used radiographs (x-rays) to measure vocal tract length in rhesus macaques (*Macaca mulatta*) and found a strong correlation between vocal tract length (from the glottis to the lips) and both body mass and length. Second, he measured formant frequencies using a spectral-estimation algorithm called linear prediction, which finds the optimal all-pole (all-formant) model to fit a particular spectrum (Markel and Gray 1976). Fitch found a strong negative correlation between formant dispersion and body size in these monkeys. Using similar techniques, Riede and Fitch (1999) also found strong correlations between body size, vocal tract length, and formant dispersion in domestic dogs (*Canis familiaris*). In both cases, restriction of the analysis to adults still yielded significant positive correlations between body size and vocal tract length, indicating that formant frequencies can provide an honest cue to adult body size in these two species. Finally, Fitch and Giedd (1999) found strong positive correlations between body size and vocal tract length in humans despite the fact that the human male vocal tract elongates slightly during puberty, causing an increase in vocal tract length. In this study, the sample size of fully adult humans of each sex was inadequate to evaluate within-sex adult vocal tract allometry. The correlation between body size and vocal tract length, and its acoustic correlates, provides a good example of honest, internally referential communication that results directly from the anatomy of the vocal-production system combined with basic acoustics. This honest signal does not require the invocation of any special selective forces or additional costs to the animal. Formant cues to body size thus appear to be an example of cheap, honest communication, at least in monkeys, humans, and dogs.

In birds, the situation is quite different. Because the voice source lies at the base of the trachea, the vocal tract includes not just the oral and nasal cavities but also the entire trachea. This means that the vocal tract of a bird of a given body size is much longer than that of an equivalent mammal or reptile. More importantly, it suggests that the multiuse constraints described above, which hinder change in the mammalian vocal tract, are not applicable to birds. Fitch (1999) suggested that this provides an explanation for the phenomenon of tracheal elongation in birds. An overall correlation between body size and vocal tract length was probably the primitive state for birds and has indeed been documented across species by Hinds and Calder (1971). Therefore, given an appropriately broadband source, formant frequencies would provide an indication of the vocalizer's body size. Once perceivers had evolved to take advantage of this information, it provided an opportunity for vocal subterfuge: a bird with an elongated trachea could duplicate the formant dispersion of a larger conspecific and thus exaggerate its own apparent size. Unlike the case in other vertebrates, where selection for vocal tract elongation would face stiff opposing selection from multiuse constraints, tracheal elongation in birds would be opposed only by a decrease in respiratory efficiency due to increased tracheal dead space (Hinds and Calder 1971; Clench 1978). However, due to the one-way, flow-through nature of the avian respiratory tract (Lasiewski 1972; Liem 1985; Schmidt-Nielsen 1997) and the small volume of the trachea relative to the extensive respiratory air sacs system, this physiological effect may be negligible (Prange et al. 1985). Thus, little stood in the way of the acoustic exaggeration of size via the evolution of tracheal elongation in birds, which would explain its repeated independent evolution in many orders of birds and virtual absence in all other taxa; the only other example of tracheal elongation of which we are aware is in the tortoise *Geochelone pardalis* (Crumly 1984) and is of uncertain acoustic or behavioral significance. If this hypothesis is correct, tracheal elongation in birds is a good example of size exaggeration via vocal tract elongation (Fitch 1999).

There are a number of other possible examples of deceptive elongation of the vocal tract. Weishampel (1981) suggested that the prominent crest of many lambeosaurine dinosaurs, which contained an elongated nasal passageway, functioned to lengthen the vocal tract and thus to decrease formant frequencies. The proboscises found in many nonhuman mammals (e.g., elephants, elephant shrews, various pinnipeds, and proboscis monkeys, as well as oreodonts and other extinct taxa) have the inevitable result of lengthening the nasal vocal tract and thus lowering the frequencies of nasal formants. Whether this serves the function of exaggerating acoustically conveyed size remains an untested hypothesis but seems plausible in the case of species such as elephant seals and proboscis monkeys where the proboscis is a sexually dimorphic trait. Finally, the vocal tract elongation resulting from the descent of the human larynx may have some size-exaggerating effect. This hypothesis is supported both by the fact that formant dispersion is known to be used as a cue to body size by human observers (Fitch 1994) and that an additional descent of the larynx occurs at puberty in males simultaneously with (but anatomically and functionally independent of) the growth of the male larynx (Fitch and Giedd 1999).

We have focused in this section on acoustic cues to body size mainly because body size is easily measured, is an extremely important variable in many species, and has a direct and obvious effect on the production of acoustic signals. However, we would like to stress that the approach outlined above is likely to be applicable to many other types of information in animal signals as well. For example, individual differences in vocal tract anatomy may provide robust cues to individual identity (Rendall et al. 1996), and differences in the use of nasal versus oral vocal tracts might serve as a cue to group membership (Hauser 1992). Sex hormones can bind preferentially to laryngeal tissues (Tuohima et al. 1981), suggesting that certain aspects of the voice source may provide cues to sexual readiness or other endocrinological information (Yamaguchi and Kelley, Chapter 6). A possible example is oestrous-related calling in gelada baboons (Moos-Heilen and Sossinka 1990): could steroid-related changes in tissue hydration over a female's cycle lead to vocal cues to ovulation? Finally, there may be vocal cues to age in some species. The histological composition of the vocal folds changes with age (Titze 1994), potentially resulting in vocal cues to a caller's age and experience. Similarly, the vocal tract in male plain chachalacas (*Ortalis vetula*) elongates with age (Marion 1977), presumably lowering formant frequencies. Could male rivals use such cues to avoid more experienced rivals, or might females use formants in mate choice? All of these questions are highly relevant to the evolution and structure of acoustic communication systems but demand advances in our knowledge of proximate mechanisms before they can be adequately addressed. If the preceding review spurs research along these lines, it will have achieved its goal.

2.4. Conclusion

To summarize and conclude Section 2 of this chapter, we have seen that physical and physiological constraints play a fundamental role in shaping the signaling systems of terrestrial vertebrates, interacting with multiple selective forces in various ways to produce an impressive variety of morphological adaptations in tetrapod vocal-production systems. Physical constraints, by creating nonarbitrary mappings between behaviorally relevant parameters (such as body size) and aspects of acoustic signals (such as frequency) can provide a starting point for the use of a certain parameter in a species' communication system. Physiological constraints (such as the multiple functions of the mammalian larynx or the restriction of most tetrapod vocal tracts to the skull) can play an important role in maintaining signal honesty in lieu of any specific selection "for" honesty. Knowledge of these constraints can also provide a principled starting point for scientific analysis of a species' vocal repertoire, allowing us to identify precisely acoustic parameters that might play a role in signaling. Finally, "key innovations," such as the syrinx in birds, can allow a species to evade such constraints in evolutionary time, opening up new vistas in the adaptive landscape. More often than not, however, new physical or physiological constraints will probably exist, even in this new adaptive space, which again will influence the evolutionary trajectory of a particular species' communication system. We now turn to the role of cognitive mechanisms and behavioral flexibility in dishonest signaling and its detection.

3. Perceptual and Cognitive Constraints on Skepticism: The Behavior of Deception

3.1. Lies of Commission

When President William Clinton was asked about his alleged relationship with Monica Lewinsky, he claimed that he never had an illicit affair with her. As the world now knows, he lied. The public perceived Clinton's statement as a lie because they detected a mismatch between what Clinton said he did and what he actually did.

Humans are not alone in their ability to create lies of commission—actions that actively falsify information. In fact, a wide variety of animal species appear to be comparably endowed. In order for such lies to work, however, three conditions must hold. First, the species must have a signal that is tightly correlated with a particular context. Thus, for example, when an animal gives an alarm call, it must signal the presence of a predator on a significant proportion of occasions. At present, we cannot say precisely how tight this correlation must be, but the signal must have relatively high predictive value with respect to the sequelae of signal and response. Second, when individuals hear such signals, they must respond in a relatively stereotyped or consistent way and must do so on a statistically significant number of occasions. Thus, when animals hear an alarm call, they must consistently flee. Third, individuals must have the flexibility to manipulate the behavior of other group members by producing a species-typical signal in a novel context; in this sense, there must be some level of independence between signal and context. Thus, while competing over food or a mate, one animal might give an alarm call causing the competitor to dive under the bushes.

Falsifying information should theoretically be a rare event, or at least have a low cost for the deceived, in order to maintain the effectiveness of the lie. Thus, individuals might be expected to produce false alarm calls infrequently in order to avoid generating a completely ineffectual signal—"the boy who cried wolf." However, there may be considerable variability within taxa in how sensitive individuals are to being deceived as well as in the cost of deception. To flesh out these ideas, we explore a set of observations and experiments on insects, birds, and primates designed to reveal how lies of commission are enacted and sometimes foiled by skeptical receivers. In each case, we evaluate the evidence in light of the three conditions discussed and, where possible, lay out a series of experiments that might take our understanding further.

Lies of commission appear relatively often during interspecific interactions. Thus, in the nonvocal domain, we know of plovers that dupe their predators by performing the injury-feigning display, predatory *Photuris* fireflies that mimic the mating-flash patterns of their congeners, thereby providing the mimic with a meal, snakes that play dead in order to avoid being eaten, fish, birds, and mammals that enlarge some portion of their body in order to look bigger, and insects and frogs that evolve coloration patterns that resemble a sympatric but poisonous species (Lloyd 1984; Mitchell and Thompson 1986; Burghardt 1991; Ristau 1991; Hauser 1996, 2000). In the vocal domain, there are fewer examples, but Charles Munn's (1986a, 1986b) study of a mixed-species flock in Peru is perhaps one of the more compelling examples.

In a Peruvian rainforest, Munn noted that among the members of a mixed-species flock of birds, some species appeared to be responsible for finding food, whereas other species appeared to be responsible for alerting the flock to danger. The fluidity with which these species interacted was spectacular, but perhaps more intriguing was the fact that the alarm-calling species—the bluish-slate antshrike and the white-winged shrike tanager—sometimes produced alarm calls when there were no predators in view. These were not mistakes. Rather, the alarm calls were given almost exclusively when the antshrike or tanager was in direct competition with the food-finding species over insect prey. As they approached the insect, the antshrike and tanager produced an alarm call, causing the food-finding species to look up and thereby forfeit its access to the insect. Surprisingly perhaps, this was not a rare event. Out of 104 alarm calls recorded from the tanager, Munn found that 55% were false alarms. Although this rate is high, we cannot conclude that it accurately represents the rate of deceptive alarm calls. It is entirely possible that in some cases the bird detected an animal it considered to be (or confused with) a predator when there was no predator at all. The alarm call is a false alarm in the signal-detection sense—a perceptual error that is likely to occur when sensitivities are set high, as are the costs of a miss (i.e., of failing to detect the predator when it is present).

Is there any evidence that antshrikes and tanagers are capable of creating lies of commission? Let us return to our three conditions. First, the antshrike and tanager alarm calls are often given during encounters with predatory birds. Thus, there is a correlation between the signal and a specific context. Further, playback experiments of alarm calls given to actual predators as opposed to virtual predators revealed no differences with respect to the food-finding species' responses; in both situations, they looked up and fled. This shows that the false alarm call sounds like the true alarm call and thus should be equally evocative. Although there may be other acoustic cues that have been overlooked in the analysis, these possible differences do not appear to be perceptually salient to listeners.

Second, when the food-finding species hears the alarm call, it responds by looking up and fleeing. Thus, the alarm call reliably elicits a response from the target receiver. Finally, the antshrike and tanager have the flexibility to produce the alarm in the absence of a predator. Thus, our conditions for a lie of commission have been met.

False alarm calls by antshrikes and tanagers are relatively common. Thus, there is no support for the prediction that lies of commission must be rare. One explanation for this high rate of deception comes from an economic analysis of the interaction. For both species, the benefit comes from capturing an insect. For the food-finding species, the potential cost comes from ignoring the alarm call. Looking at the trade-offs, it never pays the food-finding species to ignore the alarm call because the benefit of eating an insect is greatly outweighed by the costs of being eaten by a predatory bird. Because of this imbalance, the antshrike and tanager can give false alarm calls at high rates. Within this snapshot of an evolutionary arms race, the antshrikes and tanagers have the upper hand.

Several questions emerge from Munn's studies for which we have no answers. For example, how often do antshrikes and tanagers make mistakes in terms of detecting a predator? Establishing the error rate is important because it sets up a more accurate measure of the rate of deceptive alarm calls. When a false alarm call is sounded, how often do tanagers and antshrikes manage to capture the insect? How do other antshrikes and tanagers respond when they hear a deceptive as opposed to an honest alarm call? Can they detect a difference? Or do they go along with the prank in order to preserve the trick on a subsequent occasion when it is their turn? How often do antshrikes and tanagers produce deceptive alarm calls to the same individuals? Although the overall rate of deceptive alarm calls is quite high, one might expect that a sufficiently high rate with one individual, over a short period of time, would cause the receiver to begin ignoring the call. Finally, how do antshrikes and tanagers acquire the ability to deceive by producing deceptive alarm calls? Do young birds make mistakes, giving deceptive alarm calls to other members of their species? Do they give deceptive alarm calls in contexts outside of food competition? Answers to these questions are certainly attainable by manipulating the contexts in which each species encounters the other, by exploring the acoustics of deceptive and honest alarm calls in greater detail, and by selectively playing back deceptive alarm calls under controlled conditions. For example, one could play back deceptive alarm calls at higher rates than what is given naturally in order to determine when members of the food-finding species habituate. Given the level of description already provided by Munn, the mixed-species flocks in Peru provide an ideal situation for looking at the dynamics of interspecific deception.

The dynamics of inter- and intra-specific interactions may be quite different with respect to the necessary and sufficient conditions for evolving the capacity to generate lies of commission. Most studies of intraspecific deception have focused on the use of false alarm calls or food calls to exploit the behavior of other group members.

For many avian species, the winter months are difficult due to the relative scarcity of food. As a result, competition over food is more intense. Anders Møller (1988a) noticed that great tits regularly produced alarm calls in the absence of predators, suggesting that they might use such signals to gain access to limited resources. To test this possibility, Møller collected observations of alarm calling by great tits at feeding stations where food was either concentrated or dispersed. Out of the total number of alarm calls recorded, 63% were given in the absence of a predator. Great tits produced such false alarm calls when the feeding stations were occupied either by other great tits or other birds (e.g., house and tree sparrows). Specifically, the nonfeeding great tit gave an alarm call and then flew straight toward the feeding station. The feeding birds flew away upon hearing the alarm call, thereby yielding access to the food station.

To determine whether both conspecifics and heterospecifics perceived the false alarm calls as similar to the real alarm calls, playbacks were conducted. Both great tits and sparrows responded to the playbacks of real and false alarm calls in the same way: they fled the feeding station and headed for shelter. This suggested that real and false alarm calls carry the same message.

Møller also found that the use of false alarm calls was contingent on weather conditions as well as the relative dominance rank of the bird at the feeding station. Thus, great tits produced more false alarm calls during adverse weather conditions (e.g., snowstorms) as well as when the bird feeding at the station was dominant; when subordinates were present at the feeding station, dominants did not use false alarm calls but rather approached and quietly displaced the subordinate. Further, great tits were more likely to give false alarm calls when sparrows were present at a concentrated spread of food than at a dispersed spread of food and when the heterospecifics were from a flocking rather than a nonflocking species; similar results have been presented by Matsuoka (1980) working on marsh tits and willow tits.

With respect to our definitional conditions, Møller's results indicate that the great tit's alarm call is commonly given during predator detection. We do not know, however, how often great tits make errors of predator detection, and thus we cannot assess whether the documented level of false alarm calls is accurate. Second, the alarm call elicits a reliable flight response in both conspecifics and heterospecifics. This claim is supported by both the natural observations as well as the playback experiments with real and false alarm calls. Third, individuals clearly have the flexibility to exploit the manipulative power of the false alarm call, as evidenced by the contexts in which they use them. Great tits certainly do not use false alarm calls reflexively. Rather, their use of false alarm calls appears to be under voluntary control, as revealed by their sensitivity to the dominance rank of conspecifics, current weather conditions, and whether heterospecifics are flock or nonflock feeders.

Paralleling Munn's results, Møller's findings also violate the intuition that for deception to be effective, the deceptive act must be rare: great tits produce false alarm calls at extremely high rates. However, the great tit data are a bit less clear than those collected on antshrikes and tanagers because Møller only presents the overall rate of false alarm calls. Some of these calls are given in the presence of heterospecifics and some in the presence of conspecifics. To assess whether the rate of false alarm calls differs for conspecifics and heterospecifics, it would be necessary to break down the pooled data.

Møller's results raise many fascinating questions, several paralleling those raised for Munn's work on antshrikes and tanagers. Specific to the biology of great tits, however, it would be interesting to determine how often individuals produce false alarm calls during the breeding season, when resources are more abundant, and whether individuals are less likely to produce such calls prior to the mating season given that they might be deceiving a potential mate. One could test this hypothesis by making a male extremely deceptive, playing back his alarm calls at high rates when no predator is present, and contrasting this situation with one in which a male is made extremely honest—play back his alarm call and simultaneously present a hawk. Given these two male types, one could then look at differences in mate choice by females during the breeding season.

Møller (1990) followed up on his great tit work by looking at a comparable problem in barn swallows, a species that has been carefully studied with respect to its breeding biology and the selective forces operating on male–male competition and female choice (Møller 1988b, 1989, 1993). Like great tits, barn swallows also give alarm calls in the absence of predators, leading to the hypothesis that they are generating lies of commission. In contrast to great tits, barn swallows most often produced false alarm calls when their fertile mates left the nest area, apparently in search of extra-pair copulations. Observations revealed that females engaged in extra-pair copulations stopped upon hearing their mate's alarm calls.

To determine whether males produce alarm calls deceptively, Møller conducted two experiments. In the first experiment, he chased females away from their nests in order to determine whether such departures elicited false alarm calls in males. Females were chased away at the start of nest building, during egg laying, and during the incubation period. When males returned to the nest and detected the female's absence, they rarely gave alarm calls during the nest-building and incubation periods but produced false alarm calls on about 95% of all experimental trials in the egg-laying period. Observations revealed that the false alarm calls were functionally equivalent to the real alarm calls in that other swallows either flew away or gave alarm calls. Further, whereas solitarily breeding swallows produced a constant, low rate of false alarm calls across the breeding period, colonially breeding swallows—which are more vulnerable to being cuckolded—produced high rates of alarm calls almost exclusively during the egg-laying period, when females are most likely to engage in extra-pair matings.

To determine whether the difference between solitarily and colonially breeding swallows reflects a behavioral polymorphism, Møller conducted a second experiment involving the presentation of a model male swallow. Males were more likely to produce false alarm calls to a model male swallow during the nest-building and egg-laying periods than in the incubation period and were more likely to produce false alarm calls to the model swallow than to the control, a model willow warbler. This shows that solitarily breeding male swallows are sensitive to the risks of extra-pair copulations and are most responsive to this risk when their mates are fertile.

Once again, Møller's observations and experiments on barn swallows fit our three definitional conditions but appear inconsistent with the prediction of rarity. The barn swallows' alarm call is generally given in the context of predator detection and elicits a flight response. Males have evolved the capacity to use this signal to manipulate the responses of their mates, thereby fending off the threat of extra-pair copulations. False alarm calls therefore provide barn swallows with a mechanism to decrease paternity uncertainty. Like Munn's antshrikes and tanagers, it appears that the cost of ignoring the alarm call is high relative to the benefit of an extra-pair mating. Even if the male has made an error, falsely signaling the presence of a predator, it is to the female's advantage to flee and then return at a later time to mate. This economic imbalance may enable males to produce false alarm calls at high rates.

Møller's experiments reveal that barn swallows are not acting reflexively. The use of false alarm calls appears to be under facultative control, sensitive to the risks of extra-pair copulations and the female's reproductive cycle. Several questions remain, however. For example, although the male's false alarm call temporarily breaks up a covert mating, does the female in fact obtain fewer extra-pair matings? If a male produces a false alarm call and the female fails to return to the nest, does he try again, perhaps even more frenetically? Ristau (1991), in her work on the broken-wing distraction display in plovers, has noted that when a predator ignores the plover's first try with an injury-feigning display, the plover tries again, and does so more dramatically, swooping at the predator in order to grab its attention. Although male barn swallows distinguish between a model swallow and a model warbler, do they distinguish between a model male swallow who is in the company of a model female swallow? Do they perceive a potentially mated pair as a lower risk? What about an anesthetized swallow who looks dead? There is clearly no risk, but only if swallows make a clean distinction between living and dead. If we artificially escalate the rate of false alarm calls, and do so in a situation where the female can see her mate, will she abandon him in search of a more honest mate? What are the acoustic cues to individuality and context, and can they be perturbed so that false alarm calls are no longer effective?

All of the work described thus far focuses on animals using alarm calls to deceive others during competitive interactions over food or mates. Domestic chickens, however, deceive each other in the context of mating opportunities by producing food calls in the absence of food. Marler and his colleagues (Marler et al. 1986a, 1986b) first showed that roosters produce characteristic vocalizations when they discover food, with the rate of call production positively correlated with food quality. They further observed that roosters give food calls in the absence of food and are most likely to do so when a female is present; roosters are silent when another rooster is nearby, regardless of the presence or absence of food.

One explanation for the chicken's calling behavior is that rather than providing external reference by calling attention to food, the call reflects the signaler's willingness to engage in social interactions. In other words, chickens might often call in the context of food, but the call does not refer to food but to something more general. To explore whether food-associated calls refer to food or to the intent to engage in social interactions, Marler and his colleagues (Marler et al. 1986a, 1986b, 1991; Gyger and Marler 1988; Evans and Marler 1994, 1995) carried out experiments with chickens living in a seminaturalistic environment. If food-associated calls refer to food, then calls produced in the absence of food would represent lies of commission. Results showed that 45% of all calls were produced with no identifiable object present. When calls were produced in the presence of food, changes in call rate were related to food quality but not to the distance between mates or the probability of performing the waltzing display, a behavior used by males as an invitation to mate. In an experiment using an operant procedure, males pecked for food most when a light indicated that food was available. Food calling was considerably lower when the light was off, even when a receptive female was present; waltzing was highest when the female was first introduced. These two experiments suggest that the call refers to food even though it is mediated by social context.

Given that the call refers to food, its production in the absence of food appears to represent a case of deception. Support for this claim comes from looking at the relationship between food-call production and interindividual distance. Males were more likely to produce food calls in the absence of food when the females were far away than when they were close. This makes sense if a male's vocal behavior is sensitive to the female's visual perspective and, in particular, the probability that she will notice the absence of food. Thus, males should call honestly if females are sufficiently close that they can see whether the male does or does not have food. In contrast, they should act deceptively when females are sufficiently far away to prevent a clear view of the potential feeding area.

To function as a deceptive signal, calls produced in the absence of food must sound like those produced in the presence of food and must have an equal probability of eliciting an approach from females in hearing range. More specifically, for the act of deception to work, females must recognize the call as a food call, must perceive the call as an indication that a male has discovered food, and must then approach the male. Results show that females approached males 86% of the time when they called in the presence of food but only approached 35% of the time when males called in the absence of food. Further, females were more likely to approach males who called in the absence of food when their call rates were high than when they were low, and were more likely to approach when they were close to the male than when they were far. When females failed to approach in response to a male calling in the absence of food, males often approached females. These results suggest that males attempt to use food calls to attract females and that females assess the veridicality of the signal by using the rate of calling as well as contextual information. It may be that the lower rate of approach to deceptive calls is due to perceptible differences in the acoustic morphology of the calls, but no acoustical analyses of honest versus deceptive calls has yet been performed.

The chicken food-call system satisfies our definitional conditions. Once again, however, it appears to violate the prediction of rarity. Specifically, the food call is primarily given in the context of food and, as recent production and perception experiments suggest, it functionally refers to food rather than to a more generic event or context such as the willingness to engage in social interactions (Evans and Marler 1994; Evans and Evans 1999). When chickens produce food calls, they elicit characteristic responses that are distinct from the responses elicited by contact or alarm calls. Chickens apparently take advantage of the referential properties of the food call as well as the behavior it elicits in females to produce such calls in the absence of food. For reasons that are currently unclear, males appear to get away with such lies at relatively high rates. Almost 50% of all food calls are given in the absence of food and, when given, elicit female approach approximately 33% of the time.

In terms of a mismatch between signal and context, the chicken study provides an example of a lie of commission. What is unclear, especially when contrasted with the previous examples of avian deception, are the costs and benefits of this putative case of deception. Thus, males presumably gain some benefit by eliciting an approach from a female. However, Marler and his colleagues have yet to demonstrate that the female's approach translates into a reproductive advantage for the male. In terms of costs, females lose by disrupting their current activity and by traveling a distance to the male. At present, it is unclear whether there are costs that would constrain or limit the frequency with which males give false food calls. For example, is it the case that females are less likely to mate with a male who has given a food call when no food is available? One could test this possibility with a design that we have already mentioned. Specifically, make one male completely dishonest (100% of his calls are produced in the absence of food) and one male honest (100% of his calls are produced in the presence of food). Once a female has been exposed to these two males, set up a mate-choice experiment and record the female's preferences. These experiments, accompanied by others that focus more specifically on the costs and benefits of honest as opposed to deceptive food calls, will help us understand how male chickens can get away with such high levels of deceptive behavior.

The piece of this story that has been neglected is the skepticism of the receiver. Most of our discussion has focused on how individuals can manipulate the behavior of receivers by using functionally deceptive acoustic signals. However, selection will favor both mechanisms that facilitate deception and those that enable accurate skepticism. Is there evidence of skepticism? The honey bee dance language represents an exquisite example of a functionally referential signal. As decades of research have revealed, when honey bees dance, attentive listeners extract information about the location, quality, and distance to food using visual, acoustic, olfactory, and tactile cues (von Frisch 1967; Gould and Towne 1987; Gould and Gould 1988; Dyer and Seeley 1989; Michelsen et al. 1992; Seeley 1992). No other signaling system in the animal kingdom is this precise, with the exception of human language (Hauser 1996). To assess whether individuals are ever skeptical of the information conveyed in a dance, Gould (1990) conducted an ingenious experiment. Using a hive with a long history of experience in one location, he removed a group of foragers and trained them to move back and forth from a point on land to a second point on land where a pollen-filled boat was located. Over time, he increased the distance between these two points and also moved the boat from a position on land to a position out in the middle of a lake; throughout the training period, Gould prevented the bees from returning to their hive. Once the bees reliably traveled to the boat and back, he allowed them to return to the hive and dance.

When the trained foragers returned to the hive, they danced, indicating that a rich food source could be found out in the middle of the lake. Although the bees watched the foragers dance, relatively few of them flew off to the boat. Why? Because, as Gould argued, food has never been found out in the middle of this lake, or presumably any lake, and thus the information in the signal was unreliable and inaccurate. The hive members refused to move, treating the signal skeptically. This interpretation is quite reasonable when one takes into account the results of a control experiment. Specifically, Gould trained a second group of foragers to find food in a boat located on the water but along the edge of the lake. When the foragers returned and danced, other individuals immediately left the hive and flew to the pollen-filled boat; presumably, the edge of the lake represents a more likely place to find bee food.

What we do not learn from Gould's work is the nature of the information stored in the bee's brain—the extent to which an individual's own knowledge of pollen location can override the social message. For example, if a bee knows that a field of flowers has been burned down, leaving no pollen behind, would it accept or reject a dance indicating pollen at this location a week after the burn? What about one month after the burn, giving time for new growth? If an experimenter brings the bees to the lake and allows them to feed from the boat, would they then follow the dancer to this location? If a bee repeatedly lies about the location of pollen, does it lose respect? Is it punished for falsely "crying" pollen?

We do not have answers to these questions. However, the critical aspect of Gould's work for the present discussion is that bees, and perhaps other animals, can check on the veracity of a piece of information by comparing what they are told with what they have experienced or are currently experiencing. If this interpretation is correct, then we should be able to turn reliable animals into unreliable ones, as we have already suggested.

Over the past ten years, primatologists have accumulated a large number of anecdotes of potentially deceptive behavior. These observations are, as pointed out by Andrew Whiten and Dick Byrne (Whiten and Byrne 1988; Byrne and Whiten 1990), strikingly different from other cases of deception in the animal kingdom in that they are rare events—they satisfy the prediction of rarity. Among the many examples cataloged, several are of the form of a monkey or ape using a false alarm call to gain access to a resource (food or mate) or to deflect an aggressive attack. These observations suggest that nonhuman primates may have the capacity to create lies of commission. To address the problem of skepticism, Dorothy Cheney and Robert Seyfarth designed an experiment with vervet monkeys, adopting a classic technique from cognitive psychology—the habituation–dishabituation paradigm. Taking advantage of detailed acoustic analyses of the vervets' vocal repertoire, the general procedure started with a habituation series involving repeated playbacks of a single call type from one individual. Following a fixed number of habituation trials, Cheney and Seyfarth then played back either a different call type from the same individual or the same call type from a different individual. Although these experiments were primarily designed to examine the problem of referentiality, the nature of the design was ideally suited to exploring the problem of deception and skepticism. Specifically, the habituation series represents a case of experimental slander: repeatedly play back an individual's call in the absence of an appropriate context. For example, in one set of experiments, Cheney and Seyfarth contrasted the vervets' response to two calls given in the context of intergroup encounters, the "wrr" and "chutter." Although these calls are acoustically distinctive, they are given in the same general context and thus convey the same general message: that a competitive group is nearby. In one condition, subjects were repeatedly played A's wrrs and then tested with A's chutter. In another condition, subjects were played A's wrrs and then tested with B's wrr or chutter. Subjects habituated to repeated exposure to A's wrr, showing less response with successive playbacks. When they were then tested with A's chutter, they transferred the level of habituation. In contrast, when they were tested with B's wrr or chutter, they dishabituated, showing a renewed, strong response. In other words, if A is unreliable with respect to the information conveyed by wrrs, she is also unreliable when she produces chutters because these two calls are produced in the same general context. However, the fact that A is unreliable about intergroup encounters does not mean B is unreliable, as revealed by the vervets' undiminished response to B's chutter or wrr.

These results show that the attribute "unreliable" is assigned to individuals, not contexts. Moreover, experimenters can create unreliable individuals simply by playing back their vocalizations over and over again in the absence of the relevant context. Because individuals habituate to repeated vocalizations in the absence of a relevant context, a mechanism—"skepticism"—is in place for challenging lies of commission. Comparable evidence has been found for vervet and diana monkey alarm calls as well as rhesus monkey food and contact calls (Cheney and Seyfarth 1988; Seyfarth and Cheney 1990; Rendall et al. 1996; Zuberbühler, Noe, and Seyfarth 1997; Hauser 1998).

The habituation–dishabituation paradigm is ideally suited to pushing the issue of skepticism further. For example, all of the playback experiments take place over the course of less than an hour. What is lacking from all of these experiments is the extent to which such experimental slandering affects the individual's subsequent social interactions and for how long. Thus, if we make A unreliable about the presence of an eagle by repeatedly playing back his eagle alarm call, will others continue to ignore his eagle alarm call if it is played back after an hour of silence? How about a day later? Two weeks? If we play A's eagle alarm call and then pair it with the presence of an eagle, will this reinstate his reliability? If not, how many times do we have to play back A's alarm call in the presence of a predator before the rest of the group trusts A? Moreover, we do not yet understand the acoustic basis for distinguishing between call types or between individuals. An understanding of these factors (Section 2) will put us in a much stronger position regarding design and implementation of playback experiments. These types of experiments are needed if we are to better understand the nature of deceptive interactions, the factors that lead to skepticism, and ultimately how individuals acquire reputations as reliable or unreliable signalers.

3.2. Lies of Omission

Before President Clinton's affair with Monica Lewinsky emerged into the public's awareness, it was a secret. By not mentioning their affair to other interested parties (e.g., the First Lady), both the President and Ms. Lewinsky committed lies of omission.

Nonhuman animals are also capable of committing lies of omission and in fact may commit them more often than lies of commission (Cheney and

Seyfarth 1985; Mitchell and Thompson 1986; Marler et al. 1991; Hauser 1996, 1997). One reason for this difference in frequency is that it is more difficult to catch someone who simply remains silent than someone who actively falsifies information. From an empirical perspective, however, researchers working on lies of omission are faced with a far more difficult problem, for they are forced to interpret the absence of a response or behavior. To show that an animal has committed a lie of omission, the same conditions as discussed earlier must hold, plus an extra one. Thus, as for lies of commission, there must be a reliable association between a call type and some event internal or external to the organism; receivers should show a reliable response to such calls, and there must be some behavioral flexibility on the part of callers such that call production is not rigidly triggered by the relevant stimulus. The fourth condition for a lie of omission is that situations that reliably lead to calling in some circumstances fail to do so in a different social context. Given this, studies of withholding information are necessarily tied to studies of what Marler and colleagues (Marler et al. 1986a, 1991) have described as "audience effects," analyses of the social conditions mediating call production as opposed to suppression.

An important goal in studying lies of omission is to document the costs associated with withholding information. If animals have mechanisms for punishing those who are caught withholding information, this would indicate that lies of omission are perceived as such by conspecifics. Thus, if an animal fails to announce the discovery of food and is then caught by another group member, there should be a penalty imposed on the silent discoverer. For example, a male caught withholding information about food might be attacked by a dominant animal, denied shared food in the future, or rejected by a female in future mating attempts. Although the first possibility is nearly immediate, and thus easily observed, the latter requires long-term studies of social reciprocation.

To flesh out these ideas, we return to some of the themes discussed above, exploring when animals attempt to withhold information, how they benefit from such lies, and the costs incurred if caught. We focus on cases where animals withhold food and alarm calls because the absence of production can be contrasted with the many contexts in which such calls are typically given. To broaden the diversity of contexts, we also discuss calls given by animals during mating and the contexts in which animals are selectively silent about their sexual behavior.

When domestic chickens find food, they give a distinctive food call. When domestic chickens detect an aerial predator, they give a different-sounding alarm call than when they detect a ground predator. Such calls are functionally referential in that they appear to be highly correlated with particular contexts, and when receivers hear these calls, they respond in highly specific ways, approaching food calls, looking up to aerial predator alarm calls, and scanning the horizon in response to ground predator alarm calls. Playbacks of these calls show that they reliably elicit different, adaptive responses (Gyger et al. 1986, 1987; Karakashian et al. 1988; Marler et al. 1991, 1992). As Marler and his colleagues have documented over the past ten years, whether a chicken produces a food or alarm call depends on who is around—its audience (Marler et al. 1986a, 1986b, 1991; Karakashian et al. 1988; Evans and Marler 1995).

When cockerels were presented with food (e.g., mealworms), they announced their discovery 100% of the time in the presence of a familiar female, 95% of the time in the presence of an unfamiliar female, 75% of the time when alone, and never called in the presence of another cockerel. Thus, the presence of a hen potentiated food calling, whereas the presence of a cockerel apparently suppressed food calling. The pattern of calling and suppression is, however, slightly different when cockerels are presented with a nonfood item such as a peanut shell. When peanut shells were present, cockerels produced food calls about 50% of the time in the presence of a strange female but only about 20% of the time in the presence of familiar females. Cockerels almost never called when they were alone or in the presence of another male.

The food-calling system of the chicken provides evidence for the definitional conditions set out above. Several studies have shown that when chickens find food, they give a characteristic call that functionally refers to food; the rate at which chickens give food calls appears to reflect the individual's preference for the particular food type. Given the fact that the signal functionally refers to food, we must then ask why chickens sometimes suppress their calls, committing a lie of omission. The results obtained by Marler and his colleagues suggest that chickens assess the costs and benefits of withholding information by attending to the composition of the audience. When another male is present, cockerels are silent because calling would increase the costs of food competition while bringing no benefits. In contrast, when females are present, calling to food increases the costs of food competition but returns the benefits associated with sexual access or opportunity. Importantly, however, cockerels fail to call to nonfood items if a familiar female is nearby but often call if an unfamiliar female is present. Marler and colleagues offer the intriguing speculation that this pattern results from the fact that cockerels can tolerate the potential costs of a lie of commission—giving a food call to a nonfood item—when the female is unfamiliar but cannot afford such costs when the female is familiar.

The data that Marler and his colleagues have collected clearly show that chickens are sensitive to the contexts in which a lie of omission pays. At present, we have no understanding of the costs associated with being caught committing a lie of omission. For example, do dominant cockerels attack subordinate cockerels who have withheld information about the presence of food? Do unfamiliar females reject the mating advances of cockerels who have withheld information about food? In addition to addressing these questions, studies of the chicken food-call system might also profitably explore additional manipulations of the audience, food, and the signaler's motivational state. Thus, for example, all of the studies conducted to date have used audiences consisting of a single individual. Presumably, this is a relatively rare situation in nature, where several individuals are likely to be in view of a cockerel finding food; minimally, there will be several individuals within hearing range. How would a cockerel's food-calling behavior be affected by the presence of its mate and an unfamiliar female, or its mate and a cockerel? In addition to audience composition, it would be useful to look at the interaction between the cockerel's hunger level and the composition of the audience. If a cockerel is extremely hungry, and only a limited amount of food is present, does it remain silent even if a female is nearby? Finally, does the probability of remaining silent change as a function of whether the food is shareable as opposed to nonshareable? Are cockerels more likely to call if the food is spread out, thereby reducing the potential costs of competition should other individuals approach?

Showing that animals are sensitive to an audience is a critical component in investigations of lies of omission. Cockerels are not only sensitive to the presence or absence of another chicken but are also sensitive to whether the audience is male or female and, if female, whether they are familiar or unfamiliar. A crucial question then is whether such sensitivity is preserved in kind across contexts. If it is, then the system is quite rigid. Marler, Evans, and their colleagues have investigated this problem in considerable detail, and the results indicate considerable flexibility rather than rigidity. In striking contrast to the effects of an audience on food-calling behavior, alarm calls are potentiated equally by hens and cockerels. That is, the rate of alarm-call production is the same for male and female audiences, and this is true of real audiences as well as audiences simulated by video playbacks (Evans and Marler 1991, 1995). The rate of alarm-call production is higher in the presence of either a cockerel or hen than it is when there is no audience present or when the audience is comprised of a different species, such as a bobwhite quail. The decrease in alarm-call rate in the presence of bobwhite quail is not due to their smaller size because chickens produce a higher rate of alarm calls to chicks who are even smaller than quail. Finally, cockerels call more to a sexually receptive mate than to a broody hen with and without chicks and also call more when testosterone levels are elevated.

In summary, chickens have the capacity to withhold information in the context of food and predation. Chickens are capable of committing lies of omission, but the social consequences of such deception remain unclear.

Cheney and Seyfarth's work on the alarm-call system of vervet monkeys in Amboseli National Park, Kenya, represents one of the best-studied functionally referential call systems (Struhsaker 1967; Seyfarth et al. 1980a, 1980b; Cheney and Seyfarth 1981, 1990; Marler 1985; Hauser 1996). These vervets produce a suite of acoustically distinctive alarm calls in the context of predator encounters. Of the set produced, the best-studied are those given to snakes, eagles, and leopards. These three predator types exhibit different hunting strategies, and such differences appear to have led to the evolution of different alarm calls and escape strategies in vervets. Thus, in terms of the first condition for exploring lies of omission, vervets are ideally suited. These three alarm calls are primarily heard in the context of predator encounters, and, in particular, snake alarm calls are given to snakes, eagle alarm calls to eagles, and leopard alarm calls to leopards. When such calls are heard, listeners respond in highly specific ways that are suited to the style of predation. Two field observations suggest that, as in chickens, the social context also mediates alarm-call production in vervets. First, Cheney and Seyfarth (1988) noted that when lone vervets detect a predator, they remain silent. Second, low-ranking animals produce far fewer alarm calls than do high-ranking animals. The difference in alarm-call rate between high- and low-ranking animals is not due to differences in the number of kin or in the probability of detecting a predator. Rather, the observations suggest that low-ranking animals may actively suppress their alarm calls, committing lies of omission.

To test the prediction that vervet monkey alarm calls are mediated by social context, Cheney and Seyfarth (1985) conducted an experiment with captive vervet monkeys. Using a human dressed up as a predator (a graduate student in a monkey mask and lab coat, carrying a net), adult females were tested in the presence of their offspring or an unrelated but age-matched infant, whereas adult males were tested in the presence of an adult female or another adult male. Results demonstrated flexibility in alarm calling in vervets, with both males and females producing calls in some circumstances and suppressing them in others. Females produced significantly more alarm calls in the presence of their offspring than in the presence of unrelated infants, and males produced more alarm calls in the presence of adult females than in the presence of adult males; adult males were virtually silent in the presence of other males. For females, therefore, kinship appears to play an important role in the mediation of alarm calls, whereas for males, mating opportunities and intrasexual competition play a role.

As with studies of chickens, these experiments clearly show that both wild and captive vervets have the capacity to withhold information. They do not, however, allow us to assess the costs and benefits of such omissions. To better assess the economics of withholding information, one would need to perform similar experiments over a longer time course, assessing how individuals respond to those who have concealed information about predators, food, and so forth. We now turn to such a study.

Many primate species produce distinctive calls in the context of food (Wrangham 1977; Dittus 1984; Hauser and Wrangham 1987; Elowson et al. 1991; Benz 1993; Hauser and Marler 1993a and b). Like chickens, food commonly elicits such calls, and, in many species, the rate of call production covaries with the quality of the food or the individual's preference for a particular food type. When individuals hear food-associated calls, they typically orient and then approach the caller and sometimes call back with the same call type.

Hauser and Marler (1993) investigated the food-associated calls of rhesus monkeys living on the island of Cayo Santiago off the coast of Puerto Rico. When individuals discover food, they give one or more of five acoustically distinct vocalizations. Three of these call types are given when high-quality, rare food is discovered, whereas the two other call types are given when lower-quality, common food is discovered. Playback experiments reveal that these call types are classified on the basis of their putative referents and not on the basis of acoustic morphology (Hauser 1998).

Given that rhesus monkeys are sometimes silent when they discover food, experiments were conducted to provide a more precise quantification of the necessary and sufficient conditions for call production as opposed to suppression. A lone individual was presented with the same quantity of either monkey chow (low-quality/common food) or coconut (high-quality/rare food) and its behavior recorded. On approximately 50% of all trials with individuals who were members of a social group, the food was consumed, but there were no vocalizations. Paralleling our naturalistic observations, males called less often in the context of a food discovery than did females. However, there were no differences in call rate between high- and low-ranking discoverers. Independently of rank or sex, individuals who called obtained more food, and received less aggression, than individuals who were silent and caught with the food by other group members. That is, silent discoverers were chased and physically attacked, and, as a result, obtained less food. However, individuals who were silent and never caught at the food source obtained more food than anyone else. These results suggest that rhesus monkeys have the capacity to commit lies of omission and that there are measurable social costs and benefits to such deception.

What makes the rhesus case particularly interesting is that the dynamics of this kind of deception change as a function of the discoverer's group status. All of the results reported above were obtained from individuals resident within a social group. When the same experiment is conducted on peripheral males—individuals who have yet to join a social group—such males never call and when they are caught with food are never attacked (Hauser 1997). Rather, when members of a social group catch a peripheral male, they supplant him from the food or chase him away without making physical contact. Thus, peripheral males can get away with lies of omission without paying the costs. One explanation for this difference between resident and peripheral males may be that only resident males are involved in subsequent social interactions.

The rhesus food-call system satisfies the conditions for an analysis of lies of omission and furthermore provides some evidence of an immediate punishment for those who commit them. Food calls are generally produced in the context of finding food, and when individuals hear them, they approach. Thus, there is a strong association between call and context, and the call elicits a predictable response. When animals withhold information about the discovery of food, they incur the costs of targeted aggression if caught but otherwise benefit if no one discovers them. However, like the other studies, research on rhesus monkey food calls fails to address the long-term consequences of withholding information. For example, when silent peripheral males are caught at a food source, how does their silence affect the odds of joining a social group? If an estrous female finds a silent male at a food source, is she less likely to initiate or accept a sexual consortship with him? Are males less likely to withhold information about food in the presence of an estrous female as opposed to a nonestrous female? Are males more likely to remain quiet when no one is around or when an adult male is nearby?

One last example illustrates that lies of omission are possible outside the contexts of food and alarm and provides suggestive evidence of more long-term costs and benefits of deception. Like many other primates (Hauser 1996), rhesus monkeys produce distinctive vocalizations during copulation (Hauser 1993). These copulation screams are among the loudest calls in the repertoire, are acoustically different from all other call types, and are highly distinctive by individual. Thus, when a rhesus monkey gives a copulation call, listeners know what the caller is doing and who he is. Focal animal samples of 47 adult males and 59 estrous females during the mating season indicated that males were more likely to call during copulation when the number of estrous females was high than when it was low. Thus, males often copulated in silence. Interestingly, silent males generally produced the facial expressions that accompany the copulation call, suggesting that they can inhibit the vocalization but not the facial gesture. When the data are divided according to male dominance rank, high-ranking males called more often than expected, low-ranking males called less often than expected, and middle-ranking males called at the average rate.

To assess the consequences of calling as opposed to withholding information about mating behavior, we carried out analyses of the relationship between mating success and call frequency. Results showed that for a given female, the male who obtained the most copulations was a male who called during copulation. Moreover, calling males copulated more overall than did silent males. Thus, there may be a long-term cost to withholding information, at least in terms of behavioral measures of mating success. The benefits obtained from calling were, however, associated with costs. Analyses revealed a statistically significant, positive relationship between the number of copulation calls produced and the number of aggressive attacks received. Withholding information about mating is therefore associated with lower benefits than calling but is also associated with lower costs. It would seem that the benefits of increased mating opportunities outweigh the costs of intrasexual competition.

Thus, studies of rhesus copulation calls also satisfy the conditions for lies of omission and provide suggestive evidence for punishment. Copulation calls are strictly associated with the context of mating, and males often produce such calls when they copulate. Given that males were more likely to remain silent when competition for estrous females was high, and that both high- and low-ranking males have the capacity to remain silent during copulation, these data indicate the ability to withhold information about mating. Although such lies of omission are associated with reduced aggression, they are also associated with increased costs due to the fact that silent males obtain fewer opportunities to mate than do vocal males. It would be interesting to follow up this work with studies in captivity that explore more carefully the mechanisms underlying mating decisions and whether females copy the mating preferences of others. Thus, one could set up a situation in which one female watched a male copulate and vocalize with female A and a second male copulate and remain silent with female B. Given a choice, which male will the observer female select? On Cayo Santiago, some females cycle in and out of estrous several times during the mating season. Sometimes females mate with the same male on each cycle, and sometimes they mate with different males. When females switch males, it would be interesting to determine whether they switched from a silent to a vocal male. Such studies would help our understanding of why rhesus monkeys sometimes choose to vocalize when they mate and sometimes choose to remain silent.

4. Summary and Conclusions

In this paper, we have argued that the study of the proximate mechanisms underlying vocal behavior, both physiological and cognitive, is a necessary part of the study of the evolution of communication and in particular for analyzing honesty in communication. In the first section, we surveyed basic principles of vocal production in terrestrial vertebrates and the morphological diversity of their production systems. We then provided some examples of the interactions between acoustics and anatomy that can enforce honesty or subvert it. In the second section, we examined the evidence for cognitive mechanisms that allow animals to produce deceptive calls as well as "retaliatory" perceptual mechanisms that allow perceivers to accurately identify and ignore (and in some cases even punish) the deceivers. Both vocal-production mechanisms and cognitive mechanisms controlling vocalization play a crucial role in determining what is possible or impossible in a particular species' communication system. A better understanding of these mechanisms can lead to rich insights into the evolution of acoustic communication.

For the reader already interested in mechanism, the chapter also provided illustrations of the value of an ultimate evolutionary viewpoint. An evolutionary perspective proves valuable both for identifying functional problems that are solved by communicators and for using phylogenies and the comparative method as tools to identify and understand widespread selective pressures and functional constraints. The species we observe today are the outcome of a long, dynamic process of coevolution and interaction.

Signalers' ability to avoid, repel, or attract predators, competitors, and potential mates has played a critical role in the evolution of their acoustic signals, including the mechanisms that produce them. A comprehensive answer to the question "why do birds sing?" or "why do deer roar?" will always go beyond the proximate mechanisms to the ultimate function, the selective value that allowed singing or roaring animals to outreproduce their mute conspecifics. As pointed out long ago by Tinbergen (1963), these two perspectives, proximate and ultimate, are complementary. Each provides a rich source of insights and testable hypotheses that the other does not. We believe that vertebrate acoustic communication provides numerous model systems that are ideally suited to integrating these two perspectives and that such integration will prove vital in understanding the remarkable diversity of acoustic signals and the mechanisms that produce them.

Acknowledgments. This work was supported by NIH/NIDCD Grant T32 DC00038 to WTF and NSF Grants SBR-9602858 and SBR-9357976 to MDH. We gratefully acknowledge the comments of the editors, Asif Ghazanfar, Hanspeter Herzel, Philip Lieberman, Tobias Riede, David Reby, Mike Ryan, and Brad Story on an earlier version of the manuscript.

References

Alcantara M, Diaz M, Pulido FJP (1991) Variabilidad en las relaciones alométricas entre el peso y las medidas craneales en el raton de campo *Apodemus sylvaticus* L. Efectos sobre su utilidad en estudios de ecologia trofica de aves rapaces. Doñana, Acta Vertebrata 18:205–216.

Amadon D (1969) Variation in the trachea of the Cracidae (Galliformes) in relation to their classification. Nat Hist Bull Siam Soc 23:239–248.

Andrew RJ (1976) Use of formants in the grunts of baboons and other nonhuman primates. Ann N Y Acad Sci 280:673–693.

Asquith A, Altig R (1990) Male call frequency as a criterion for female choice in *Hyla cinerea*. J Herpetol 24:198–201.

August PV, Anderson JGT (1987) Mammal sounds and motivation-structural rules: A test of the hypothesis. J Mammal 68:1–9.

Baken RJ (1987) Clinical measurement of speech and voice. Boston: Little, Brown and Co.

Bass AH (1989) Evolution of vertebrate motor systems for acoustic and electric communication: Peripheral and central elements. Brain Behav Evol 33:237–247.

Bass AH, Baker R (1997) Phenotypic specification of hindbrain rhombomeres and the origins of rhythmic circuits in vertebrates. Brain Behav Evol 50:3–16.

Bauer HR (1987) Frequency code: Orofacial correlates of fundamental frequency. Phonetica 44:173–191.

Beckford NS, Rood SR, Schaid D (1985) Androgen stimulation and laryngeal development. Ann Otol Rhinol Laryngol 94:634–640.

Beil RG (1962) Frequency analysis of vowels produced in a helium-rich atmosphere. J Acoust Soc Am 34:347–349.

Benz JJ (1993) Food-elicited vocalizations in golden lion tamarins: Design features for representational communication. Anim Behav 45:443–455.
Beranek LL (1954) Acoustics. New York: McGraw–Hill.
Bickley C, Stevens K (1986) Effect of a vocal tract constriction on the glottal source: Experimental and modeling studies. J Phonetics 14:373–382.
Bradbury JW (1977) Lek behavior in the hammer-headed bat. Z Tierpsychol 45:225–255.
Bradbury JW, Vehrencamp SL (1998) Principles of Animal Communication. Sunderland, MA: Sinauer Associates.
Brittan-Powell EF, Dooling RJ, Larsen OH, Heaton JT (1997) Mechanisms of vocal production in budgerigars (*Melopsittacus undulatus*). J Acoust Soc Am 101: 578–589.
Brown CH, Cannito MP (1995) Modes of vocal variation in Syke's monkey (*Cercopithecus albogularis*) squeals. J Comp Psychol 109:398–415.
Brown CH, Gomez R (1992) Functional design features in primate vocal signals: The acoustic habitat and sound distortion. In: Nishida T, McGrew WC, Marler P, Pickford M, Waal FD (eds) Topics in Primatology, Vol 1, Human Origins. Tokyo: Tokyo University Press, pp. 177–198.
Burghardt GM (1991) Cognitive ethology and critical anthropomorphism: A snake with two heads and hognose snakes that play dead. In: Ristau CA (ed) Cognitive Ethology: The Minds of other Animals. Hillsdale, NJ: Lawrence Erlbaum Associates, pp. 53–91.
Byrne R, Whiten A (1990) Tactical deception in primates: The 1990 database. Primate Rep 27:1–101.
Calder WA (1970) Respiration during song in the canary (*Serinus canaria*). Comp Biochem Physiol 32:251–258.
Calder WA (1984) Size, Function, and Life History. Cambridge, MA: Harvard University Press.
Capranica RR, Moffat AJM (1983) Neurobehavioral correlates of sound communication in anurans. In: Ewert JP, Capranica RR, Ingle DJ (eds) Advances in Vertebrate Neuroethology. New York: Plenum Press, pp. 701–730.
Carroll RL (1997) Patterns and Processes of Vertebrate Evolution. New York: Cambridge University Press.
Catchpole CK, Slater PLB (1995) Bird Song: Themes and Variations. New York: Cambridge University Press.
Chapin JP (1922) The function of the oesophagus in the bittern's booming. Auk 39:196–202.
Cheney DL, Seyfarth RM (1981) Selective forces affecting the predator alarm calls of vervet monkeys. Behaviour 76:25–61.
Cheney DL, Seyfarth RM (1985) Vervet monkey alarm calls: Manipulation through shared information? Behaviour 94:150–166.
Cheney DL, Seyfarth RM (1988) Assessment of meaning and the detection of unreliable signals by vervet monkeys. Anim Behav 36:477–486.
Cheney DL, Seyfarth RM (1990) How Monkeys See the World: Inside the mind of another species. Chicago: University of Chicago Press.
Clench MH (1978) Tracheal elongation in birds-of-paradise. Condor 80:423–430.
Clutton-Brock TH, Albon SD (1979) The roaring of red deer and the evolution of honest advertising. Behaviour 69:145–170.

Clutton-Brock TH, Harvey PH, Rudder B (1977) Sexual dimorphism, socionomic sex ratio and body weight in primates. Nature 269:797–800.
Cohen JR, Crystal TH, House AS, Neuburg EP (1980) Weighty voices and shaky evidence: A critique. J Acoust Soc Am 68:1884–1886.
Crelin E (1987) The Human Vocal Tract. New York: Vantage Press.
Crumly CR (1984) Evolution of land tortoises. Ph. D. Thesis, Rutgers University, Piscataway, NJ.
Cullen JM (1966) Reduction of ambiguity through ritualization. Philos Trans R Soc Lond B Biol Sci 251:363–374.
Darling FF (1937) A Herd of Red Deer. Oxford: Oxford University Press.
Darwin C (1859) On the Origin of Species. London: John Murray.
Darwin C (1871) The Descent of Man and Selection in Relation to Sex. London: John Murray.
Davies NB, Halliday TR (1978) Deep croaks and fighting assessment in toads *Bufo bufo*. Nature 274:683–685.
Dawkins R, Krebs JR (1978) Animal signals: Information or manipulation? In: Krebs JR, Davies NB (eds) Behavioural Ecology. Oxford: Blackwell Scientific Publications, pp. 282–309.
de Waal FBM (1988) The communicative repertoire of captive bonobos (*Pan paniscus*), compared to that of chimpanzees. Behaviour 106:183–251.
Dechow PC (1983) Estimation of body weights from craniometric variables in baboons. Am J Phys Anthropol 60:113–123.
Delacour J, Amadon D (1973) Curassows and Related Birds. New York: American Museum of Natural History.
Demski LS, Gerald JW, Popper AN (1973) Central and peripheral mechanisms of teleost sound production. Am Zool 13:1141–1167.
Dittus WPG (1984) Toque macaque food calls: Semantic communication concerning food distribution in the environment. Anim Behav 32:470–477.
Drewry GE, Heyer WR, Rand AS (1982) A functional analysis of the complex call of the frog *Physalaemus pustulosus*. Copeia 1982:636–645.
Dudley R, Rand AS (1991) Sound production and vocal sac inflation in the Túngara frog, *Physalaemus pustulosus* (Leptodactylidae). Copeia 1991:460–470.
Dyer FC, Seeley TD (1989) On the evolution of the dance language. Am Nat 133:580–590.
Elowson AM, Tannenbaum PL, Snowdon CT (1991) Food-associated calls correlate with food preferences in cotton-top tamarins. Anim Behav 42:931–937.
Evans CS, Evans L (1999) Chicken food calls are functionally referential. Anim Behav 58:307–319.
Evans CS, Marler P (1991) On the use of video images as social stimuli in birds: Audience effects on alarm calling. Anim Behav 41:17–26.
Evans CS, Marler P (1994) Food-calling and audience effects in male chickens. *Gallus gallus*: Their relationships to food availability, courtship and social facilitation. Anim Behav 47:1159–1170.
Evans CS, Marler P (1995) Language and Animal Communication: Parallels and Contrasts. In: Roitblatt H (ed) Comparative Approaches to Cognitive Science. Cambridge, MA: MIT Press, pp. 241–382.
Fant G (1960) Acoustic Theory of Speech Production. The Hague: Mouton and Co.
Fay FH (1960) Structure and function of the pharyngeal pouches of the walrus (*Odobenus rosmarus* L.). Mammalia 24:361–371.

Fitch WT (1994) Vocal tract length perception and the evolution of language. Ph.D. Thesis, Brown University, Providence, RI.
Fitch WT (1997) Vocal tract length and formant frequency dispersion correlate with body size in rhesus macaques. J Acoust Soc Am 102:1213–1222.
Fitch WT (1999) Acoustic exaggeration of size in birds by tracheal elongation: comparative and theoretical analyses. J Zool (London), 248:31–49.
Fitch WT (2000a) The phonetic potential of nonhuman vocal tracts: Comparative cineradiographic observations of vocalizing animals. Phonetica 57:205–218.
Fitch WT (2000b) The evolution of speech: A comparative review. Trends Cogn Sci 4:258–267.
Fitch WT (2000c) Skull dimensions in relation to body size in nonhuman mammals: The causal bases for acoustic allometry. Zoology 103:40–58.
Fitch WT, Giedd J (1999) Morphology and development of the human vocal tract: A study using magnetic resonance imaging. J Acoust Soc Am 106:1511–1522.
Fitch WT, Hauser MD (1995) Vocal production in nonhuman primates: Acoustics, physiology, and functional constraints on "honest" advertisement. Am J Primatol 37:191–219.
Fitch WT, Kelley JP (2000) Perception of vocal tract resonances by whooping cranes, *Grus americana*. Ethology 106:559–574.
Fitch WT, Reby D (2001) The descended larynx is not uniquely human. Proc R Soc Lond B Biol Sci 268:1669–1675.
Fujimura O, Lindqvist J (1970) Sweep-tone measurements of vocal tract characteristics. J Acoust Soc Am 49:541–558.
Gans C (1970) Respiration in early tetrapods—the frog is a red herring. Evolution 24:723–734.
Gans C (1973) Sound production in the Salientia: Mechanism and evolution of the emitter. Am Zool 13:1179–1194.
Gans C, Maderson PFA (1973) Sound producing mechanisms in recent reptiles: Review and comment. Am Zool 13:1195–1203.
Gaunt A, Nowicki S (1998) Sound production in birds: Acoustics and physiology revisited. In: Hopp SL, Owren MJ, Evans CS (eds) Animal acoustic communication: Sound analysis and research methods. New York: Springer, pp. 291–321.
Gaunt AS, Gaunt SLL (1985) Syringeal structure and avian phonation. In: Johnson RF (ed) Current Ornithology. New York: Plenum Press, pp. 213–245.
Gaunt AS, Wells MK (1973) Models of syringeal mechanisms. Am Zool 13:1227–1247.
Gaunt AS, Gaunt SLL, Prange HD, Wasser JS (1987) The effects of tracheal coiling on the vocalizations of cranes (Aves: Gruidae). J Comp Physiol A 161:43–58.
Gautier JP (1971) Etude morphologique et fonctionnelle des annexes extra-laryngées des cercopithecinae; liaison avec les cris d'espacement. Biol Gabonica 7:230–267.
Gerhardt HC (1982) Sound pattern recognition in some North American tree-frogs (Anura: Hylidae): Implications for mate choice. Am Zool 22:581–595.
Gerhardt HC (1991) Female mate choice in treefrogs: Static and dynamic acoustic criteria. Anim Behav 42:615–636.
Goldstein UG (1980) An articulatory model for the vocal tracts of growing children. D. Sc. Thesis, Massachusetts Institute of Technology, Cambridge, MA.
Goller F, Larsen ON (1997a) *In situ* biomechanics of the syrinx and sound generation in pigeons. J Exp Biol 200:2165–2176.

Goller F, Larsen ON (1997b) A new mechanism of sound generation in songbirds. Proc Nat Acad Sci USA 94:14787.
Gould JL (1990) Honey bee cognition. Cognition 37:83–103.
Gould JL, Gould CG (1988) The Honey Bee. New York: Freeman Press.
Gould JL, Towne WF (1987) Evolution of the dance language. Am Nat 130:317–338.
Grafen A (1990) Biological signals as handicaps. J Theor Biol 144:475–546.
Greenewalt CH (1968) Bird Song: Acoustics and Physiology. Washington, DC: Smithsonian Institution Press.
Griffin DR (1958) Listening in the Dark. New Haven, CT: Yale University Press.
Gyger M, Marler P (1988) Food calling in the domestic fowl (*Gallus gallus*): The role of external referents and deception. Anim Behav 36:358–365.
Gyger M, Karakashian S, Marler P (1986) Avian alarm calling: Is there an audience effect? Anim Behav 34:1570–1572.
Gyger M, Marler P, Pickert R (1987) Semantics of an avian alarm calling system: The male domestic fowl, *Gallus domesticus*. Behaviour 102:15–40.
Harrison DFN (1995) The anatomy and physiology of the mammalian larynx. New York: Cambridge University Press.
Hartley DJ, Suthers RA (1988) The acoustics of the vocal tract in the horseshoe bat, *Rhinolophus hildebrandti*. J Acoust Soc Am 84:1201–1213.
Hartley RS, Suthers RA (1989) Airflow and pressure during canary song: Direct evidence for mini-breaths. J Comp Physiol A 165:15–26.
Hast M (1989) The larynx of roaring and non-roaring cats. J Anat 163:117–121.
Hausberger M, Black JM, Richard J-P (1991) Bill opening and sound spectrum in barnacle goose loud calls: Individuals with "wide mouths" have higher pitched voices. Anim Behav 42:319–322.
Hauser MD (1992a) Articulatory and social factors influence the acoustic structure of rhesus monkey vocalizations: A learned mode of production? J Acoust Soc Am 91:2175–2179.
Hauser MD (1992b) Costs of deception: Cheaters are punished in rhesus monkeys. Proc Nat Acad Sci USA 89:12137–12139.
Hauser MD (1993) Rhesus monkey (*Macaca mulatta*) copulation calls: Honest signals for female choice? Proc R Soc Lond 254:93–96.
Hauser MD (1996) The Evolution of Communication. Cambridge, MA: MIT Press.
Hauser MD (1997) Minding the behavior of deception. In: Whiten A, Byrne RW (eds) Machiavellian Intelligence II. Cambridge, UK: Cambridge University Press, pp. 112–143.
Hauser MD (1998) Functional referents and acoustic similarity: Field playback experiments with rhesus monkeys. Anim Behav 55:1647–1658.
Hauser MD (2000) Wild Minds: What Animals Really Think. New York: Henry Holt.
Hauser MD, Fowler C (1991) Declination in fundamental frequency is not unique to human speech: Evidence from nonhuman primates. J Acoust Soc Am 91: 363–369.
Hauser MD, Marler P (1993a) Food-associated calls in rhesus macaques (*Macaca mulatta*). I. Socioecological factors influencing call production. Behav Ecol 4:194–205.
Hauser MD, Marler P (1993b) Food-associated calls in rhesus macaques (*Macaca mulatta*). II. Costs and benefits of call production and suppression. Behav Ecol 4:206–212.

Hauser MD, Schön-Ybarra M (1994) The role of lip configuration in monkey vocalizations: Experiments using xylocaine as a nerve block. Brain Lang 46:232–244.
Hauser MD, Wrangham RW (1987) Manipulation of food calls in captive chimpanzees: A preliminary report. Folia Primatol 48:24–35.
Hauser MD, Evans CS, Marler P (1993) The role of articulation in the production of rhesus monkey (*Macaca mulatta*) vocalizations. Anim Behav 45:423–433.
Hayes MP, Krempels DM (1986) Vocal sac variation among frogs of the genus *Rana* from western North America. Copeia 1986:927–936.
Henton C (1992) The abnormality of male speech. In: Wolf G (ed) New Departures in Linguistics. New York: Garland Publishing.
Hersch GL (1966) Bird voices and resonant tuning in helium air mixtures. Ph.D. Thesis, University of California, Berkeley.
Herzel H (1996) Possible mechanisms of vocal instabilities. In: Davis PJ, Fletcher NH (eds) Vocal Fold Physiology: Controlling Complexity and Chaos. London: Singular Publishers, Inc.
Herzel H, Wendler J (1991) Evidence of chaos in phonatory samples. Proceedings of the 2nd European Conference on Speech Technology. Eurospeech 91: Genoa, Italy, pp. 263–266.
Hildebrand M (1974) Analysis of Vertebrate Structure, 1st ed. New York: John Wiley and Sons.
Hinde RA (1981) Animal signals: Ethological and games-theory approaches are not incompatible. Anim Behav 29:535–542.
Hinds DS, Calder WA (1971) Tracheal dead space in the respiration of birds. Evolution 25:429–440.
Hirano M (1991) Phonosurgical anatomy of the larynx. In: Ford CN, Bless DM (eds) Phonosurgery: Assessment and Surgical Management of Voice Disorders. New York: Raven Press, Ltd.
Hirano M, Kurita S, Yukizane K, Hibi S (1989) Asymmetry of the laryngeal framework: A morphological study of cadaver larynges. Ann Otol Rhinol Laryngol 98:135–140.
Hollien H (1960) Some laryngeal correlates of vocal pitch. J Speech Hear Res 3:52–58.
Hollien H, Green R, Massey K (1994) Longitudinal research on adolescent voice change in males. J Acoust Soc Am 96:2646–2653.
Hosokawa H (1950) On the Cetacean larynx, with special remarks on the laryngeal sack of the Sei whale and the aryteno-epiglottideal tube of the Sperm whale. Sci Rep Whale Res Inst 3:23–62.
Houle D (1998) High enthusiasm and low r-squared. Evolution 52:1872–1876.
Ishizaka K, Flanagan JL (1972) Synthesis of voiced sounds from a two-mass model of the vocal cords. Bell Syst Tech J 51:1233–1268.
Isshiki N, Tanabe M, Ishizaka K, Broad D (1977) Clinical significance of asymmetrical vocal cord tension. Ann Otol Rhinol Laryngol 86:58–66.
Johnsgard PA (1961) Tracheal anatomy of the Anatidae and its taxonomic significance. Wildfowl Trust 12th Annual Report, pp. 58–69.
Johnsgard PA (1971) Observations on sound production in the Anatidae. Wildfowl 22:46–59.
Johnsgard PA (1983) Cranes of the World. Bloomington, IA: Indiana University Press.

Karakashian SJ, Gyger M, Marler P (1988) Audience effects on alarm calling in chickens (*Gallus gallus*). J Comp Psychol 102:129–135.
Kimura D (1983) Sex differences in cerebral organization for speech and praxic functions. Can J Psychol 37:19–35.
Klump GM, Gerhardt HC (1987) Use of non-arbitrary acoustic criteria in mate choice by female gray tree frogs. Nature 326:286–288.
Koenigsknecht RA, Friedman P (1976) Syntax development in boys and girls. Child Dev 47:1109–1115.
Krebs JR, Davies NB (1997) Behavioural Ecology: An Evolutionary Approach, 4th ed. Oxford: Blackwell Scientific Publications.
Krebs JR, Dawkins R (1984) Animal signals: Mind reading and manipulation. In: Krebs JR, Davies NB (eds) Behavioural Ecology. Sunderland, MA: Sinauer Associates, pp. 380–402.
Krogman WM (1941) Growth of Man. The Hague: W. Junk.
Künzel HJ (1989) How well does average fundamental frequency correlate with speaker height and weight? Phonetica 46:117–125.
Larsen ON, Goller F (1999) Role of syringeal vibrations in bird vocalizations. Proc R Soc Lond B Biol Sci 266:1609–1615.
Lasiewski RC (1972) Respiratory function in birds. In: Farner DS, King JR (eds) Avian Biology, Vol 2. New York: Academic Press, pp. 287–342.
Lass NJ, Brown WS (1978) Correlational study of speakers' heights, weights, body surface areas, and speaking fundamental frequencies. J Acoust Soc Am 63: 1218–1220.
Lieberman P (1961) Perturbations in vocal pitch. J Acoust Soc Am 33:597–603.
Lieberman P (1968) Primate vocalization and human linguistic ability. J Acoust Soc Am 44:1574–1584.
Lieberman P (1984) The Biology and Evolution of Language. Cambridge, MA: Harvard University Press.
Lieberman P, Blumstein SE (1988) Speech Physiology, Speech Perception, and Acoustic Phonetics. Cambridge, UK: Cambridge University Press.
Lieberman P, Klatt DH, Wilson WH (1969) Vocal tract limitations on the vowel repertoires of rhesus monkeys and other nonhuman primates. Science 164: 1185–1187.
Liem K (1973) Evolutionary strategies and morphological innovations: Cichlid pharyngeal jaws. Syst Zool 22:425–441.
Liem KF (1985) Ventilation. In: Hildebrand M (ed) Function Vertebrate Morphology. Cambridge, MA: Belknap Press of Harvard University Press, pp. 185–209.
Lloyd JE (1984) On deception, a way of all flesh, and firefly signalling and systematics. In: Dawkins R, Ridley M (eds) Oxford Surveys in Evolutionary Biology, Vol 1. New York: Oxford University Press, pp. 48–54.
MacLarnon A, Hewitt G (1999) The evolution of human speech: The role of enhanced breathing control. Am J Phy Anthropol 109:341–363.
Marion WR (1977) Growth and development of the plain chachalaca in south Texas. Wilson Bull 89:47–56.
Markel JD, Gray AH (1976) Linear Prediction of Speech. New York: Springer-Verlag.
Marler P (1955) Characteristics of some animal calls. Nature 176:6–7.
Marler P (1969) Vocalizations of wild chimpanzees. Recent Adv Primatol 1:94–100.

Marler P (1985) Representational vocal signals of primates. In: Hölldobler B, Lindauer M (eds) Experimental Behavioral Ecology and Sociobiology. Stuttgart: Gustav Fischer Verlag, pp. 211–221.
Marler P, Tenaza R (1977) Signalling behavior of apes with special reference to vocalization. In: Sebeok TA (ed) How Animals Communicate. Bloomington, IN: Indiana University Press, pp. 965–1033.
Marler P, Dufty A, Pickert R (1986a) Vocal communication in the domestic chicken. I. Does a sender communicate information about the quality of a food reterent to a receiver? Anim Behav 34:188–193.
Marler P, Dufty A, Pickert R (1986b) Vocal communication in the domestic chicken. II. Is a sender sensitive to the presence and nature of a receiver? Anim Behav 34:194–198.
Marler P, Evans CS, Hauser MD (1992) Animal signals? Reference, motivation or both? In: Papoucek H, Jürgens U, Papoucek M (eds) Nonverbal Vocal Communication: Comparative and Developmental Approaches. Cambridge, UK: Cambridge University Press, pp. 66–86.
Marler P, Karakashian S, Gyger M (1991) Do animals have the option of withholding signals when communication is inappropriate? The audience effect. In: Ristau C (ed) Cognitive Ethology: The Minds of other Animals. Hillsdale, NJ: Lawrence Erlbaum Associates, pp. 135–186.
Martin BGH, Bellairs ADA (1977) The narial excresence and pterygoid bulla of the gharial, *Gavialis gangeticus* (Crocodilia). J Zool Lond 182:541–558.
Martin WF (1972) Evolution of vocalizations in the genus *Bufo*. In: Blair WF (ed) Evolution in the genus *Bufo*. Austin, TX: University of Texas Press, pp. 279–309.
Matsuoka S (1980) Pseudo warning call in titmice. Tori 29:87–90.
Maynard Smith J (1976) Sexual selection and the handicap principle. J Theor Biol 57:239–242.
Maynard Smith J, Price GR (1973) The logic of animal conflict. Nature 246:15–18.
Maynard Smith J, Burian R, Kauffman S, Alberch P, Campbell J, Goodwin B, Lande R, Raup D, Wolpert L (1985) Developmental constraints and evolution. Q Rev Biol 60:265–287.
McComb KE (1991) Female choice for high roaring rates in red deer, *Cervus elaphus*. Anim Behav 41:79–88.
McLelland J (1989) Larynx & trachea. In: King AS, McLelland J (eds) Form and Function in Birds, Vol 4. New York: Academic Press, pp. 69–103.
Mende W, Herzel H, Wermke K (1990) Bifurcations and chaos in newborn infant cries. Phys Lett A 145:418–424.
Mergell P, Herzel H (1997) Modelling biphonation—The role of the vocal tract. Speech Commun 22:141–154.
Mergell P, Fitch WT, Herzel H (1999) Modeling the role of non-human vocal membranes in phonation. J Acoust Soc Am 105:2020–2028.
Michelsen A, Andersen BB, Storm J, Kirchner WH, Lindauer M (1992) How honeybees perceive communication dances, studied by means of a mechanical model. Behav Ecol Sociobiol 30:143–150.
Miller AH (1934) The vocal apparatus of some North American owls. Wilson Bull 36:204–213.
Miskimen M (1951) Sound production in passerine birds. Auk 68:493–504.
Mitchell RW, Thompson NS (1986) Deception: Perspectives on Human and Nonhuman Deceit. Albany, NY: State University of New York Press.

Modig AO (1996) Effects of body size and harem size on male reproductive behaviour in the southern elephant seal. Anim Behav 51:1295–1306.
Møller AP (1988a) False alarm calls as a means of resource usurpation in the great tit *Parus major*. Ethology 79:25–30.
Møller AP (1988b) Female choice selects for male sexual tail ornaments in the monogamous swallow. Nature 332:640–642.
Møller AP (1989) Viability costs of male tail ornaments in a swallow. Nature 339:132–135.
Møller AP (1990) Deceptive use of alarm calls by male swallows, *Hirundo rustica*. Behav Ecol 1:1–6.
Møller AP (1993) Morphology and sexual selection in the barn swallow *Hirundo rustica* in Chernobyl, Ukraine. Proc R Soc Lond 252:51–57.
Møller AP, Thornhill R (1998) Bilateral symmetry and sexual selection: A meta-analysis. Am Nat 151:174–192.
Moos-Heilen R, Sossinka R (199O) The influence of oestrus on the vocalization of female gelada baboons (*Theropithecus gelada*). Ethology 84:35–46.
Morita S, Ohtsuki F (1973) Secular changes of the main head dimensions in Japanese. Hum Biol 45:151–165.
Morton ES (1975) Ecological sources of selection on avian sounds. Am Nat 109:17–34.
Morton ES (1977) On the occurrence and significance of motivation-structural rules in some birds and mammal sounds. Am Nat 111:855–869.
Müller J (1848) The Physiology of the Senses, Voice and Muscular Motion with Mental Faculties (W Baly, translator). London: Walton and Maberly.
Munn C (1986a) Birds that 'cry wolf'. Nature 319:143–145.
Munn CA (1986b) The deceptive use of alarm calls by sentinel species in mixed species flocks of neotropical birds. In: Mitchell RW, Thompson NS (eds) Deception: Perspectives on Human and Nonhuman Deceit. Albany, NY: State University of New York Press, pp. 169–175.
Napier JR, Napier PH (1985) The natural history of the primates. Cambridge, MA: MIT Press.
Negus VE (1949) The Comparative Anatomy and Physiology of the Larynx. New York: Hafner Publishing Company.
Niemeier MM (1979) Structural and functional aspects of vocal ontogeny in *Grus canadensis* (Gruidae: Aves). Ph.D. Thesis, University of Nebraska, Lincoln, NE.
Nottebohm F (1971) Neural lateralization of vocal control in a passerine bird. J Exp Zool 179:35–49.
Nowicki S (1987) Vocal tract resonances in oscine bird sound production: Evidence from birdsongs in a helium atmosphere. Nature 325:53–55.
Nowicki S, Marler P (1988) How do birds sing? Music Percep 5:391–426.
Owren MJ (1990) Acoustic classification of alarm calls by vervet monkeys (*Cercopithecus aethiops*) and humans: I. Natural calls. J Comp Psychol 104:20–28.
Owren MJ, Bernacki R (1988) The acoustic features of vervet monkey (*Cercopithecus aethiops*) alarm calls. J Acoust Soc Am 83:1927–1935.
Owren MJ, Bernacki RH (1998) Applying linear predictive coding (LPC) to frequency-spectrum analysis of animal acoustic signals. In: Hopp SL, Owren MJ, Evans CS (eds) Animal Acoustic Communication: Sound Analysis and Research Methods. New York: Springer, pp. 130–162.

Owren MJ, Seyfarth RM, Cheney DL (1997) The acoustic features of vowel-like *grunt* calls in chacma baboons (*Papio cyncephalus ursinus*): Implications for production processes and functions. J Acoust Soc Am 101:2951–2963.
Parker GA (1974) Assessment strategy and the evolution of fighting behavior. J Theor Biol 47:223–243.
Perry SF (1989) Mainstreams in the evolution of vertebrate respiratory structures. In: King AS, McLelland J (eds) Form and Function in Birds, Vol 4. New York: Academic Press, pp. 1–67.
Peters RH (1983) The ecological implications of body size. New York: Cambridge University Press.
Prange HD, Wasser JS, Gaunt AS, Gaunt SLL (1985) Respiratory responses to acute heat stress in cranes (Gruidae): The effects of tracheal coiling. Respir Physiol 62:95–103.
Pye JD (1967) Synthesizing the waveforms of bat's pulses. In: Busnel R-G (ed) Animal Sonar Systems. Jouy-en-Josas, France: Laboratoire de Physiologie Acoustique, pp. 69–83.
Quayle CJ (1991) A dissection of the larynx of a humpback whale calf with a review of its functional morphology. Mem Queensland Mus 30:351–354.
Raikow RJ (1986) Why are there so many kinds of Passerine birds? Syst Zool 35:255–259.
Rand AS, Dudley R (1993) Frogs in helium: The anuran vocal sac is not a cavity resonator. Physiol Zool 66:793–806.
Remmers JE, Gautier H (1972) Neural & mechanical mechanisms of feline purring. Respir Physiol 16:351–361.
Rendall CA (1996) Social communication and vocal recognition in free-ranging rhesus monkeys (*Macaca mulatta*). Ph.D. Thesis, University of California, Davis.
Rendall D, Rodman PS, Emond RE (1996) Vocal recognition of individuals and kin in free-ranging rhesus monkeys. Anim Behav 51:1007–1015.
Riede T, Fitch WT (1999) Vocal tract length and acoustics of vocalization in the domestic dog *Canis familiaris*. J Exp Biol 202:2859–2867.
Ristau C (1991) Aspects of the cognitive ethology of an injury-feigning bird, the piping plover. In: Ristau C (ed) Cognitive Ethology: The Minds of other Animals. Hillsdale, NJ: Erlbaum, pp. 91–126.
Rüppell W (1933) Physiologie und Akustik der Vogelstimme. J Ornithol 81:433–542.
Ryan MJ (1985) The Tungara Frog, A Study in Sexual Selection and Communication. Chicago: University of Chicago Press.
Ryan MJ (1988) Constraints and patterns in the evolution of anuran acoustic communication. In: Fritzch B, Ryan MJ, Wilczynski W, Hetherington TE, Walkowiak W (eds) The Evolution of the Amphibian Auditory System. New York: Wiley, pp. 637–677.
Ryan MJ, Brenowitz EA (1985) The role of body size, phylogeny, and ambient noise in the evolution of bird song. Am Nat 126:87–100.
Scammon RE (1927) The developmental anatomy of the chest and thoracic organs. In: Myers JA (ed) The normal chest of the adult and child. Baltimore: Williams & Wilkins, pp. 300–335.
Schevill WE, Watkins WA, Ray C (1966) Analysis of underwater *Odobenus* calls with remarks on the development and function of the pharyngeal pouches. Zoologica (NY) 51:103–106.

Schmidt RS (1965) Larynx control and call production in frogs. Copeia 1965: 143–147.
Schmidt-Nielsen K (1972) How Animals Work. New York: Cambridge University Press.
Schmidt-Nielsen K (1984) Scaling: Why Is Animal Size so Important? New York: Cambridge University Press.
Schmidt-Nielsen K (1997) Animal Physiology: Adaptation and Environment, 5th ed. New York: Cambridge University Press.
Schneider H (1988) Peripheral and central mechanisms of vocalization. In: Fritzsch B, Ryan MJ, Wilczynski W, Hetherington TE, Walkowiak W (eds) The Evolution of the Amphibian Auditory System. New York: John Wiley and Sons, pp. 537–558.
Schneider R (1964) Der Larynx der Saugetiere. Handb Zool 5:1–128.
Schneider R, Kuhn H-J, Kelemen G (1967) Der Larynx des männlichen *Hypsignathus monstrosus* Allen, 1861 (Pteropodidae, Megachiroptera, Mammalia). Z Wiss Zool 175:1–53.
Schön M (1971) The anatomy of the resonating mechanism in howling monkeys. Folia Primatol 15:117–132.
Schön-Ybarra M (1986) Loud calls of adult male red howling monkeys (*Alouatta seniculus*). Folia Primatol 34:204–216.
Schön-Ybarra M (1988) Morphological adaptations for loud phonation in the vocal organ of howling monkeys. Primate Rep 22:19–24.
Schön-Ybarra M (1995) A comparative approach to the non-human primate vocal tract: Implications for sound production. In: Zimmerman E, Newman JD (eds) Current Topics in Primate Vocal Communication. New York: Plenum Press, pp. 185–198.
Schuett GW (1997) Body size and agonistic experience affect dominance and mating success in male copperheads. Anim Behav 54:213–224.
Seeley TD (1992) The tremble dance of the honey bee: Message and meanings. Behav Ecol Sociobiol 31:375–383.
Seyfarth RM, Cheney DL (1990) The assessment by vervet monkeys of their own and another species' alarm calls. Anim Behav 40:754–764.
Seyfarth RM, Cheney DL, Marler P (1980a) Monkey responses to three different alarm calls: Evidence of predator classification and semantic communication. Science 210:801–803.
Seyfarth RM, Cheney DL, Marler P (1980b) Vervet monkey alarm calls: Semantic communication in a free-ranging primate. Anim Behav 28:1070–1094.
Siebenrock F (1899) über den Kehlkopf und die Luftröhre der Schildkröten. Sitzungsberichten der kaiserl. Akad Wiss Wien 108:581–595.
Siller S (1998) A note on errors in Grafen's strategic handicap models. J Theor Biol 195:413–417.
Slavitt DH, Lipton RJ, McCaffrey TV (1990) Glottographic analysis of phonation in the excised canine larynx. Ann Otol Rhinol Laryngol 99:396–402.
Sleptsov MM (1940) On adaptations of pinnipeds to swimming. Zool J 19:379–386.
Smith-Gill SJ, Berven KA (1980) In vitro fertilization and assessment of male reproductive potential using mammalian gonadotropin-releasing hormone to induce spermiation in *Rana sylvatica*. Copeia 1980:723–728.
Steinecke I, Herzel H (1995) Bifurcations in an asymmetric vocal-fold model. J Acoust Soc Am 97:1874–1884.

Strote J, Nowicki S (1996) Responses to songs with altered tonal quality by adult song sparrow (*Melospiza melodia*). Behaviour 133:161–172.
Struhsaker TT (1967) Auditory communication among vervet monkeys (*Cercopithecus aethiops*). In: Altmann SA (ed) Social Communication among Primates. Chicago: University of Chicago Press, pp. 281–324.
Sullivan BK (1984) Advertisement call variation and observations on breeding behavior of *Bufo debilis* and *B. punctatus*. J Herpetol 18:406–411.
Sutherland CA, McChesney DS (1965) Sound production in two species of geese. Living Bird 4:99–106.
Suthers RA (1988) The production of echolocation signals by bats and birds. In: Nachtigall PE, Moore PWB (eds) Animal Sonar: Processes and Performance. New York: Plenum Press, pp. 23–45.
Suthers RA (1990) Contributions to birdsong from the left and right sides of the intact syrinx. Nature 347:473–477.
Suthers RA (1994) Variable asymmetry and resonance in the avian vocal tract: A structural basis for individually distinct vocalizations. J Comp Physiol 175:457–466.
Suthers RA, Fattu JM (1973) Mechanisms of sound production in echolocating bats. Am Zool 13:1215–1226.
Suthers RA, Hector DH (1988) Individual variation in vocal tract resonance may assist oilbirds in recognizing echoes of their own sonar clicks. In: Nachtigall PE, Moore PWB (eds) Animal Sonar: Processes and Performances. New York: Plenum Press, pp. 87–91.
Suthers RA, Hartley DJ, Wenstrup JJ (1988) The acoustic role of tracheal chambers and nasal cavities in the production of sonar pulses by the horseshoe bat, *Rhilophus hildebrandti*. J Comp Physiol A 162:799–813.
Tavolga WN (1964) Sonic characteristics and mechanisms in marine fishes. In: Tavolga WN (ed) Marine Bio-acoustics. Oxford: Pergamon Press, pp. 195–211.
Tigges M, Mergell P, Herzel H, Wittenberg T, Eysholdt U (1997) Observation and modelling of glottal biphonation. Acustica 83:707–714.
Tinbergen N (1963) On aims and methods of ethology. Z Tierpsychol 20:410–433.
Titze IR (1976) On the mechanics of vocal fold-vibration. J Acoust Soc Am 60: 1366–1380.
Titze IR (1980) Comments on the myoelastic-aerodynamic theory of phonation. J Speech Hear Res 23:495–510.
Titze IR (1994) Principles of voice production. Englewood Cliffs, NJ: Prentice–Hall.
Truby HM, Lind J (1965) Cry sounds of the newborn infant. In: Lind J (ed) Newborn Infant Cry. Uppsala: Almquist & Wiksells Boktryckeri, pp. 7–59.
Tuohimaa PT, Kallio S, Heinijoki J (1981) Androgen receptors in laryngeal cancer. Acta Otolaryngol 91:159–164.
Tyack P, Miller EH (in press) Vocal anatomy, acoustic communication, and echolocation in marine mammals. In: Hoelzel R (ed) Evolutionary Biology of Marine Mammals. New York: Blackwell Scientific Publications.
van den Berg J (1958) Myoelastic-aerodynamic theory of voice production. J Speech Hear Res 1:227–244.
van den Berg J (1968) Sound production in isolated human larynges. Ann NY Acad Sci 155:18–27.
van Dommelen WA (1993) Speaker height and weight identification: A reevaluation of some old data. J Phonetics 21:337–341.

von Frisch K (1967) The Dance Language and Orientation of Bees. Cambridge, MA: Harvard University Press.
Warner RW (1972a) The syrinx in family Columbidae. J Zool Lond 166:385–390.
Warner RW (1972b) The anatomy of the syrinx in passerine birds. J Zool Lond 168:381–393.
Watkins WA, Baylor ER, Bowen AT (1970) The call of *Eleutherodactylus johnstonei*, the whistling frog of Bermuda. Copeia 1970:558–561.
Weishampel DB (1981) Acoustic analysis of potential vocalization in lambeosaurine dinosaurs (Reptilia: Ornithischia). Paleobiology 7:252–261.
Welch AM, Semlitsch RD, Gerhardt HC (1998) Call duration as an indicator of genetic quality in male gray tree frogs. Science 280:1928–1930.
Westneat MW, Long JHJ, Hoese W, Nowicki S (1993) Kinematics of birdsong: Functional correlation of cranial movements and acoustic features in sparrows. J Exp Biol 182:147–171.
White SS (1968) Movements of the larynx during crowing in the domestic cock. J Anat 103:390–392.
Whiten A, Byrne RW (1988) Tactical deception in primates. Behav Brain Sci 11: 233–273.
Wilden I, Herzel H, Peters G, Tembrock G (1998) Subharmonics, Biphonation, and Deterministic Chaos in Mammal Vocalization. Bioacoustics 9:171–196.
Wiley RH, Richards DG (1982) Adaptations for acoustic communication in birds: Sound propagation and signal detection. In: Kroodsma DE, Miller EH (eds) Acoustic Communication in Birds, Vol 1. New York: Academic Press, pp. 131–181.
Williams H, Crane LA, Hale TK, Esposito MA, Nottebohm F (1992) Right-side dominance for song control in the zebra finch. J Neurobiol 23:1006–1020.
Wolff PH (1969) The natural history of crying and other vocalizations in early infancy. In: Foss BM (ed) Determinants of Infant Behaviour IV. London: Methuen, pp. 81–109.
Wrangham RW (1977) Feeding behaviour of chimpanzees in Gombe National Park, Tanzania. In: Clutton-Brock TH (ed) Primate Ecology: Studies or Feeding and Ranging Behaviour in Lemurs, Monkeys and Apes. London: Academic Press, pp. 504–538.
Wunderlich L (1886) Beiträge zur vergliechenden Anatomic und entwicklungsgeschichte des unteren Kehlkopfes der Vögel. Nova Acta Acad Caesareae Leopoldino-Carolinae Germanicae Nat Curiosorum 48:1–80.
Young BA (1992) Tracheal diverticula in snakes: Possible functions and evolution. J Zool Lond 227:567–583.
Zahavi A (1975) Mate selection: A selection for a handicap. J Theor Biol 53:205–214.
Ziswiler V, Farner DS (1972) Digestion and the digestive system. In: Farner DS, King JR (eds) Avian Biology, Vol 2. New York: Academic Press, pp. 343–430.
Zuberbühler K, Cheney DL, Seyfarth RM (1997) Conceptual semantics in a nonhuman primate. J Comp Psychol 113:33–42.
Zuberbühler K, Noe R, Seyfarth RM (1997) Diana monkey long-distance calls: Messages for conspecifics and predators. Anim Behav 53:589–604.

4
Social Sounds: Vocal Learning and Development of Mammal and Bird Calls

Janette Wenrick Boughman and Cynthia F. Moss

1. Introduction

First, we hear a rustle in the shrubs, and then a bird whose call sounds like "where are you?" From a short distance away, another bird gives a slightly different call: "oh, where are you?" For simplicity, we will call these WAY calls. In answer, several small gray and white wrens fly toward the first caller, giving WAY calls that sound virtually identical to the first caller's. At the same time, a small flock flies toward the second caller, giving their own WAY calls. A skirmish follows, with much calling on either side. Careful listening picks up distinct WAY calls throughout and another staccato call, which turns out to be a duet. When the dust settles, the small flocks move off in different directions, territory boundaries apparently stable for the day.

The birds are stripe-backed wrens, and this species lives in stable groups composed of a dominant pair and their offspring, who cooperate to care for the pair's brood (Rabenold 1984). Group mates also cooperate to defend their joint territory. The dominant pair advertises boundaries by singing duets, but when a neighbor intrudes, WAY calls provide a means of identifying group mates. All males in a group share a repertoire of about 12 WAY calls that they use both to maintain contact among the group and to identify themselves in territory disputes. Female group members have their own shared repertoire of WAY calls (Price 1998a). WAY calls indicate group membership—a kind of badge to indicate "I belong to this group." All stripe-backed wrens live in social groups, so belonging to a group is critical. But, how do group members come to share calls? Most group mates are closely related, so call similarity based on genetic similarity is one possibility. Yet, calls are passed down along sex-specific lines, and a genetic mechanism for this is difficult to envision. Instead, it seems that calls are learned—sons copy their fathers and daughters copy their mothers, and do so with remarkable precision. Like a surname passed on from father to son, or mother to daughter (Price 1998a), calls convey family identity (Price 1999). The patterns observed suggest that call sharing is essential to call function. Although most males in a group are closely related, occasionally juvenile males immigrate (Piper et al. 1995) and then learn their new group's call repertoire (Price 1998b). For the occasional immigrants to be able to match the calls of a new social group, call repertoires cannot be hardwired. Learning vocalizations allows both the incorporation of unrelated individuals into groups (which is a way for pairs to increase group size) and sex-specific transmission of calls.

1.1. Detecting Vocal Learning—What Constitutes Good Evidence?

Vocal learning has a fairly clear function in the example above, yet what exactly is vocal learning? Vocal learning refers to changes in vocalizations influenced by social interaction. Here, we focus on vocalizations given by species that live in social groups and organize our discussion around the function these vocalizations serve. This functional focus rests on the premise that vocal learning has evolved to facilitate the social function of some calls. Learning is likely to occur for a specific subset of the vocalizations in a species' repertoire whose function is enhanced by the degree of acoustic similarity to social partners who may or may not be related (e.g., members of a social group, territorial neighbors, competitors, and kin). We discuss the evidence for learning of calls that serve a variety of functions, including affiliative vocalizations, contact calls, food vocalizations, vocalizations used in defending a group or individual territory, and vocalizations that convey individual or group identity. We present evidence for several families of birds and mammals. The taxonomic distribution of various forms of vocal learning supports the hypothesis that it has evolved independently in several taxa. Our review is intended to help develop hypotheses about the social and ecological factors that favor its evolution and to stimulate additional research to test the hypotheses we present.

We distinguish two types of vocal learning: *learned acquisition* and *social modification*. These may differ in the learning process involved, and each type may be the outcome of different forms of selection and rely on different neural mechanisms. However, both are true vocal learning.

Learned acquisition is the adding of new vocalizations to an individual's repertoire. Acquisition is frequently, but not always, age-dependent, and changes are influenced by social interactions. Typically, juveniles require auditory experience to produce adult-form vocalizations. Social interactions are often essential and can be important in identifying who serves as the tutor in this developmental process. Subtle forms of interaction can reinforce the acquisition and retention of specific vocalizations during development (e.g., West and King 1988). The production of abnormal vocalizations in acoustically or socially isolated individuals has long been considered strong evidence for vocal learning. In contrast, vocalizations that are produced normally by deafened, or acoustically or socially isolated individuals are good evidence against learned acquisition. They do not;

however, preclude social modification (see below). Learned acquisition is not restricted to the juvenile phase in all species; adults can acquire new vocalizations, and this acquisition is often guided by social partners (e.g., Slater 1989). This category of vocal learning includes age-dependent learning (Payne and Payne 1997) and selective attrition (Nelson and Marler 1994; Nelson et al. 1996).

Song learning in oscine passerines is an excellent and well-studied example of learned acquisition. We elect not to describe this process here because there are a number of excellent recent reviews (e.g., Catchpole and Slater 1995; Kroodsma and Miller 1996). We refer the reader to these. Instead, our review focuses on call learning in oscines and on less well-studied groups.

Social modification occurs when existing vocalizations are altered in response to social interactions. Frequently, although not necessarily, this results in increased acoustic similarity between the social partners' and target individuals' vocalizations. Occasionally, vocalizations can become more distinct, which may facilitate identification of individuals (Janik and Slater 1997; Janik 1999). In many species, social modification is not restricted to a particular developmental stage. Vocalizations that are acquired developmentally can be subject to later social modification. But even vocalizations that do not require auditory experience for normal production can be socially modified. This category of vocal learning includes social-dependent learning (Payne and Payne 1997).

Changing frequency characteristics of vocalizations through learning may require more complex perceptual and motor mechanisms than changing temporal characteristics. We discuss both frequency and temporal changes because we consider both to be interesting for understanding vocal development. Other authors have limited their definition of vocal learning to frequency changes or learned production (e.g., Janik and Slater 1997, 2000). We elect not to restrict ourselves in this way because we do not think this is the most critical distinction, although we do differentiate frequency from temporal changes in our discussion of evidence for vocal learning, and data on frequency changes are more abundant.

Here, we focus on behavioral and acoustic data. Data on the biophysics of vocal production are reviewed by Fitch and Hauser (Chapter 3), data on ecological constraints are discussed by Bass and Clark (Chapter 2), and data on neural mechanisms underlying call detection are reviewed by Gentner and Margoliash (Chapter 7).

1.2. What Does Vocal Learning Teach us about Acoustic Communication?

Vocal learning involves modification of acoustic signals produced in a social context and requires the coordinated operation of both perceptual and motor systems. For vocal learning to take place, two things must happen.

An animal's perceptual system must support discrimination among conspecific acoustic signals, and its vocal system must support flexible signal production. A well-developed perceptual system that operates independently from a flexible vocal-production system is insufficient. A flexible vocal system that is not guided by socially relevant acoustic input is also inadequate. There must be an audio–vocal interface that allows for adjustments in signal production with changing acoustic input. Without this neural interface, there is no way for acoustic input to influence subsequent vocal production.

Feedback between the study of call function and vocal learning can be productive. Knowing call function can help to focus research on those vocalizations that are likely to be learned. For example, when acoustic similarity among group members facilitates recognition and individuals are not closely related, vocal learning is likely. Similarly, studying the structural features influenced by the social environment can reveal additional subtleties of call function and where vocal flexibility is adaptive. In addition, the extent of modification through vocal learning—whether learned acquisition or social modification is involved—gives us insight into the direct forces that shape vocalizations and points to the perceptual, motor, and neural mechanisms necessary for such vocal flexibility.

Learned vocalizations are by no means the only interesting ones. Study of acoustic structure and usage of genetically determined vocalizations can give great insight into many aspects of a species' social behavior and cognition. Vocal learning is but one way to achieve communicative complexity. Contrasting learned with hardwired vocalizations can tell us much about the conditions favoring cultural inheritance and those favoring genetic inheritance.

1.3. Testing the Vocal Learning Hypothesis

Observational data can provide evidence in support of vocal learning, but only carefully controlled experiments can unambiguously demonstrate it, and experiments are necessary to identify mechanisms and rule out alternative explanations. Important baseline observational data should include descriptions of normal ontogenetic changes in vocalizations and the social contexts in which they occur (e.g., Moss et al. 1997). When ontogenetic change is absent, learned acquisition is unlikely. In addition, interpreting results from many experimental manipulations requires data on normal vocal development. Acoustic data accompanied by data on morphological development are especially valuable, because they facilitate exploration of how morphological changes affect acoustic structure (e.g., Jones et al. 1991). Comparisons of acoustic structure among individuals, territorial neighbors, social groups, and geographic regions help to characterize call function and reveal when call similarity is favored. When individuals copy neighbors or other social partners, both microgeographic and macrogeographic patterns can emerge. Thus, the occurrence of call sharing at any of these scales implicates vocal learning. However, alternative mechanisms can cause similar patterns (see below), necessitating experimental tests to demonstrate unequivocally that vocal learning does occur. Without this experimental proof, vocal learning can only be conjectured.

A number of experimental protocols have been used to test for both forms of vocal learning and to identify the mechanisms involved. Perhaps the most common and most convincing for learned acquisition are acoustic isolation experiments and deafening (e.g., Winter et al. 1973; Heaton and Brauth 1999). Social isolation experiments provide evidence of when social interaction is essential to normal call acquisition (e.g., Hughes et al. 1998). Many experiments on song learning provide either recorded or live call tutors and then quantify the extent of acoustic similarity between tutor and pupil, which can test either learned acquisition or social modification, depending on the call repertoire at the start of the experimental manipulation (see Kroodsma and Miller 1996); unfortunately, this protocol has seldom been used to study call learning. Data from experiments where pupils copied call tutors who produced aberrant calls (including humans) clearly demonstrate the power of social interactions (e.g., Brittan-Powell et al. 1997). An important experimental protocol to test social modification is to alter the social environment (i.e., create new social groups) and quantify acoustic changes. When convergence or divergence occur in novel social groups, social modification is strongly supported (e.g., Farabaugh et al. 1994; Boughman 1998).

In all of these observational and experimental studies, careful characterization of vocalizations is required to ensure that the acoustic parameters measured capture the relevant variation. Determining the appropriate measurements to make can be complex. Ideally, prior experiments have identified the acoustic features that the animals use to discriminate among individuals, groups, or dialects.

1.4. What Alternative Mechanisms Can Produce Patterns Similar to those Found in Learned Vocalizations?

Alternative mechanisms such as maturation of the vocal tract, body size, genetics, and local ecology can affect the structure and development of vocalizations and can produce patterns similar to those found for learned vocalizations, although these mechanisms do not involve vocal learning. Thus, these mechanisms need to be ruled out when tests of vocal learning are being done.

As animals mature, vocal tract morphology changes and motor control improves. These maturational changes can result in subtle to striking changes in vocalizations (e.g., Scherrer and Wilkinson 1993). However, these changes have a purely biophysical basis and are not influenced by social interactions and thus are not due to vocal learning. Age-related changes in vocalizations can provide substantial information to listeners about the caller's age and thus may have important functions, even though learning is not involved.

Body size is an important determinant of vocalization frequency in a number of taxa (e.g., Hauser 1993; Fitch and Hauser, Chapter 3). Individuals of similar body size often produce vocalizations of similar frequency; individuals of different size produce vocalizations that differ in frequency. Larger individuals typically produce lower-frequency vocalizations. The effects of body size on call frequency can profoundly influence consequent function of vocalizations and can contribute to individual distinctiveness and to signature function. Size dimorphism can be part of the basis for sex differences in vocalization frequency (e.g., Gouzoules and Gouzoules 1989b, 1990). Combined with maturational effects, body size can contribute to age-specific patterns in vocalizations (Gouzoules and Gouzoules 1995). Dialects can occur as a consequence of populations differing in body size because of adaptation to local environments or genetic differences (Barclay et al. 1999). These differences might be very important for identifying individuals, for classifying social partners and opponents, and even for deciding who to mate with, but vocal learning through social interaction need not be involved. Body size can have large effects on individual fitness; thus, finding that calls indicate body size is an important discovery, even when calls are not learned.

Social groups are often organized along kinship lines. Genetic variation among social groups can generate parallel variation in vocalization characteristics, facilitating recognition of kin (e.g., Rendall et al. 1996). Kin recognition based on vocalizations may function importantly in many aspects of social behavior. However, vocal learning is not necessary to produce these patterns and may even be costly. Learning can introduce copy errors that can decrease similarity among kin. Additional error can result if the individuals who serve as the tutors are unknowingly unrelated. Vocalizations open to learning might be more easily copied by nonkin who usurp a group's resources through their cheating. Selection against cheaters can favor vocalizations that directly reflect underlying genetics—a type of relatedness marker that is uncheatable. Therefore, when groups consist of relatives, genetic explanations need to be carefully considered and tested before vocal learning can be inferred. This is especially true when there is no previous experimental evidence demonstrating that the species under consideration learns vocalizations. In these situations, vocal learning might be a viable hypothesis, but we cannot be sure it occurs without direct experiments that control for the influence of relatedness. Even when vocalizations are not learned, finding that individuals use heritable vocalizations to recognize kin would be extremely important (e.g., Grafen 1990; Medvin et al. 1992).

Dialects can also result from processes that do not involve vocal learning. For instance, species distributions can be disjunct because of geographic barriers that reduce matings between populations, creating population substructure. Sustained isolation can give rise to sufficient genetic differentiation to produce vocalizations that differ systematically among populations. This can be independent of any possible advantages to vocal variation and need not involve vocal learning. In this case, morphological and behavioral traits will also reflect population substructure. This expectation provides one way to test the influence of population substructure on vocal differences. Finding vocal variation without concomitant variation in other traits would refute the population substructure hypothesis.

The same kinds of dialect patterns can result from local adaptation, which can also result in genetic variation among populations. Where population substructure can result from genetic drift generating random differences between isolated populations, the local adaptation hypothesis argues that the action of divergent selection is important. Populations that inhabit areas differing in physical features may produce vocalizations whose characteristics allow for efficient transmission in their respective habitat (Slobidchikoff and Coast 1980; Rydell 1993; Barclay et al. 1999; Bass and Clark, Chapter 2; Ryan and Kime, Chapter 5). Characteristics can include frequency shifts, changes in the proportion of tonal, trilled, and noisy elements, and temporal pattern (Wiley and Richards 1982). In captive studies, the physical characteristics of housing facilities can influence frequency and temporal characteristics of vocalizations. Individuals or groups housed in different facilities can produce vocalizations that differ systematically, whereas those housed in similar facilities can sound similar, independent of actual vocal learning. Changes in housing can induce changes in vocalizations, again independent of social influences and vocal learning. Predation can also exert selection on signal structure by selecting for vocalizations that are hard for predators to detect, localize, or recognize. Consequently, vocalizations can vary among areas that vary in predation risk or predator identity.

To unequivocally demonstrate vocal learning, the alternative mechanisms described above must be ruled out. When a species' biology implicates one or more as possible explanations for observed patterns, observational studies are not conclusive. Descriptions of dialects are especially vulnerable to this. Demonstrating that vocal learning produces the patterns observed is especially challenging in animal systems where experimental manipulations are difficult and direct observations of behavioral interactions and context are not possible. However, we suggest possible approaches and point out those that have been successfully used. Experiments are necessary, and these should be carefully designed to control or directly test the likely alternatives. Determining which of these alternatives is involved is difficult because they can predict the same kinds of patterns—in many cases only the process differs.

In the next sections we begin by describing vocal learning in the context of individual signatures, group signatures, and dialects. We specifically focus on expected patterns of vocal variation at each of these levels and with respect to call function to highlight potential sources of selection that may favor vocal learning. We follow with case studies of birds and mammals. These in-depth case studies are mostly in the bird groups oscines, psittacines, and trochilids and in the mammalian groups chiropterans, cetaceans, and primates because that is where research has focused or evidence suggests vocal learning may occur. We recognize that different sources of selection may favor vocal learning at each level and so divide the following sections into three functional levels: individual signatures, group signatures, and dialects. However, in some taxa, data at multiple levels indicate that vocal learning occurs. In these cases, it may be difficult to determine where the primary function for vocal learning resides. For instance, dialects may be a by-product of vocal learning that has evolved to serve a function on a smaller scale. We include studies at the level we believe is most important or where the system has been best studied, and we also describe evidence at other levels.

2. Individual Signatures

Individually distinctive vocalizations have been described in many birds and mammals. In fact, individual distinctiveness is ubiquitous and has been found in almost every taxon in which it has been explored. Individual distinctiveness is enhanced when differences between individuals are large and variation within individuals is minimal, so that each individual produces a consistently unique vocalization. Distinctly different call types are the most noticeable individually distinctive vocalizations (e.g., whistle signatures in bottlenose dolphins, *Tursiops truncatus*; Caldwell and Caldwell 1965). However, large differences are not essential; even subtle acoustic differences can generate sufficiently distinctive voices for individual identification in some species (e.g., Weary and Krebs 1992). This is especially true when individuals differ on several acoustic dimensions, but subtle differences on a single acoustic dimension can be sufficient for individual identification. Nor does variation within individuals necessarily obliterate distinctive voice characteristics. Differences in several acoustic dimensions are more likely in complex, frequency-modulated, or noisy calls, whereas tonal calls may differ in a single acoustic dimension.

Individual distinctiveness is necessary but insufficient for vocalizations to function as individual signatures. Conspecifics must be able to use vocalizations to discriminate between individuals. This is a more demanding criterion. Although many taxa have individually distinctive calls, far fewer cases of individual signatures have been demonstrated, in part because recognition is difficult to test. Yet, individual signatures are defined both by their structure and their function. Individual signatures need not be learned to serve their function. Individual distinctiveness can arise in two ways: by *designed individual signatures* or *by-product distinctiveness*. Only designed individual signatures are likely to rely on learning.

Vocalizations can be specifically designed by learning or by natural selection to enhance individual distinctiveness and facilitate individual recognition. These *designed individual signatures* appear to be relatively rare. They include the cases where individuals produce distinct call types. The more common mechanism is for individual distinctiveness to arise as a by-product of individual variation in vocal tract morphology and body size. Such variation has a purely biophysical basis and appears to be sufficient in many cases to generate a large number of unique signatures or distinct voices. Such *by-product distinctiveness* can serve the signature function quite well, although it is unlikely to involve learned acquisition. Individual distinctiveness is a simple consequence of morphological variation that may have other functions when by-product distinctiveness is involved. In contrast, selection must directly favor individual distinctiveness, or learning must occur specifically to enhance it to produce designed signatures. Designed individual signatures are likely to be found primarily when by-product distinctiveness is insufficient or unreliable. Even taxa that have repertoires of call types may rely, in part, on by-product distinctiveness. For example, call types usually fall into classes so that some individuals sound roughly similar. Individuals that share call types may be differentiated on the basis of by-product distinctiveness (e.g., Weary and Krebs 1992).

Understanding how selection has shaped vocalizations to serve as individual signatures requires knowledge of the potential benefits to conveying individual identity offset against the costs that might arise. From the signalers' point of view, indicating identity could be beneficial to facilitate interactions between social partners. Good examples are isolation calls that infants give to facilitate reunions with caregivers in many species of bats (e.g., Scherrer and Wilkinson 1993). However, indicating identity might not always be favored. A signaler might also benefit from concealing identity or from imitating another individual. Vocal learning can facilitate both of these kinds of cheating, making it costly from the receivers' point of view. Vocal learning will only be favored if its benefits outweigh its costs. In developing and testing hypotheses for the evolution of vocal learning these potential costs should not be overlooked. Costs can be a powerful force shaping signal design and development (see Fitch and Hauser, Chapter 3).

Vocal learning is one mechanism to enhance differences between individuals (e.g., Janik 1999), and this mechanism operates on the time scale of individual lifetimes. However, individual signatures need not be learned, and in fact several demonstrated cases have a strong heritable component (e.g., Scherrer and Wilkinson 1993). Heritable individual signatures reflect the action of selection over evolutionary time to enhance differences between individuals. As an example, learning plays almost no role in structuring squirrel monkey calls, but individuals vary in voice (Winter et al. 1973; Symmes et al. 1979) and are likely to be recognized by their calls (Snowdon et al. 1985; Boinski and Mitchell 1997), so the calls function as individual signatures. Studies of vocalizations in several taxa have shown significant heritabilities for multiple features of calls that serve as individual signatures, especially in bats (Jones and Ransome 1993; Scherrer and Wilkinson 1993; G.S. Wilkinson and D. Lill unpublished data). The distinctiveness in these species' calls has a strong genetic basis. Indeed, Scherrer and Wilkinson (1993) calculated that sufficient heritable variation exists in the isolation calls of evening bat pups (*Nycticeius humeralis*) for more than 1,800 pups to produce unique calls; the largest colonies consist of 1,000 bats. This is a huge amount of variation, none of which is due to learning, and these calls indicate individual identity (100% correct classification in a discriminant analysis) and relatedness (significant heritability) quite well. A strong inherited component will lead to acoustic similarity among relatives; thus, when relatives need to identify each other, heritable individual signatures should be favored.

We expect vocal learning of individual signatures to play a role in the following cases: (1) determining which call types to include in an individual's repertoire, and (2) increasing distinctiveness of voices among a specific set of individuals (for instance, a social group, set of relatives, or territorial neighbors) to allow for greater precision in individual identification within that group. In both cases, mechanisms to detect cheaters should be built in. Determining which call types to produce, especially when the calls are designed signatures, is likely to rely on learned acquisition. In contrast, social modification is likely to be the predominant form of vocal learning for increasing distinctiveness of individual voices. Group mates provide a benchmark against which to gauge distinctiveness. Learned acquisition is unlikely here because group mates are the most likely models, and copying call types from them makes it difficult to achieve the desired outcome of sounding different from social partners.

Although the specific function of repertoires is not clearly known for many taxa, learning repertoires may provide different benefits than learning a single, variable call type and may rely on different neural and behavioral mechanisms. The form of selection that has molded designed signatures is also likely to differ from that underlying individual voices arising from by-product distinctiveness. Identifying the selective factors that generate these different vocal systems promises to be a difficult but exciting area of research.

In the case studies described below, we begin by describing the social biology and the context in which calls are produced and identification of individuals favored for each species. We present information on call patterns to demonstrate that calls are individually distinctive and are used by conspecifics in identification. We follow with a discussion of the evidence that vocal learning produces individual signatures and conclude with a consideration of alternative mechanisms. Little information is available on the process of vocal learning for individual signatures, but we include it where we can. In addition to these case studies, we briefly review other taxa where vocal learning has been studied for calls that are individually distinctive (Tables 4.1 and 4.2).

TABLE 4.1. Selected bird species for which vocal learning of calls has been studied.

Bird group	Species	Social structure	Call	Function	Genetics and relatedness	Vocal learning evidence	LA SM*	Alternative explanations	Exp‡	References
Oscines	*Campylorhynchus nuchalis* (stripe-backed wren)	stable family groups that defend long-term territories	family- and sex-specific call repertoires	maintain contact among group mates; territorial defense; identify group members	primarily related males; females disperse from neighboring territories; some males disperse to neighboring territories	all males share sex-specific repertoire; females share sex-specific repertoire; unrelated juvenile males learn new groups' repertoire	U	repertoires reflect patrilineal or matrilineal kinship; therefore, genetics could contribute	N	Price 1998a, 1998b, 1999
	Cacicus cela vitellinus (yellow-rumped caciques)		colony-specific songs	unknown	unknown	shared song types among males in colony; differences between colonies; distance effect	U		N	Feekes 1977, 1982; Trainer 1987, 1988, 1989

	Gymnorhina tibicen (Australian magpie)	social groups	duets; communal chorus; carol	territorial defense; within-group cohesion	unknown	song sharing within group; sex differences; individual differences; mimicry of human whistles	SM		N	Brown et al. 1988; Farabaugh et al. 1988; Brown and Farabaugh 1991; Farabaugh et al. 1992b
	Taeniopygia guttata (zebra finch)	breeding colony and pairs	distance call	maintain contact with pair mate	fathers and sons; unknown for pairs	individual differences; sex differences; no dialects; sons copy fathers or other adults	LA		Y	Zann 1984, 1985, 1990
Phasianids	*Colinus virginianus* (bobwhite quail)	family groups	hoypoo call	contact calls	family groups	coveys of unrelated birds differ in hoypoo calls; dialects based on temporal characters; individuals differ in frequency	SM	coveys differ only in temporal variables	Y	Bailey 1978; Goldstein 1978; Bailey and Baker 1982; Baker and Bailey 1987

* LA = learned modification; SM = social modification; U = unknown.

‡ This column indicates whether an experiment has been conducted to test vocal learning.

TABLE 4.2. Selected mammalian taxa for which vocal learning has been studied.

Mammal group	Species	Social structure	Call	Function	Genetics and relatedness	Vocal learning evidence	LA SM*	Alternative explanations	Exp‡	References
	Individual signatures									
Bat	*Rhinolophus ferremuquinem* (horseshoe bat)	maternity colony; mother–pup	CF echolocation pulse	navigation	mother–offspring	age changes in mother's resting frequency mirrored by pups; heritability greater than 1 suggests possibility of learning	SM	body size effects potentially confound analysis	N	Jones and Ransome 1993
	Nycticeius humeralis (evening bat)	maternity colony; mother–pup	pup isolation calls	facilitate reunions	mother–offspring; half sibs	none	N	strong heritable component suggests calls are genetically coded	N	Scherrer and Wilkinson 1993
	Eptesicus fuscus (big brown bat)	maternity colony; mother–pup	FM echolocation pulse	navigation	mother–offspring	calls vary by age, sex, and individual; geographic differences	U	probably genetic and maturational; geographic differences reflect different foraging strategies	N	Moss 1988; Rydell 1993; Masters et al. 1995

	Myotis lucifugus (little brown bat)	maternity colony; mother–pup	FM echolocation pulse	navigation	maternity colonies; mother–offspring	colony differences	U	primarily ontogenetic changes	N	Pearl and Fenton 1996; Moss et al. 1997
Primate	*Cercopithecus aethiops* (vervet monkey)	social group	alarm calls; social calls	predator warning; intergroup encounter	group mates; variable relatedness including parent–offspring and matrilineal	correct usage learned	N	no evidence for learning of acoustic structure	Y	Seyfarth and Cheney 1980, 1986
	Pan troglodytes (Chimpanzee)	fission–fusion social groups	pant–hoot	contact call; recruit social partners; intergroup spacing; mating	group mates; territorial neighbors; potential mates; unknown relatedness	populations differ; call matching among chorusing partners; novel element included in one group's calls	SM ?	genetic differences between populations; calls recorded in different contexts in each population	N	Marler and Hobbett 1975; Mitani et al. 1992; Mitani and Nishida 1993; Clark and Wrangham 1999 1993; Mitani and Brandt 1994; Mitani and Gros-Louis 1995; Marshall et al. 1999

TABLE 4.2. *Continued*

Mammal group	Species	Social structure	Call	Function	Genetics and relatedness	Vocal learning evidence	LA SM*	Alternative explanations	Exp‡	References
	Individual signatures									
Rodent	*Dipodyms* spp. (kangaroo rat)	territorial neighbors; parent–offspring	foot drums (nonvocal)	territorial advertisement	unknown relatedness between neighbors; parent–offspring	greater than expected distinctiveness between neighbors	SM		N	Randall 1989a, 1989b, 1994, 1995
	Group signatures									
Cetacean	*Physeter macrocephalus* (sperm whale)	stable matrilineal groups of 10–12; temporary association with other groups and individuals	coda	not known	matrilineal	codas are specific to matrilineal groups	U	groups are not very stable; some group members are related, which could indicate that genetic similarity underlies vocal similarity	N	Whitehead et al. 1991; Weilgart and Whitehead 1993, 1997; Richard et al. 1996; Lyrholm and Gyllensten 1998; Christal et al. 1998
Primate	*Macaca nemestrina* (pigtail macaque)	social groups	screams	recruit allies in dominance interactions	matrilines within groups	screams indicate maternal lineage; learned usage suggested by increasing accuracy of contextual use with age	N	analysis to test contextual usage confounded by body size effects, which are known to affect vocalization frequency	N	Gouzoules and Gouzoules 1989a, 1989b, 1990, 1995

	Dialects									
Bat	*Saccopteryx bilineata* (white-lined bat)	harems	male tonal calls	attract females; advertise and defend territory	males within a colony are likely to be related	individual differences; differences between colonies	U	relatedness of males within colonies could mean colony differences have a genetic basis	N	Davidson 1999
Primate	*Hylobates* sp. (gibbon)	territorial social groups	great call; mother–daughter chorus; female–male duet	advertise and defend territory; pair-bond maintenance	mother–daughter; mated pairs; primarily monogamous	groups respond to own group's call; daughters sing with mothers, which could indicate that practice is necessary for call development; hybrids give calls intermediate between parental species	N	intermediacy of hybrids suggests genetic basis; bulk of evidence suggests calls are not learned	N	Marshall and Marshall 1976; Chivers and Raemaekers 1980; Brockelman and Srikosamatara 1980; Deputte 1982; Brockelman and Schilling 1984; Raemaekers et al. 1984
	Microcebus murinus (grey mouse lemur)	family groups	trilled loud calls	territorial maintenance; mate attraction	males within groups related; females within group likely related	individual differences; differences between captive groups	U	individually distinctive calls of few males in a group would give appearance of dialects; males within groups related, making genetic basis likely	N	Zimmermann and Lerch 1993; Zimmermann 1995; Hafen et al. 1998

TABLE 4.2. *Continued*

Mammal group	Species	Social structure	Call	Function	Genetics and relatedness	Vocal learning evidence	LA SM*	Alternative explanations	Exp[‡]	References
	Dialects									
Seals	*Leptonychotes weddeli* (Weddell seals)	population	underwater calls	unknown; possibly mating	probable discrete populations	differences between geographic regions in repertoires; occurrence of "unique" call types; differences between populations in frequency characteristics of shared calls	U	genetic differences between populations could give rise to differences; incomplete sampling or repertoires makes designation of "unique" call types premature	N	Thomas and Stirling 1983; Green and Burton 1988; Thomas et al. 1988; Morrice et al. 1994
	Erignathus barbatus (bearded seal)	population	underwater calls	unknown; possibly mating	probable discrete populations	differences between geographic regions	U	genetic differences between populations could give rise to differences	N	Cleator et al. 1989
	Mirounga angustirostris (Northern elephant seal)	population	threat calls	deter rivals	breeding populations	differences between breeding colonies; individual bulls differ	U	differences between populations deteriorate over time	N	Le Boeuf and Peterson 1969; Le Boeuf and Petrinovich 1974; Shipley et al. 1981, 1986

Rodent	*Cynomys gunnisoni* (prairie dog)	population	alarm calls	predator warning	unknown	variation in temporal characteristics	N	divergent characteristics correlated with local habitat; possible genetic differences	N	Slobidchikoff and Coast 1980
Lagomorph	*Ochotona princeps* (pika)	population	alarm calls; songs	predator warning; territorial maintenance and mate attraction	unknown	variation among populations in duration and maximum frequency of fundamental frequency	N	no evidence of learning; geographic variation correlates with pelage and body size variation, suggesting genetic basis	N	Somers 1973

* LA = learned modification; SM = social modification; U = unknown.

‡ This column indicates whether an experiment has been conducted to test vocal learning.

2.1. Mammals

2.1.1. Bats—Lesser Spear-Nosed Bat Pup Isolation Calls

Lesser spear-nosed bats (*Phyllostomus discolor*) roost in hollow trees in large colonies of 50–300 bats. These colonies are composed of several smaller groups of 4–14 females and a single male. Colonies move from roost tree to roost tree as a unit and thus appear to be relatively stable, whereas the composition of groups within colonies is dynamic (C. Kagarise, J. Bradbury, and L. Emmons unpublished data). Adult females leave their infants in the roost each night when they go out foraging and must find their pup when they return. Finding their mothers has immediate fitness consequences for pups who depend on their mothers for sustenance, grooming, thermoregulation, and protection. Identifying her own pup affects adult female fitness as well. Females produce a single pup each year, so losing a pup can seriously depress reproductive success. Misdirecting maternal care is also costly because it decreases the amount of care a mother can provide her own offspring. These costs should favor a mechanism for mothers to identify offspring and for offspring to find mothers.

Like many bat species, when separated from their mothers, *P. discolor* infants give sinusoidally frequency-modulated (FM), multiharmonic calls, termed isolation calls (Fig. 4.1A). Mothers produce maternal directive calls whose frequency structure is quite similar to infant isolation calls. Isolation calls and maternal directive calls vary individually in the pattern of sinusoidal FM and duration (Esser and Schmidt 1989). Adults can discriminate modulation frequencies that differ by as little as 2 or 3 Hz, more than sufficient to discriminate individuals whose modulation frequencies differ by as much as 49–100 Hz (Esser and Lud 1997). Infants appear to recognize the directive call of their own mother (Schubert and Esser 1997), and mothers respond to their own pup's calls. Reciprocal calling facilitates reunions between pups and their mothers when pups fall or fly from the roost site or when mothers return from foraging (Gould 1983). We focus here on data concerning vocal learning of pup isolation and maternal directive calls.

Research on vocal learning in *P. discolor* includes observational information on normal development (Esser and Schmidt 1989), descriptive work on geographic variation (Esser and Schubert 1998), and experimental tests of social modification (Esser 1994). In addition, psychophysical work has described auditory thresholds in infants (Esser and Schmidt 1990) and adults (Esser and Daucher 1996), and detection of frequency modulation (Esser and Kiefer 1996), which may be important for individual discrimination of these FM calls.

Newborns produce isolation calls at birth, suggesting that learned acquisition is not necessary for normal production. However, pup isolation calls do undergo progressive changes during development (Fig. 4.1A). Calls given by newborns show little FM and the pattern of sinusoidal FM increases with age. In addition, some calls given by newborns drop out of

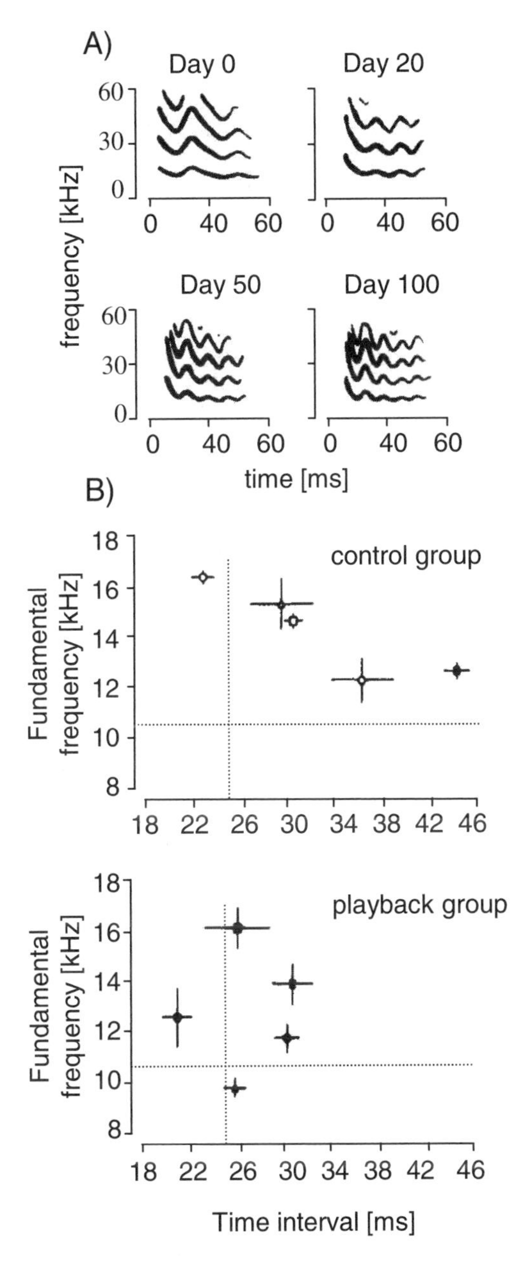

FIGURE 4.1. Lesser spear-nosed bat infant isolation calls. (**A**) Sonograms showing changes during development for one infant. (**B**) Sinusoidal frequency-modulation rate at day 50 for control and playback groups. Time interval between the first and third frequency minima on the x-axis, fundamental frequency on the y-axis. Values of maternal reference call are indicated with dotted lines. (From Esser 1994. Reprinted with permission from Lippincott, Williams & Wilkins.)

the repertoire during the first weeks of life (Esser and Schmidt 1989; Esser 1994). Esser and Schmidt (1989) suggest that a female infant's isolation calls develop into maternal directive calls. Social modification is possible, but these developmental changes could simply be due to maturation of the vocal tract and increasing motor and respiratory control.

One piece of evidence that the fine structure of vocalizations is modified by learning is the suggestion of dialects in maternal directive calls (Esser and Schubert 1998). Captive adult females from two regions show differences in carrier frequency, modulation frequency, and the number of FM peaks, which separates them into two partially overlapping multivariate clusters in a multidimensional scaling analysis. The authors present no direct data that dialects result from learning; rather, they draw parallels between their findings and dialects in oscine birds to suggest learning. They also cite the experimental evidence for social modification described below. Alternative explanations have not been ruled out. Groups from the two regions are likely to have been genetically isolated, so genetic differences could contribute to the differences described. Variation among captive groups in body size or age could also contribute to differences in the frequency measures. Testing for the influence of these factors should be straightforward and would greatly strengthen claims for learning-based dialects.

One experiment supports the importance of social modification in this species' vocalizations (Esser 1994). Four individually housed, hand-reared bats were presented with one maternal directive call as a reference vocalization, and changes in the isolation calls of these pups over the first 100 days of life were monitored. The playback group was compared with a control group of hand-reared pups that heard no maternal directive calls (Fig. 4.1B). Both groups increased the amount of sinusoidal FM in their calls during development; however, the playback group showed a larger amount of FM at 100 days, coming close to the value of the maternal directive call. The rate of sinusoidal FM in the playback group's calls showed some convergence with the maternal directive call as compared with the control group, although the extent of convergence was not great. No apparent convergence in fundamental frequency of the first FM peak was found. Bandwidth of both control and playback groups overlapped the distribution of the reference call (Esser 1994). The observed convergence was interpreted as evidence for vocal learning; clearly, social modification is the relevant form. For *P. discolor*, vocal learning functions to increase vocal similarity to promote mutual recognition and facilitate reunions.

There may be two opposing forces at work shaping isolation calls. Pups that sound similar to their mother may facilitate recognition and reunion. But pups that sound different from infant roost mates should also facilitate recognition. The extent of convergence on the reference call that Esser (1994) observed may have been constrained by the need for pups to distinguish themselves from other infants. This theme runs through other studies on learned individual signatures—similarity to some individuals may be favored simultaneously with increased distinction from other individuals.

2.1.2. Cetaceans—Bottlenose Dolphin Whistles

Bottlenose dolphin (*Tursiops truncatus*) social structure is characterized as fluid, meaning individuals interact with many others during their long lives. However, two types of long-term associations occur: mother–calf pairs and male coalitions (Wells et al. 1987; Smolker et al. 1992). Presumably, identifying individuals in these long-term associations is important, although little data are available to test this hypothesis. Calves are dependent on their mothers for three to five years and may remain with mothers for up to ten years, and females of reproductive age associate with other females in their matriline (Wells et al. 1987). Benefits of long-term associations vary. Female matrilines may receive aid in caring for and protecting calves against predators and may gain feeding benefits from associating with conspecifics (Hoese 1971). Males form long-term coalitions (Smolker et al. 1992; Connor and Smolker 1995) that aggressively herd and consort with females in reproductive condition (Connor et al. 1992b) and compete with other males and coalitions for access to these females. Long-term and large coalitions tend to be more successful in these competitions (Connor et al. 1992b).

Bottlenose dolphins produce a variety of social calls in addition to echolocation calls, including pulsed, tonal, and whistled sounds. The specific function of pulsed and tonal sounds has not been thoroughly investigated, but whistles have been studied in some depth. Adults produce a repertoire of three to nine FM whistles (McCowan and Reiss 1995a). Usually one of these (but up to three) is unique to an individual and is termed that individual's signature whistle (Caldwell and Caldwell 1965). Signature whistles can make up from 5% (McCowan and Reiss 1995a) to 95% (Caldwell and Caldwell 1965) of an individual's call production. The primary function of signature whistles is thought to be individual identification (Caldwell and Caldwell 1965; Caldwell et al. 1990). The combination of small, stable groups such as mother–calf pairs and male coalitions, and the fluid and repeated association of these groups with other individuals over many years is thought to favor individual identification (Janik 1999). Whistles are somewhat variable within individuals (Sayigh et al. 1990), but tend to have a greater amount of variation between individuals (P.L. Tyack unpublished data), so their structure is appropriate to serve as individual signatures. Statistical analysis can classify calls to the correct individual (Buck and Tyack 1993), and dolphins have sufficiently good discrimination abilities to differentiate among whistle variants, although no study has tested for recognition of individuals by whistles (Tyack 1997). This is a case where data on signal structure have been taken as proof of signal function. Calves who are separated from their mothers whistle as they initiate reunions (McBride and Kritzler 1951; Smolker et al. 1993). The exact function of whistling in this context is unclear. It seems doubtful that whistles facilitate reunions in a manner similar to isolation calls given by bat pups because the calf approaches the mother and mothers rarely call in response to calves. Also, calves have usually oriented to the mother prior to calling.

The signature hypothesis has been reassessed recently (Caldwell et al. 1990; McCowan and Reiss 1995a, 2001). McCowan and Reiss (1995a) argue that repertoires include many whistles shared among the members of a social group or even across social groups. These whistles are not likely to be individual signatures; rather, hypotheses suggest they may be referential (Tyack 1997) or affiliative (Smolker 1993; Tyack 1997). McCowan and Reiss (1995a) also provide data indicating that the proportion of whistles unique to an individual can be quite low. They suggest that the preponderance of signature whistles documented by some researchers is due to recording context. When restrained and isolated from others, both adults and infants produce many signature whistles (McCowan and Reiss 1995a; Janik and Slater 1998). When freely interacting with others, dolphins produce many shared whistles (McCowan and Reiss 1995a, 1995b). Thus, defining a signature as the whistle most commonly produced is problematic.

Much of the evidence for vocal learning in *Tursiops* comes from observational studies of captive animals (Tyack 1986; Reiss and McCowan 1993; McCowan and Reiss 1995a, 1995b). Observational studies of communication are being incorporated into long-term studies of two wild populations, one in Sarasota, Florida (Wells et al. 1987; Sayigh et al. 1990, 1995; Wells 1991), and the other in Shark Bay, Australia (Connor et al. 1992a; Smolker 1993; Connor and Smolker 1995, 1996). Experimental work on vocal learning is restricted primarily to eliciting specific whistles by operant conditioning (Richards et al. 1984; Reiss and McCowan 1993). Experimental work is hampered by two factors: small sample sizes due to difficulties keeping animals in captivity, and neither social and acoustic isolation nor deafening are possible due to ethical considerations or because they induce abnormal social and acoustic behavior.

Vocal development has been studied for both captive and free-living dolphins. Infant bottlenose dolphins produce whistles when they are only a few days old (Caldwell and Caldwell 1979), suggesting that auditory experience is not required for whistle production. Like adults, calves give some whistles that are unique to each individual (73% of whistle types produced during the first year), but do not retain one particular whistle as their individual signature (McCowan and Reiss 1995b). A high proportion of whistles are shared within social groups (23%), and many whistles are shared by calves in the three groups studied (17%). Certain whistle types are produced only by very young infants, and the same type is given by all infants. The proportion of this infant whistle gradually decreases and the proportion of other whistles increases as the calves age, resulting in a turnover in the repertoire. Calves continue to acquire new whistle types even as they discontinue use of earlier types. This process continues past the first year, as repertoires of 12-month-old calves differ from adults. At any one developmental stage, calves have larger repertoires than adults and by 12 months have given as many as 22–55 different types of whistles (McCowan and Reiss 1995b). This suggests that individual repertoires can be very large. This overproduction suggests that a process of selective attrition (Nelson et al. 1996) may be involved in shaping adult repertoires of three to nine whistles. Calves' acquisition of this large number of whistle types may depend on auditory and social experience, although published data are lacking to determine the importance of either factor in the initial acquisition of whistles or the process of selecting which types to retain.

The contextual use of some whistles changes during development (McCowan and Reiss 1995b). Calves call more when separated than in socializing or other contexts until about 8 months of age. Their use of signature whistles in social contexts increases after this point. This pattern is similar for signature and shared whistle types. The function of these whistle types is uncertain, making it difficult to know whether developmental changes arise from increasing accuracy of use or from changing social relationships and behavior as calves mature.

The ontogeny of whistle repertoires and production is more difficult to study in free-living dolphins. Sayigh et al. (1990) repeatedly recorded 12 pairs of mothers and their one year or older calves during captures. Dolphins were lightly restrained in shallow water and recorded with a suction cup hydrophone attached to the melon. Within a single recording session, renditions of individual calls varied in such acoustic parameters as duration and absolute frequency even though whistle contour varied little. The most commonly produced whistle of calves (presumably their signature whistle) was stable, undergoing little systematic modification in contour over the course of their 3-year study. Slight modifications to whistle contour took place in a few individuals' calls, but for most individuals, calls did not vary more across years than within a recording session. Adult signature whistles were consistent over even longer time periods of up to 12 years. This study focused on stability of the signature whistle rather than cataloging the entire repertoire. Signature whistles were recorded almost exclusively—only 8% of calls recorded from female calves and 27% from male calves were other whistle types. This finding contradicts the finding of large repertoires (McCowan and Reiss 1995a, 1995b), perhaps because of the different recording contexts. Although McCowan and Reiss (1995a, 1995b) worked in captivity, which can itself introduce artifacts, they recorded in a variety of behavioral contexts, whereas Sayigh et al. (1990) recorded only loosely restrained dolphins during captures.

Sayigh et al. (1990) found sex differences in the apparent acoustic model used by males and females. Female calves produced signature whistles that differed from their mothers', whereas male calves were more likely to produce whistles similar to their mothers'. This sex difference was less pronounced in a later study (Sayigh et al. 1995), where many male whistles differed from their mothers'. Sayigh et al. (1990) suggested that vocal learning underlies this sex difference; however, their data cannot be used to determine directly whether learned acquisition or social modification occurred. They studied calves older than one year and found little change in whistle contour; therefore, the processes that influenced contour development had clearly taken place before the end of the first year and the start of their study, which agrees with McCowan and Reiss (1995b). Sayigh et al. (1990) hypothesized that because females may continue to associate with their mothers for many years, distinctive signature whistles may facilitate accurate individual identification. Males disperse and thus may not face the same constraints; rather, they suggested that similar whistles may facilitate kin recognition. For this latter interpretation to be plausible, all males should have imitated their mothers, and this was not the case (Sayigh et al. 1990, 1995). The occurrence of shared whistles in captive repertoires further complicates interpreting the results of these two studies. It seems quite likely that both females and males shared some whistles with mothers and produced others that were individually distinctive. Individuals need to be recorded in a large number of sessions, preferably in multiple contexts, before one can be sure that the entire repertoire is cataloged.

Data on sharing of whistles among social companions are somewhat conflicting. Most studies suggest that infants only occasionally copy their mother, who is their primary social partner during the first few years of life (Sayigh et al. 1990, 1995; Reiss and McCowan 1993; McCowan and Reiss 1995b). Yet other studies show sharing of whistles among other types of companions (Connor and Smolker 1995; McCowan et al. 1998) and even apparent imitation of a trainer's whistle (Tyack 1997). Currently, it is unclear who serves as the primary tutor and why.

Three males in a coalition increased the similarity of their whistles as the tenure of their coalition lengthened (Fig. 4.2; Smolker 1993; Smolker and Pepper 1999). Given the information on the potential size of repertoires and the extent of shared calls, it is unclear whether males socially modified existing whistles to increase similarity, acquired novel whistle types, or increased their use of a particular whistle already in their repertoire. Because of the possibility that convergence results from changing use rather than the generation of novel whistles, Smolker and Pepper (1999) did not demonstrate vocal learning in the sense outlined by Janik and Slater (1997) (e.g., acquisition of novel calls or modulation of call frequency characteristics). However, it seems quite likely that the whistle type these males used depended on their social affiliation, which meets the criterion for vocal learning that we use. The contribution of by-product distinctiveness to variation among individuals in these shared whistles should be studied.

Dolphins have been trained operantly to match a whistle of a particular contour presented as a stimulus (Richards et al. 1984; Reiss and McCowan 1993). This requires vocal flexibility and auditory–vocal feedback to achieve matching. Such work demonstrates remarkable intrinsic cognitive and vocal abilities, but does not tell us much about the context, function, or extent of vocal flexibility in nature.

The work on dialects in dolphins has produced conflicting results. One study found evidence of geographic variation at both the microgeographic

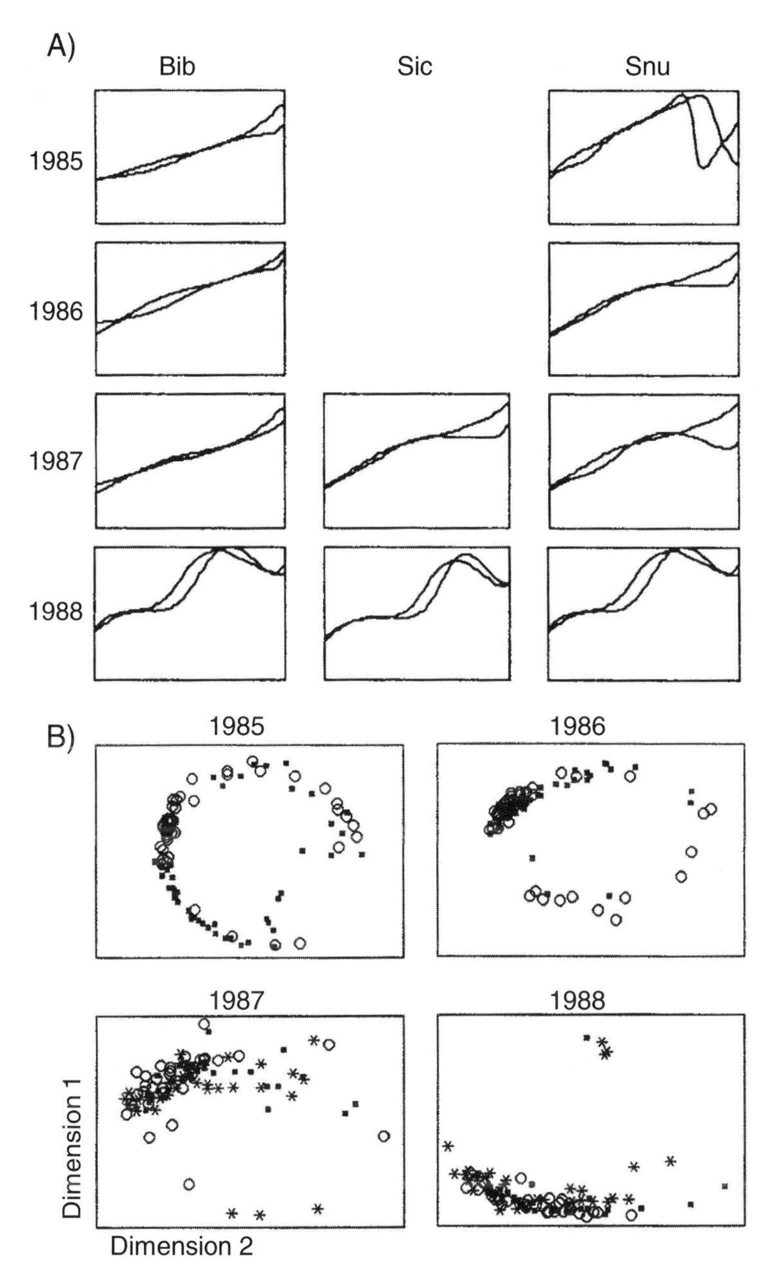

FIGURE 4.2. Whistles used by allied male bottlenose dolphins. (**A**) Average frequency contours of the two most commonly used whistles for three allied males over four years. (**B**) Multidimensional scaling of whistles for each male. Individuals noted by different symbols. Note that the scatter in multidimensional space is reduced over time, indicating acoustic convergence among individuals. (From Smolker and Pepper 1999. Reprinted with permission from Blackwell Wissenschafts-Verlag Berlin, GmgH.)

and macrogeographic scales (Ding et al. 1995). Whistles from three subpopulations in the Gulf of Mexico differed. Comparisons among populations in the Gulf of Mexico, the Gulf of California, Argentina, Japan, and Australia also indicated that these populations differed from one another. Other studies failed to find such evidence (Evans and Dreher 1962; Graycar 1976; Steiner 1981). Additional work is necessary to determine the prevalence of dialects in dolphins.

A primary factor favoring vocal learning in dolphins may be individual identification (Janik and Slater 1997; Janik 1999). In terrestrial animals, differences in vocal tract morphology and body size often result in individually distinctive vocalizations. Dolphins may not be able to rely on these by-product mechanisms because the shape of air sacs involved in whistle production is influenced by pressure, which changes when they dive. Thus, dive-induced variation may undermine the consistency of individual voice characteristics (Tyack and Sayigh 1997). Janik (1999) suggests that this constraint favored individual identification via distinctly different whistle contours. He argues that vocal learning was therefore favored—individuals could listen to social companions and develop a whistle contour that differentiated them. This idea is intriguing, but no data exist to test it directly. Some of the information described above seems consistent with this hypothesis, but the existence of shared whistles in dolphin repertoires (McCowan and Reiss 1995a, 1995b) and the occasional matching of mothers' whistles by sons (Sayigh et al. 1990, 1995) contradict it. Further work to test this hypothesis is warranted.

It is clear that dolphins can imitate other sounds and this imitation may require both forms of vocal learning—learned acquisition and social modification. What is unclear is how often they use imitation in nature, what induces them to copy a particular individual or sound, and the functions of such copying. Perhaps these questions are difficult to address because whistles from known individuals can only be recorded in limited contexts, and these contexts may not adequately elicit the full repertoire nor allow identification of who serves as tutor from the potentially large pool of individuals. More thorough sampling of free-living dolphins, although logistically difficult, would be extremely useful for clarifying the pattern of sharing among individuals in social groups, across populations, and the extent to which signature whistles are individually specific. Captive studies are likely to continue to provide important information on ontogeny, yet they cannot adequately address some important questions with respect to vocal learning because of constraints in composition of social groups, the small number of individuals in captivity, and the absence of ecological context.

2.1.3. Primates—Macaque Food and Contact Calls

The vocal behavior of several species of macaques (*Macaca* spp.) has been studied to test whether vocalizations are learned. Most macaques live in large, multimale social groups with strong dominance hierarchies among matrilines. Matrilineal groups are genetically differentiated (Olivier et al. 1981). Females remain in their natal group, whereas males disperse. Female offspring assume the dominance rank of their mothers; members of a matriline assist one another in agonistic interactions to actively defend their dominance rank. Social groups generally avoid one another rather than actively defend territories. Aggressive encounters do occur, and females usually focus their aggression on other females when defending a clumped food source. Females sometimes act aggressively toward immigrant males, and males almost always do so (Cheney 1987).

Japanese (*M. fuscata*) and rhesus (*M. mulatta*) macaques give calls, termed "coos," in various social contexts and when encountering food (Hauser 1996). Coos can be used to recognize individuals (Hansen 1976; Masataka 1985). Studies of call structure provide conflicting results with respect to vocal learning. Green (1975) argues that social transmission has produced differences in frequency modulation of coos in three troops of Japanese macaques (*M. fuscata*). Unfortunately, he provides no analysis of acoustic features and illustrates single examples from each troop, which makes quantitative assessment of troop variation impossible. The vocal differences he describes are related to the provisioning context and are the only acoustic differences among troops noted. Individuals give some coos with the "locale-specific" features and some without. Other vocalizations do not differ among troops.

Subsequent work to investigate vocal learning has used cross-fostering among two species, the Japanese macaque, *M. fuscata*, and rhesus macaque, *M. mulatta* (Masataka and Fujita 1989; Owren et al. 1992, 1993). These studies ask whether offspring raised by heterospecifics produce coos like those of their foster mothers. Masataka and Fujita (1989) cross-fostered a single Japanese and two rhesus macaques. They describe differences in peak frequency of the fundamental frequency, but not in duration, of the two species' calls, and they claim that cross-fostered infants produce coos with frequency characteristics similar to their adoptive mothers'. They also present playback data suggesting that others respond to the cross-fostered monkeys as though they are conspecifics. These data are interpreted as providing strong evidence of vocal learning.

However, Owren et al. (1992, 1993) challenge these results. Although calls of infants raised by conspecifics are individually distinctive and differ between species, there are no statistically significant differences between calls given by adults of the two species for peak frequency or duration, nor for five other frequency variables related to fundamental frequency (Owren et al. 1992). Each species shows substantial variability and overlaps the frequency range of the other. Rather than being species-specific, adult coos differ among individuals, although classification accuracy in a discriminant analysis is low. Coos are not particularly good vocalizations to use in a cross-fostering study because there are no consistent differences between species in adult calls (Owren et al. 1992). Thus, they cannot serve as distinctive models for developing species-specific characteristics. In addition, similarity between mothers and infants has not been described, and such similarity may not enhance the calls' function. If similarity is not favored, there is no reason to expect convergence between calls given by cross-fostered infants and their mothers. This lack of functional consideration may have interfered with progress in testing vocal learning in these and other primate species.

The use of cross-fostering to test vocal learning requires differences in vocalizations among species. Japanese and rhesus macaques differ in their use of several calls. Japanese infants give coos during play almost exclusively, whereas rhesus infants use coos and gruffs (Fig. 4.3). Rather than focusing on changes in acoustic structure, Owren et al. (1993) compare the proportion of coos and gruffs given by cross-fostered infants to proportions given by normally raised infants of each species. Cross-fostered Japanese infants continue to give coos almost exclusively, in proportions indistinguishable from normally raised Japanese infants. Cross-fostered rhesus infants give more coos than normally raised rhesus infants in two contexts but fewer coos than normally raised Japanese infants in two other contexts (Fig. 4.3B). Owren et al. (1993) interpret these results as very limited modification of usage. Even when interacting with others producing gruffs, cross-fostered Japanese infants almost never give these calls. Furthermore, rhesus infants give gruffs even though they almost never hear their mothers or social companions produce such calls. This result is particularly surprising because both species are capable of producing both coos and gruffs (Owren et al. 1993), so very little modification is necessary to mimic the calling behavior of social companions. New vocalizations do not need to be learned. Perhaps modification does not occur because there is no strong inducement to produce calls similar to those of social companions. Cross-fostered infants interact normally with peers, despite their unusual vocal behavior. Selection for call similarity seems to be lacking in these species and may be overridden by other social cues such as facial displays and gestures that facilitate social interactions. Thus, vocal learning may simply not be necessary.

3. Group Signatures

Group signatures are signals that indicate an individual belongs to a particular social group. They are likely to be found in species where stable social groups control access to limiting resources, such as food, mating and nesting sites, or territories. In this context, discriminating between group mates and others facilitates both cooperation within the group and exclusion of outsiders who compete for the same resources. Typically, signal structure is shared among group mates, and groups are distinct from one another, although group members can learn to recognize each others' individually distinctive calls. Groups can share call types that are distinct from the types

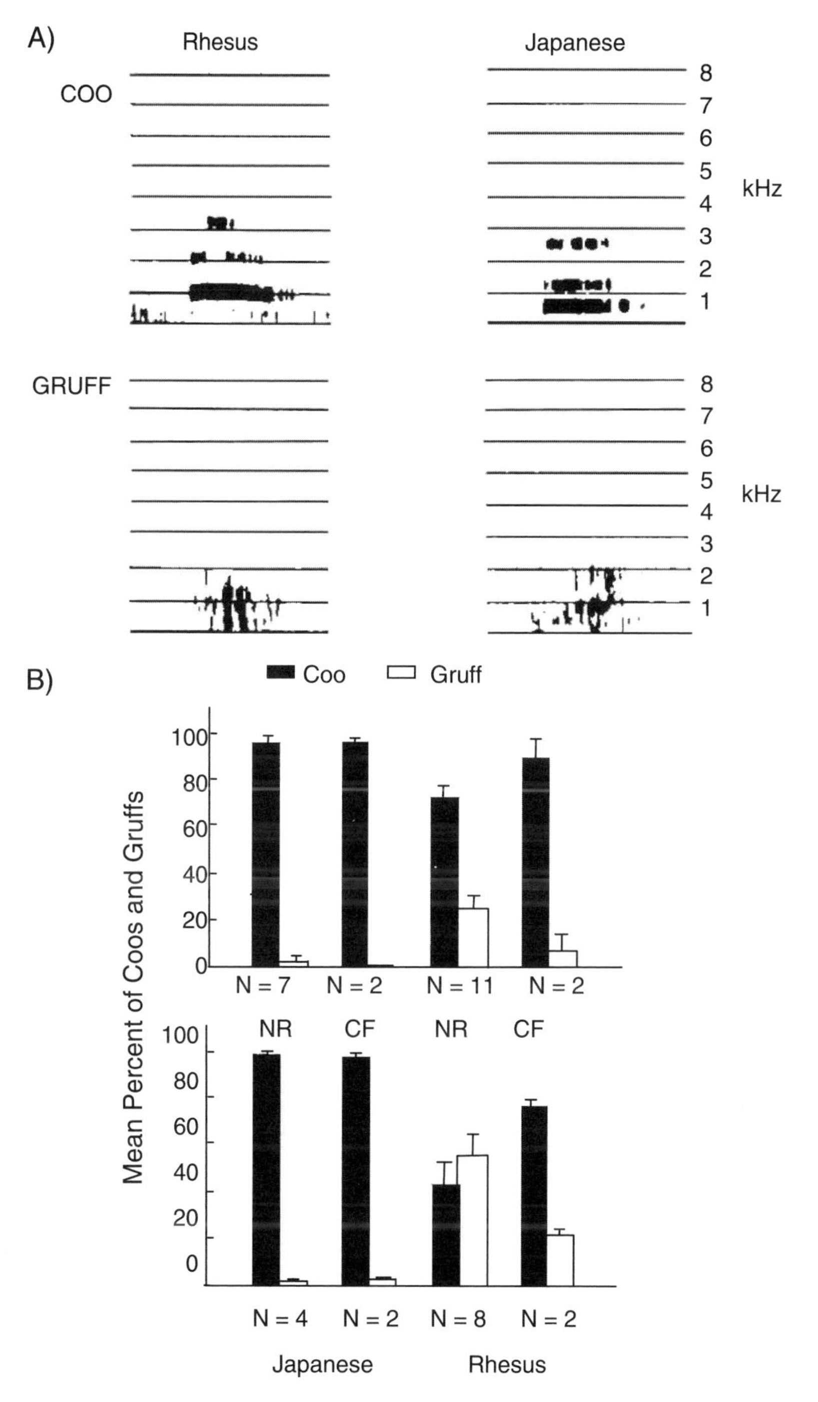

FIGURE 4.3. Coo and gruff calls for Japanese and rhesus macaques. (**A**) Sonograms of each call type for both species. (**B**) Usage by normally raised (NR) and cross-fostered (CF) juveniles during the first two years of life. (From Owren et al. 1993. Copyright © 1993 John Wiley & Sons, Inc. Reprinted with permission from Wiley-Liss, Inc., a subsidiary of John Wiley & Sons, Inc.)

of other groups (likely to involve learned acquisition), or can share more subtle acoustic features of a single call type (likely to involve social modification). As with individual signatures, to demonstrate that calls are group signatures requires demonstrating convergent call structure and recognition of group mates by call.

Using a repertoire that is partially shared with other groups is an inefficient way to convey group identity. If repertoires are shared, to unequivocally identify the group affiliation of a caller requires hearing much of the repertoire or just the few calls unique to that group. In this situation, the function of shared call types appears to be something other than group identity. That function is currently unknown.

As with individual signatures, potential benefits of indicating group identity need to be considered in conjunction with potential costs and the possibility of cheating. Imitating a group's signal can allow an outsider to gain access to group resources. Therefore, mechanisms for cheater prevention—a means of detecting imitators and preventing their access to resources—should evolve in concert with group signatures.

Group signatures need not be learned to serve their function. In species where social groups are formed of relatives, heritable signals can effectively indicate group affiliation and may also be used to determine the level of relatedness between individuals. Heritability of group signatures has not been studied directly, but studies described earlier in several taxa have shown a strong genetic basis for individual signatures (Jones and Ransome 1993; Scherrer and Wilkinson 1993). When social groups are organized along kinship lines, acoustic similarity among group mates can result from shared genetics. Thus, a pattern of convergence within groups or differences between them is insufficient to demonstrate that vocal learning occurs if group mates are close relatives.

When call similarity is favored, we expect vocal learning in two cases: when groups are composed of unrelated individuals who cannot rely on genetics to produce similarity in calls, or when group composition changes slowly over time and group members use calls to indicate group affiliation. In the latter case, the immigrants may change their vocalizations to match the new group, or the entire group may accommodate the new group composition by changing call characteristics. In both of these instances, the most relevant type of vocal learning is social modification. Learned acquisition can occur, but is unnecessary unless new call types are added to an individual's repertoire. Certainly, vocal learning can occur in groups of relatives and may increase similarity or increase distinctiveness depending on the primary function of calls. Separating the effects of heredity from those of vocal learning in such taxa is likely to be a difficult but fruitful task.

In the examples that follow, we begin by describing the species' social biology and the benefits of group living, and consider when learned group signatures are expected. Then, we describe evidence that calls are group-distinctive and present data on call function. We conclude each example by discussing the evidence for vocal learning and discuss alternative hypotheses where appropriate. We also describe evidence of the vocal learning process when that information is available. We briefly describe vocal learning of group signatures in taxa not covered in case studies in Tables 4.1 (birds) and 4.2 (mammals).

3.1. Birds

3.1.1. Oscines—Chickadee Flock Signatures

Black-capped chickadees (*Parus atricapillus*) associate in small, fairly stable flocks of four to eight birds during nonbreeding months. Breeding pairs form within these flocks and defend breeding territories (Ficken et al. 1981). Flocks defend shared feeding territories. The composition of flocks is relatively stable throughout each year, but changes from one year to the next due to adult mortality and recruitment of dispersing juveniles (Weise and Meyer 1979; Nowicki 1989). Flock mates are probably not closely related to one another, but the degree of relatedness is unknown.

Song and call distinctions are not particularly clear in chickadees, and the complexity of some call types exceeds that of song. Many call types are given by both sexes (Ficken et al. 1978). Here, we focus on two call types that have been well-studied: the gargle and the chick-a-dee. The gargle is a variable series of two to nine notes given primarily by males (Ficken et al. 1978; Ficken and Weise 1984). Gargles are given in agonistic encounters between flock mates, during territorial encounters, and prior to copulation. They are not used to advertise flock territories but rather to defend those territories against intrusion and to maintain dominance hierarchies within flocks (Ficken et al. 1978). The chick-a-dee call consists of four note types (Fig. 4.4A). Each note type is given a variable number of times in an invariant sequence: A, B, C, D. Chick-a-dee calls are given by both sexes throughout the year and function as contact calls—to coordinate movements of pairs and flocks (Ficken et al. 1978; Nowicki 1983) and facilitate recognition of flock mates (Nowicki 1983). Birds also give chick-a-dees during interflock encounters and when mobbing predators. Flocks differ in call characteristics, primarily of the D note, and calls of individuals within flocks are convergent (Mammen and Nowicki 1981; Nowicki 1989). Recognition of flock mates could be based on both flock and individual differences (Nowicki 1983).

Observational and experimental evidence for vocal learning supports the role of social modification and, to a lesser extent, learned acquisition of both gargle and chick-a-dee calls. Social modification is implicated in gargle development on the basis of observational data on dialect variation (Ficken and Weise 1984; Ficken et al. 1985; Ficken and Popp 1995; Hailman and Griswold 1996). Hand-reared birds produced calls similar to individuals with which they were raised and different in subtle ways from free-living birds (Ficken and Weise 1984). The chick-a-dee call is better studied and suggests both learned acquisition and social modification. Observational

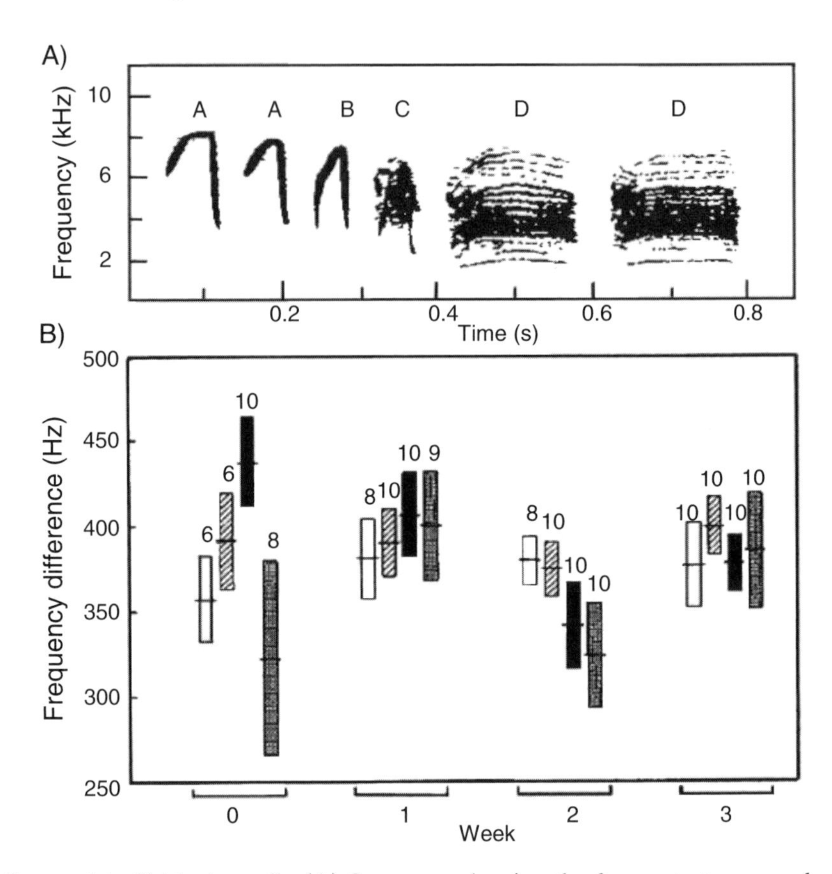

FIGURE 4.4. Chickadee calls. (**A**) Sonogram showing the four note types, marked A, B, C, D. The study focused on convergence among group mates in D notes. (**B**) Changes in frequency structure of D notes during the three-week experiment. Means are indicated by a horizontal line, SD by the vertical bars, and sample size by the numbers above bars. (From Nowicki 1989. Reprinted with permission from Academic Press.)

data (Clemmons and Howitz 1990) and one experiment (Hughes et al. 1998) support the hypothesis that vocal learning is involved in the acquisition of normal chick-a-dee calls. Clemmons and Howitz (1990) described the normal process of call development and found that approximate renditions of adult-like chick-a-dee calls are first given by fledglings at 28 days. These calls were highly variable, especially in the introductory A, B, and C notes, and did not acquire fully adult-like characteristics for another week. Acquisition of normal chick-a-dee calls required auditory and social experience (Hughes et al. 1998). The A note developed normally in isolated individuals, but birds who did not hear normal adult vocalizations produced abnormal B and C notes. Hughes et al. (1998) did not study development of the

D note. Social experience appears to be a powerful determinant of call development because birds who were socially but not acoustically isolated produced B and C notes that were at least as aberrant as those of totally isolated birds.

Social modification also occurs and produces convergence within flocks (Fig. 4.4B; Nowicki 1989). Birds who had no prior contact with one another gave calls that initially differed in several parameters of the D note (week 0). Calls converged rapidly. This is shown by the reduction in variation and increased similarity in means by week 1. No single bird appeared to serve as the model; instead, calls converged on the mean spectral characteristics of the flock. Convergence was observed in several frequency measures, including the bandwidth, maximum frequency, and difference between the minimum frequencies produced by each side of the syrinx. No convergence was observed in temporal measures or proportions of note types in calls. Calls began changing within a week, and by four weeks substantial convergence occurred. Both juveniles and adults changed calls, indicating that social modification is not limited to a sensitive phase in this species. Both types of vocal learning occur in *P. atricapillus* calls. This oscine species is a good counterpoint to studies of song learning in other oscines because it challenges some of our notions about the distinction between song and calls, and the importance of vocal learning to these two kinds of vocal signals.

3.1.2. Parrots—Budgerigar Flock Signatures

Budgerigars (*Melopsittacus undulatus*) are a highly social, nomadic parrot from inland Australia. Budgerigars live in flocks of 20–30 birds that associate seasonally with other flocks in very large assemblages. Both sexes provide parental care, and mated pairs may remain together over several breeding attempts. The arid environment has highly variable productivity both spatially and temporally, and budgerigars are opportunistic feeders that will move great distances to find productive areas. Flock living is thought to provide several benefits in food finding and predator protection (Farabaugh and Dooling 1996). Budgerigars are opportunistic breeders as well, initiating breeding when ecological conditions are favorable (Brockway 1964; Wyndham 1980a). Little is known of the composition of budgerigar flocks in regards to relatedness or stability. However, much is known of their vocal behavior.

Budgerigars give several types of calls, and complex, variably structured song. Most work on vocal learning has examined the most frequently used call, the contact call (Fig. 4.5), but all calls and song are thought to be learned. Contact calls seem to function to maintain social cohesion. Adults give them in flight and when separated from mates or other individuals in the flock (Wyndham 1980a). Contact calls also facilitate reunions between parents and their fledged offspring. Once fledged, offspring live in a creche. Parents visit the creche to feed their fledglings until they reach full independence

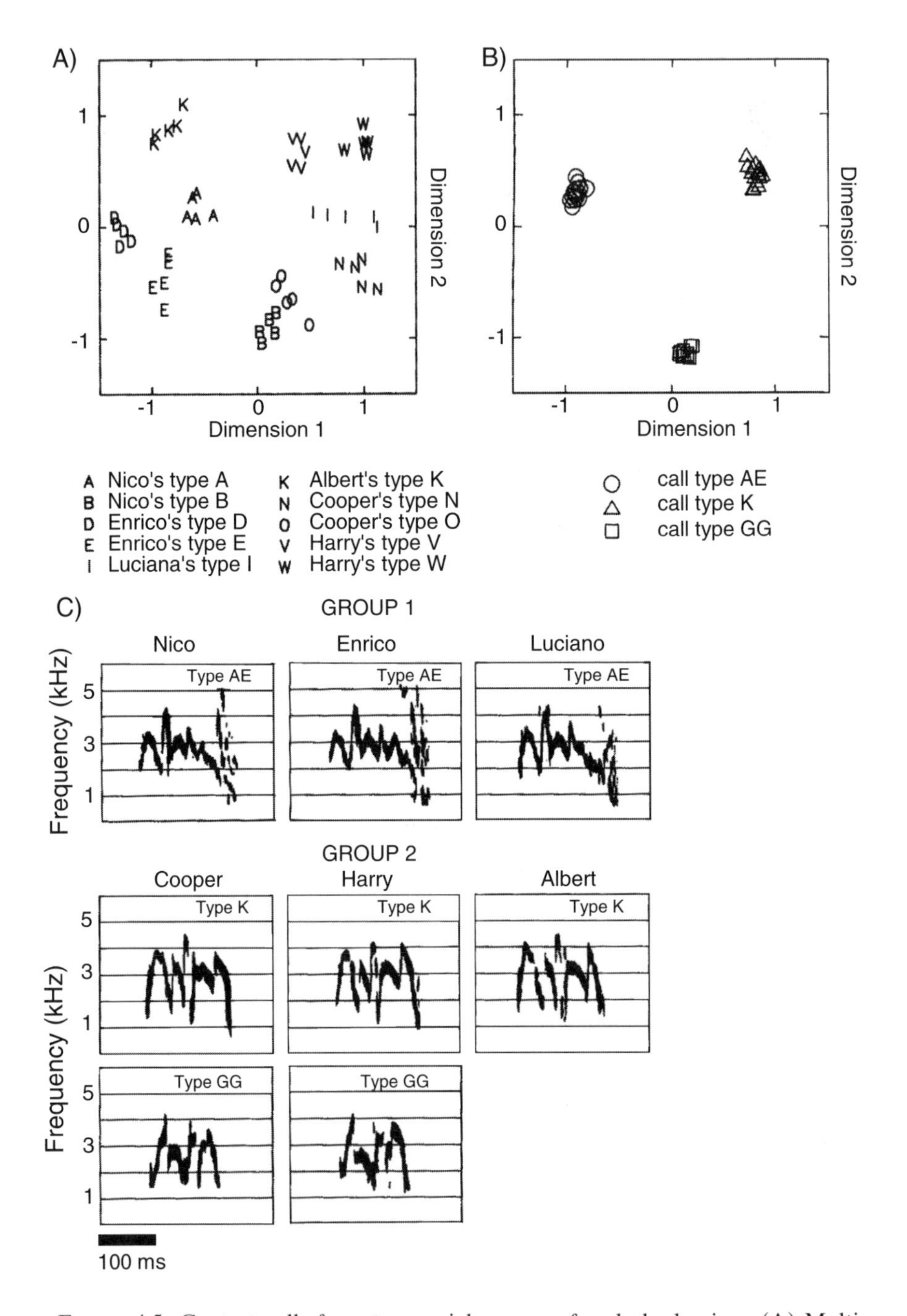

FIGURE 4.5. Contact calls from two social groups of male budgerigar. (**A**) Multidimensional scaling of five renditions of each bird's dominant contact call before contact and (**B**) after 8 weeks of social interaction. (**C**) Sonograms of dominant contact calls for each male after 8 weeks of contact have similar structure and thus show convergence within each group. (From Farabaugh et al. 1994. Copyright © 1994 by the American Psychological Association. Reprinted with permission.)

about two weeks later (Brockway 1964; Wyndham 1980b). Individual birds have a repertoire of contact calls, but use one or two most of the time (Farabaugh et al. 1994). Flock mates share call types (Farabaugh et al. 1994), resulting in flock-specific repertoires of contact calls. Budgerigars can discriminate between very similar contact calls given by their flock mates, suggesting that calls can be used to identify individuals within flocks. This perceptual discrimination appears to be learned because birds can only make these finely detailed discriminations when they are familiar with the individuals from whom the calls are recorded (Brown et al. 1988).

During normal vocal development, juveniles increase the extent and complexity of FM, decrease peak frequency, decrease bandwidth, and increase duration (Brittan-Powell et al. 1997). Contact calls appear to develop from a call given late in the nesting phase, the food-begging call. Broods show few differences and sexes develop similarly early on. However, evidence of call sharing emerges within a month of fledging (Brittan-Powell et al. 1997).

Direct evidence demonstrated that contact calls are learned and that both acquisition and social modification occur. Manipulations of auditory experience demonstrated its fundamental importance. Birds deafened at birth developed abnormal calls (Dooling et al. 1987; Heaton and Brauth 1999). Contact calls of deafened birds retained some features of the species' calls, including maximum and minimum frequency, bandwidth, and duration, but differed in tonality, number of tonal elements, and extent of frequency modulation (Dooling et al. 1987). Deafened birds produced normal food-begging calls but abnormal contact calls (Heaton and Brauth 1999), so vocal learning was confined to calls whose function was enhanced by similarity to social partners. Hand-reared birds gave contact calls that were simpler than typical budgerigar calls and that differed in bandwidth and frequency modulation (Brittan-Powell et al. 1997). Hand-reared birds had more acoustic experience than deafened birds, and their calls resembled typical budgerigar contact calls far more than those of the deafened birds. Contact calls and the song of deafened adults became abnormal within 6 months, indicating that auditory input is required to maintain normal call structure, and demonstrating the absence of a critical period. These studies showed that acoustic experience was essential for normal vocal development in budgerigars and revealed a continuum of effect.

Social experience had a powerful influence on which call types were acquired and how existing calls were modified. Learned acquisition and social modification occurred throughout development and into adulthood. Young birds learned from adults with which they were housed even when those adults produced abnormal contact calls or were heterospecific (Brittan-Powell et al. 1997). Abnormal features included bandwidth, degree of frequency modulation, and duration. Juveniles housed with zebra finches gave contact calls that resembled zebra finch calls but that showed less harmonics and more frequency modulation than the heterospecific call.

These birds also gave contact calls that were more typical of their species. Once abnormally reared birds were exposed to normal song at about 6 months of age, they developed more normal contact calls (Brittan-Powell et al. 1997), indicating that substantial ability for vocal learning persisted to this age. These data suggested that social interactions were very powerful at sculpting young birds' calls. Adults also learned the calls of the birds with which they interacted (Farabaugh et al. 1994; Bartlett and Slater 1999) and formed social bonds (Hile et al. 2000), providing further evidence that vocal learning was not limited to a sensitive phase in this species. Farabaugh et al. (1994) showed that adult birds whose contact calls were initially distinct converged on flock mates' calls within two months' time, with the first changes occurring in a week (Fig. 4.5). Mutual imitation seemed to be the mechanism because the final contact call was composed of call elements from different birds. The importance of social interaction was demonstrated because birds that could hear but not interact with each other did not share contact calls. When a single bird joined a group, that individual altered its call to match the group (Bartlett and Slater 1999). Call convergence occurs between males and females forming pair bonds due to males imitating their mates' calls (Hile et al. 2000).

Vocal learning is not limited to contact calls. Birds reared in acoustic and social isolation developed abnormal warble songs (Farabaugh et al. 1992a), indicating that acquisition of song was learned. In addition, males within social groups had similar warble songs, suggesting that social modification occurred. In support of this hypothesis, one male apparently copied syllables from a cage mate that sang abnormal songs, even though he could hear other birds whose song was typical for budgerigars (Farabaugh et al. 1992a). Birds whose tutor gave abnormal or heterospecific vocalizations continued to include some species-typical characteristics in their song and calls. This result argues that some features of vocalizations rely on underlying genetics. On top of this genetic basis is substantial vocal flexibility and social dependency.

3.2. Mammals

3.2.1. Bats—Greater Spear-Nosed Bat Group Signatures

Female greater spear-nosed bats (*Phyllostomus hastatus*) live in stable social groups that are attended by a single male who maintains nearly exclusive reproductive access to all females in the group (McCracken and Bradbury 1977). Consequently, in a given year, offspring of most females in a group are paternal half-sibs (McCracken and Bradbury 1981). However, adult females in these groups are not close relatives (McCracken and Bradbury 1981), a pattern that results from juvenile dispersal patterns. All offspring disperse from their natal group during their first year. When reproductively mature, at about two years of age, females join an existing female social group. Females obtain several benefits from group living.

Female group mates forage together (Wilkinson and Boughman 1998), which results in faster food finding and better defense of food resources (Wilkinson and Boughman 1999; Boughman unpublished). Group living and group foraging enhance reproductive success (Boughman unpublished). Reproduction is synchronous within a group (Porter and Wilkinson 2001), which may facilitate thermoregulation by young pups and allomaternal care (Porter unpublished).

Females use vocalizations, termed screech calls (Fig. 4.6A,B), to coordinate group foraging (Wilkinson and Boughman 1998), and these vocalizations differ among social groups (Boughman 1997). Screech calls differ among groups in a number of frequency and temporal features, yet calls from individuals within social groups are not statistically distinguishable. Such group-distinctive calls provide an effective mechanism for finding group mates outside the roost cave (Wilkinson and Boughman 1998). In addition to the variation among social groups within caves, variation exists between cave populations (Boughman and Wilkinson 1998). Little migration between caves occurs (McCracken et al. unpublished), yet this does not result in genetic differences between cave populations (McCracken 1987). Thus, geographical variation in screech calls is more likely to result from social modification than from population substructure.

Screech calls convey group membership (Boughman and Wilkinson 1998). Bats discriminate between calls given by group mates and other bats. Bats also differentiate between calls from their own cave and other caves. However, bats reveal no ability to discriminate among calls given by individuals within their group (Boughman and Wilkinson 1998). Screech calls function as contact calls that enable group mates to find one another outside the roost cave. Bats give them when departing on foraging flights, en route to, and at feeding sites (Wilkinson and Boughman 1998). Bats approach when hearing familiar calls but fly away when hearing unfamiliar calls (Boughman and Wilkinson 1998). This result and a lab experiment suggest that screech calls may facilitate group defense of resources. Bats increase their rate of calling when a large number of intruders are present (Boughman unpublished).

Unrelated females give acoustically similar calls (Boughman 1997). Genetics cannot explain this pattern, because group mates are not close relatives. Ontogeny of screech calls has not been studied, but because females join social groups as adults, ontogenetic change is unlikely to give rise to group-distinctive calls. Experimental evidence demonstrated that convergence within groups results from social modification (Boughman 1998). Juvenile and adult bats were transferred reciprocally between social groups that initially differed in call characteristics. Bats gradually changed calls (Fig. 4.6C–F). The first changes occurred within one month, and by five months calls given by transfers and residents showed strong convergence in both frequency and temporal characteristics. Both residents and transfers changed calls. Results showed that the social environment was the primary determinant of acoustic change. Comparisons with age-matched

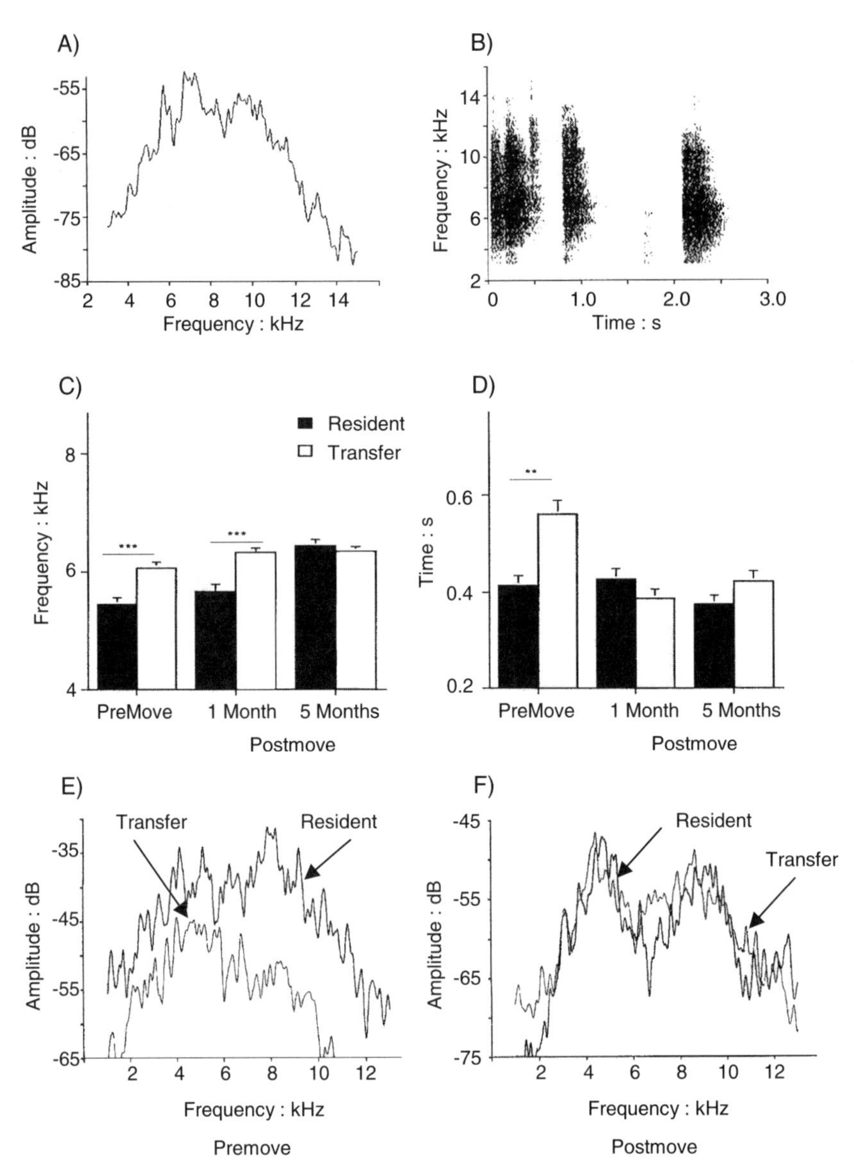

FIGURE 4.6. Greater spear-nosed bat screech calls and changes in acoustics after bats were moved between social groups. (**A**) Power spectrum and (**B**) sonogram of screech call. (**C**) Mean bandwidth and (**D**) repetition rate for group 2 residents (solid bars) and transfers (open bars) before and after social contact. Residents remained in their natal group, whereas transfers immigrated from another social group during the experiment. Premove periods are before this immigration event; postmove periods are after it. (**E**) Power spectra for group 2 residents and transfers before and (**F**) after social contact. (From Boughman 1998. Reprinted with permission from The Royal Society.)

half-sib controls demonstrated that call convergence did not result from maturational processes, acoustic adaptation to the physical environment, or heredity. Instead, half-sibs who transferred to new groups were more similar to their new, unrelated group mates than to relatives in their natal group (Boughman 1998). Call modification was not immediate, and the time required for individuals to match a new group could provide protection against outsiders who might feign identity to obtain access to food resources controlled by groups. Thus, in this species, cheater-protection mechanisms appear to have evolved in concert with group signatures.

3.2.2. Cetaceans—Killer Whale Pod Signatures

Killer whales (*Orcinus orca*) live in stable social groups called pods. Both transient and resident whales are found in the coastal waters of western North America. These types differ in foraging strategy, social structure, and vocal patterns, and they are reproductively isolated (Barrett-Lennard et al. unpublished). Here, we focus on the resident killer whales off the British Columbia (BC) and Alaska coasts, for which the bulk of work on vocal behavior has been done. Residents feed primarily on fish and live in extremely stable matrilineal groups. Neither males nor females disperse out of their natal pods (Bigg et al. 1990; Olesiuk et al. 1990; Hoelzel et al. 1998), so pods are composed of close relatives organized along matrilineal lines. Mating occurs between members of different pods (Barrett-Lennard unpublished). Pod members travel, feed, and socialize together. There is some suggestion that pod members cooperate in finding and capturing prey (Bigg et al. 1987; Hoelzel 1993), and foraging benefits may be one factor that favors living in stable groups.

Like other odontocetes, killer whales use echolocation to find prey. They produce several types of calls, but most work on vocalizations has focused on social calls, which combine tonal and pulsed elements (Fig. 4.7A). Resident pods have a repertoire of 3–16 discrete calls, with an average of nine calls (Ford and Fisher 1983; Ford 1989, 1991). Similar patterns were found for whales in coastal waters off Norway, which appear to have social structure and call patterns similar to BC residents (Strager 1995). Pods differ in their call repertoires. Although most calls are shared by several pods, some are unique to a particular pod (Fig. 4.7). The degree of call sharing between pods may depend on geographic distance, the amount of contact, or genetic similarity. Pods that rarely encounter one another share almost no calls, whereas those that interact frequently share some calls (Ford and Fisher 1983). The function of such call sharing has not been investigated. In addition to variation among pods in the type of calls given (termed repertoire variation), more subtle structural variation is present. Pods differ in frequency and temporal features of several calls, with those pods that associate least and least closely related showing the largest differences (Deecke 1998). An additional class of graded calls appears to be used at close range

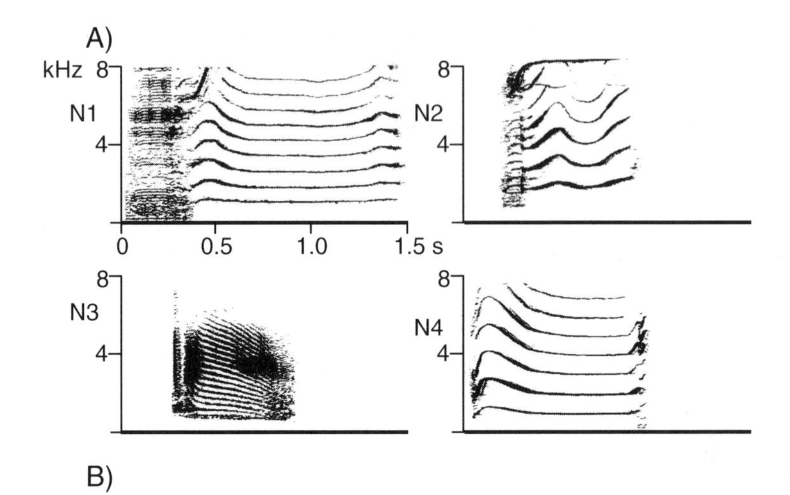

B)

Calls		Pods								
		A1	A4	A5	B	C	D	H	I1	I11
N1		X	X							
N2		X	X	X						
N3		X	X	X	X	X	X	X	X	
N4		X	X	X						
N5		X	X	X	X	X	X	X	X	
N6		X	X	X						
	N7I	X	X	X						
N7	N7II	X	X	X				X	X	
	N7III				X					
	N7IV					X	X			
	N8I	X	X	X	X				X	
N8	N8II					X	X			
	N8III							X		
N9		X	X	X						
N10		X	X	X						
N11		X	X	X	X					
	N12I	X	X	X						
N12										
	N12II				X	X	X	X	X	
N13		X	X	X						
N14					X	X	X		X	
N15					X	X	X	X	X	
	N16I				X				X	
N16	N16II					X	X			
	N16III							X		
N18					X	X				
N20					X	X	X		X	
N21					X					
N22					X				X	
N23										X
N24										X
N25										X
N26										X
N28										X
N29								X	X	
N31						X	X			
N36										X
N40										X
N41										X
Total		14	14	13	13	11	10	8	11	8

during intense socializing (Ford 1989). These calls do not differ among pods in a consistent way (Ford and Fisher 1983).

The exact function of discrete calls is unclear. They are given in various social situations, and Ford (1989) hypothesized that they function as contact calls, enabling pod mates to stay in acoustic contact while traveling and foraging. Ford speculated that calls might be important in coordinating foraging efforts during group hunts. Presumably, whales can use call differences among pods to determine the pod to which the caller belongs, although call repertoires overlap substantially (Fig. 4.7) and playback studies to test the group signature and contact-call hypotheses have not been conducted. The high degree of repertoire congruence among pods suggests that group identification might be based on structural variation between pods. Ford (1991) hypothesized that shared calls indicate the level of relatedness between pods or maternal lineage and might facilitate outbreeding (Ford and Fisher 1983). Unfortunately, there are no data to test most of these hypotheses. Testing hypotheses for call function with playback experiments is critical for establishing function and could also illuminate the selective pressures that might favor a learned or genetic basis to calls. A single call type is sufficient to identify group membership; the occurrence and function of repertoires remain unexplained, but the pattern suggests additional functions for repertoires. Genetic data indicate that mating usually occurs outside the pod and clan, so mates are likely to differ acoustically (Barrett-Lennard unpublished). This provides indirect support for the hypothesis that whales base mate choice on the degree of acoustic similarity.

No direct published evidence that calls are learned exists. One observation of captive whales is consistent with the learning hypothesis. In this case, a single Icelandic female produced calls similar to the BC whale with which she was housed for many years (Bain 1988).

Despite this lack of direct evidence. Ford (1989, 1991) proposed that calls are learned and that call sharing arises because pods that come into frequent contact imitate each others' calls. Deecke (1998) investigated this latter hypothesis with respect to structural variation, but not repertoire variation, by comparing the association patterns and vocal similarities of nine matrilineal units in three pods. He tested four call types, two of which showed some support for the hypothesis; however, the pattern was not particularly strong. Even though a significant correlation existed between acoustic similarity and association, dendrograms for acoustic similarity for

FIGURE 4.7. Killer whale social calls. (**A**) Sonograms of four call types. (**B**) Call types in the repertoire of the pods studied. Three closely related pods have almost identical calls (1, A4, A5) but produce calls that other more distantly related pods do not (B through Ill). Pods B through H have very similar repertoires. All pods except Ill produce call N3, N5, and a variant of N7 and N8. (From Ford and Fisher 1982. Reprinted with permission from the International Whaling Commission.)

each of these two calls were not congruent with dendrograms for association. The pattern does not seem to be particularly strong for repertoire variation either because Bigg et al. (1990) found no significant relationship between association patterns and repertoire variation. Thus, the call-sharing hypothesis is not strongly supported, but neither is it refuted. More data are needed. In any case, congruence between association and call repertoire or structure is insufficient to prove vocal learning because association can allow for both genetic and cultural exchange.

Even though the pattern of shared calls within pods (Ford and Fisher 1983; Ford 1989, 1991; Strager 1995) and the stable social structure are consistent with a learning hypothesis, high levels of relatedness within pods also argue for a strong genetic component. A significant correlation was found between acoustic similarity and genetic distance (Barrett-Lennard unpublished). We cannot yet determine whether this pattern results from cultural or genetic inheritance or a combination of the two modes. Genetically encoded calls would serve their presumed function quite well, indicating kinship and social affiliation simultaneously. Several authors presume that insufficient genetic variation exists to produce complex call repertoires and thus argue that pod-specific calls are unlikely to have a genetic basis. In marked contrast to this view, studies cited earlier (Jones and Ransome 1993; Scherrer and Wilkinson 1993) have shown that calls with heritable acoustic components can effectively serve as signatures, especially when the relevant information is kinship. Determining the contributions of learning and genetics to the pattern of pod-specific calls in *O. orca* requires experimental tests and additional genetic data. Analysis of both nuclear DNA and mtDNA is ongoing (Barrett-Lennard unpublished), and the results can be correlated with information on the degree of distinctiveness and overlap of calls to determine how well genetic similarity predicts call similarity for individuals and the potential role for learning. A critical test of vocal learning must control for relatedness.

3.2.3. Primates—Pygmy Marmoset Contact Calls

Pygmy marmosets (*Cebuella pygmaea*) are small primates that live in family groups composed of a breeding pair, their offspring, and an occasional unrelated adult (Soini 1993). Offspring care is communal (Kinzey 1997a), and this cooperative care may improve offspring survival although it is not essential (Rothe et al. 1993). *Cebuella* inhabit mature forests in the upper Amazon region that are seasonally flooded (Soini 1993), and they feed heavily on gum and insects. Groups defend small, exclusive territories centered around a primary feeding tree, but change their home ranges in response to food abundance (Soini 1993).

Pygmy marmosets give two types of trilled contact calls, termed the closed mouth trill (Fig. 4.8A) and the J-call. Closed mouth trills are used when animals are in close range or within visual contact, whereas J-calls are given when animals are some distance apart or unable to see one another.

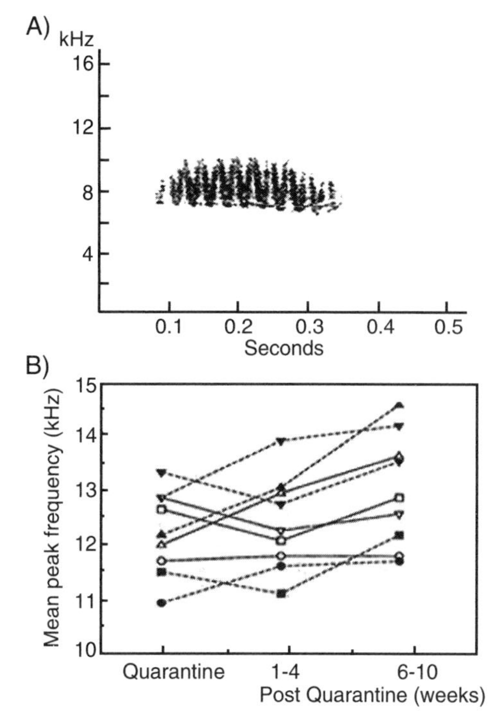

FIGURE 4.8. Pygmy marmoset trills. (**A**) Sonogram of a trill call. (From Pola and Snowdon 1975. Reprinted with permission from Academic Press.) (**B**) Changes m peak frequency of trills for residents (open symbols) and introduced animals (closed symbols) during the ten-week experiment. Groups had no auditory contact during the quarantine period. After quarantine, they had auditory but not social contact. Females (circles), subadult males (squares), juveniles (upward triangles), and infants (downward triangles). (From Elowson and Snowdon 1994. Reprinted with permission from Academic Press.)

The structure of J-calls makes them easy to localize (Snowdon and Hodun 1981). Group mates respond antiphonally to both types of contact calls and may also orient toward and approach an animal producing J-calls. The calls are used primarily in intragroup situations, but may also be used in encounters between groups. Thus, both calls are used to maintain contact between the members of a family group (Snowdon and Hodun 1981). These calls function as group signatures, yet most variation is among individuals. Individual vocalizations differ in several temporal and frequency characteristics for both types of contact calls. These differences do not appear to depend on either familial relationships or sex (Snowdon and Cleveland 1980). There is some evidence that individuals are recognized by their calls, although this conclusion is weakened because of small sample size and because statistical significance rests on responses of two males to their own calls (Snowdon and Cleveland 1980). Calls played from an unfamiliar location elicit very low responses, suggesting that a combination of vocal characteristics and context are involved in recognizing others (Snowdon and Cleveland 1980).

Infant call repertoires differ from adult repertoires both in types of calls and acoustic structure of adult-like calls (Elowson et al. 1998a). In addition, correct usage of call types appears to be learned (Elowson et al. 1998a). Several features of contact trills change with time, although stereotypy does not increase so the cause is unclear. J-calls show even less age-related modification (Elowson et al. 1992). Individuals' contact calls differ extensively in acoustic structure and the pattern of ontogenetic changes. Such high individual variance makes finding ontogenetic patterns difficult. However, one pattern that emerges is that infants go through a stage of vocal production reminiscent of human infant babbling (Elowson et al. 1998b). Similarities to human babbling include repetition of call types and use of a subset of the adult repertoire in modified infant form. Babbling infants are responded to by caregivers more than their silent counterparts, suggesting that babbling is important in the normal development of social relationships. Data on development indicate vocal flexibility exists and show that calls are not always produced in adult form by juveniles. Vocal learning may or may not play a role; vocal tract maturation and developing social relationships may be responsible for such changes. Descriptions of development cannot directly test the importance of vocal learning. Considering the current lack of evidence for most primates, alternative explanations seem likely.

The potential for social modification of contact calls was studied with family groups from two different captive colonies (Elowson and Snowdon 1994). Animals were housed in family groups initially in two separate rooms during a quarantine period. Contact calls were recorded and analyzed from this initial time period. Subsequently, family groups from both captive colonies were housed in the same room in acoustic but not visual or physical contact. Contact calls recorded after the animals were introduced to each other (postquarantine) were compared with those from the initial period (Fig. 4.8B). Peak frequency and bandwidth rose in a parallel fashion in most animals from both captive colonies, duration showed some evidence of convergence within groups, and trill rate showed normal developmental changes (Elowson and Snowdon 1994). Thus, this experiment demonstrated that vocalizations were not fixed, and the changes observed appear to be in response to a changed social context. However, neither acoustic convergence nor divergence occurred, so social modification was not demonstrated. It seems likely that the animals were stressed by the presence of unfamiliar conspecifics and may have responded by increasing the frequency of calls. Vocalizations often rise in frequency when animals are stressed, and calls that are used in agonistic contexts often are of high frequency (Morton 1975, 1982). The lack of convergence in call characters is not surprising. There is actually little reason to predict convergence between animals from the two colonies because they were not social companions, had very little social interaction, and neither individuals within family groups nor groups within colonies shared call characteristics prior to the introduction (Snowdon and Cleveland 1980). It is not clear what kind of vocal changes would demonstrate social modification in this species because the degree of similarity in calls does not appear to have functional significance. A more recent experiment investigated vocal changes in four newly formed pairs (Snowdon and Elowson 1999). No clear pattern emerged for the four variables they measured. For each variable, some pairs converged, others diverged, and others showed no change. The variable most suggestive of social modification was peak frequency because three pairs showed convergence, but in only two was it significant. Neither experiment was designed to explore call acquisition.

In 1997, Snowdon concluded that the weight of evidence does not support a role for learning in the acquisition or social modification of vocalizations in this or other species of primates. Snowdon and Elowson's latest experiment (Snowdon and Elowson 1999) is consistent with Snowdon's (1997) position. In keeping with findings for vervet monkeys and some others (Seyfarth and Cheney 1997), marmosets appear to learn the proper use of alarm calls in response to predators and may learn to recognize individually distinctive calls given by group mates (Hayes and Snowdon 1990). Learned usage is a process distinctly different from either learned acquisition or social modification, and we do not consider it vocal learning.

4. Dialects

Vocalizations often vary in structure geographically, sometimes with sharp boundaries between call types. The scale of variation can be microgeographic (variation among contiguous individuals or groups) or macrogeographic (variation among groups separated geographically) (Krebs and Kroodsma 1980). The types of geographic variation are not discrete but form a continuum (Baker and Cunningham 1985). Geographic variation has been termed dialects, although some authors prefer to reserve this term for microgeographic variation resulting from vocal learning (Conner 1985). Here, we adopt a broader definition to include geographic variation at all levels without regard to the mechanism producing such variation. We take the latter stance primarily because the causes of geographic variation are unknown for most mammalian vocalizations and bird calls, and also because dialects can be biologically important even when they arise from processes other than vocal learning. Studies of call dialects in birds and mammals have focused primarily on describing dialectical patterns. Particularly in mammals, these studies often aim at finding evidence consistent with vocal learning rather than determining the function of geographic patterns.

Dialects can emerge as a natural consequence of learning vocalizations from neighbors (Catchpole and Slater 1995). Under this scenario, geographic variation arises due to local copying and cultural drift (or the random fixation of culturally transmitted variants). Random copying errors and innovations can be transmitted horizontally (between neighbors of similar age), obliquely (from unrelated adult to juvenile), and vertically (from parent to offspring). Thus, the closer two individuals are in space, the more they sound alike. This same pattern can arise if dialects result from genetic differences among populations, making it difficult to use pattern alone to infer vocal learning. Either learned acquisition or social modification can give rise to dialects, and the calls can function as individual or group signatures, but need not have a signature function.

The function of birdsong dialects has been debated for some time (Catchpole and Slater 1995). The simplest hypothesis is that dialects themselves have no particular function but merely reflect the underlying process of vocal learning described above (Andrew 1962; Wiens 1982). Others have argued, sometimes forcefully, that dialects have evolved to preserve local adaptation—the genetic adaptation hypothesis (e.g., Baker and Cunningham 1985), to enhance sound transmission by matching local habitat—the habitat matching hypothesis (e.g., Hansen 1979; Handford 1988), or to facilitate interactions between neighbors—the social adaptation hypothesis (e.g., Payne 1981; Rothstein and Fleischer 1987). The function of call dialects in birds is often assumed to be similar to that of song dialects even though the function of the vocalizations themselves is quite different. Dialect function has infrequently been discussed in relation to mammalian dialects.

Assuming that dialects have the same function for song and calls may not be warranted because calls and song function differently. Consequently, the underlying processes might differ. Song in oscines functions to establish and maintain territories and to attract females (e.g., Kroodsma and Byers 1991). In temperate species, males sing primarily during the breeding season. A similar function is assumed for hummingbirds. Many male hummingbirds sing in assemblages, or "leks," and couple this song with elaborate courtship displays. Temperate oscine and hummingbird females rarely sing. In contrast, contact calls appear to function in various social interactions, are given by both sexes throughout the year in birds, and are usually used by both sexes in mammals. Their primary function is not to establish and defend nesting sites, although they can be used in this way. This difference in function may be reflected in different patterns of dialect variation because selection will probably act differently. The prevalence of learned acquisition and social modification may also differ between bird song and the calls of both birds and mammals.

Calls need not be learned to vary geographically. Adaptive variation in body size can produce geographic variation that correlates with habitat, latitude, or altitude. Habitat matching itself can result from local adaptation without recourse to vocal learning. Restricted gene flow between dialect areas is required to produce the genetic divergence and local adaptation responsible for both morphological and call variation. In most cases, data on gene flow required to test these alternative mechanisms are lacking, although the work on cetaceans, particularly humpback whales, has made great strides in this direction (Baker et al. 1990, 1993, 1998; Palumbi and Baker 1994; Larsen et al. 1996; Valsecchi et al. 1997).

Comparisons of geographic variation in genetic similarity and dialect boundaries can illuminate the extent to which dialects may result from vocal learning. In species with sharp dialect boundaries, a lack of congruence between dialect boundaries and the extent of sharing in molecular genetic markers (usually mtDNA haplotypes or microsatellite alleles) supports a strong role for vocal learning. Unfortunately, few study systems have such data available. We present case studies where vocal learning has been implicated and also briefly describe other birds (Table 4.1) and mammals (Table 4.2) for which vocal learning of dialects has been studied. Because population structure is important in testing alternative hypotheses, we describe this evidence when available, along with evidence consistent with vocal learning.

4.1. Birds

4.1.1. Parrots—Yellow-Naped Amazon Parrot Contact Call Dialects

Yellow-naped amazon parrots, *Amazona auropalliata*, are monogamous, yet gregarious birds (Wright 1997). Pair bonds are long-lasting, and pairs jointly defend nest sites against conspecifics. Pairs spend most of their days feeding and socializing with other pairs and at night congregate with others at large traditional roosts (Wright 1996) primarily for protection against predators. Dispersal patterns of juveniles are unknown. The species is resident throughout its range in semiarid woodland scrub and savannahs along the Pacific coast from southern Mexico to northern Costa Rica, Honduras, and several Bay Islands in the Caribbean (Juniper and Parr 1998).

Yellow-naped Amazons have a repertoire of calls that serve a variety of social functions (Wright 1997). The most ubiquitous call is the contact call, a multiharmonic, frequency-modulated call (Fig. 4.9). Both males and females give this call often near their nesting site and at communal roosts (Wright 1996). In other parrot species, contact calls function to maintain proximity between members of a flock or mated pairs (Farabaugh and Dooling 1996) and to recognize mates and other social companions, including siblings (Wanker et al. 1998). The function in *A. auropalliata* is not known, but is suspected to be similar to that of these other psittacines. Pairs also duet near nesting sites, and these duets are thought to function in territorial defense (Wright 1997).

Evidence for vocal learning in this species includes observational work describing dialects in the contact call (Wright 1996) and other calls (Wright 1997), and genetic analyses (Wright and Wilkinson 2001). Disjunct acoustic and geographic boundaries separate three dialect groups in Costa Rica: the southern, northern, and Nicaraguan. Dialects are characterized by different

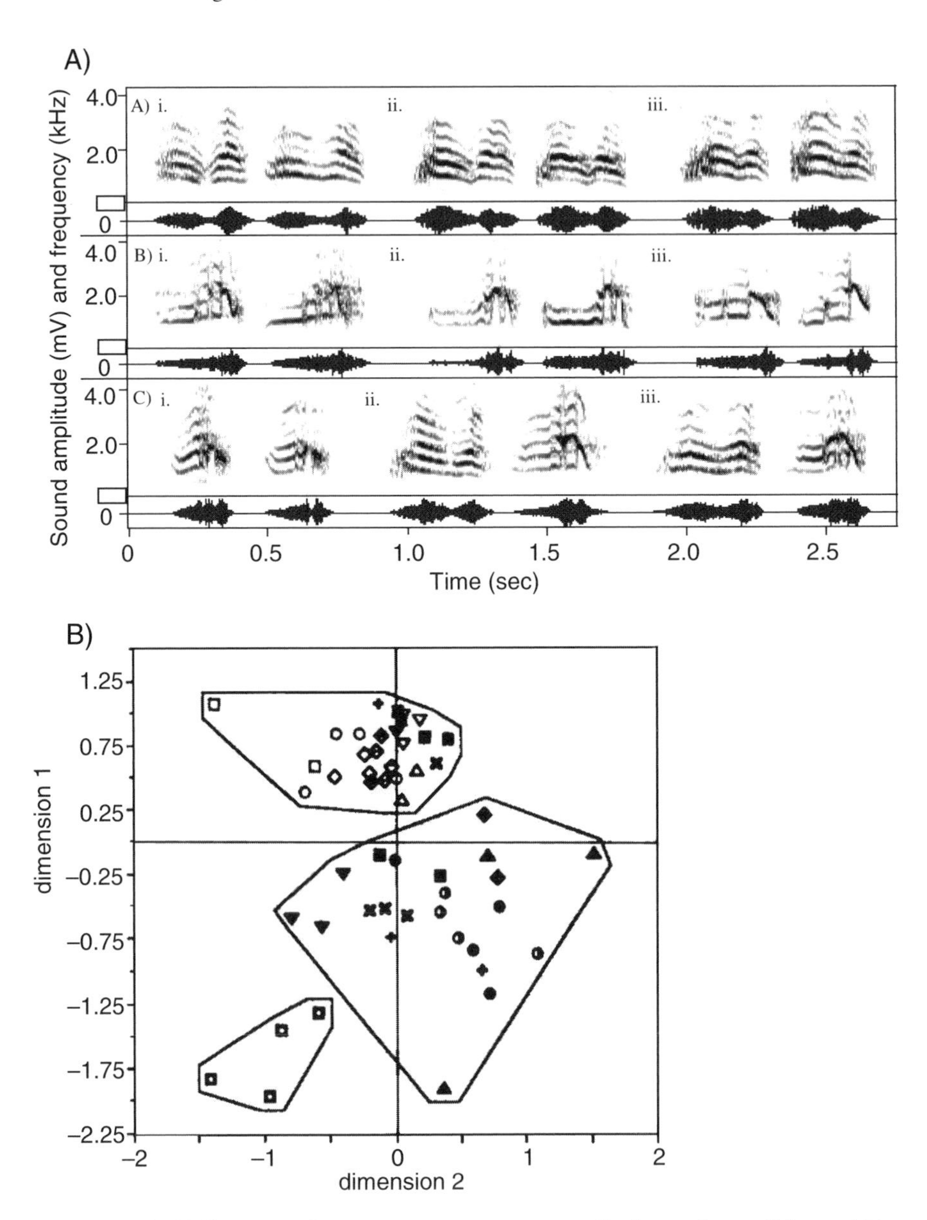

FIGURE 4.9. Yellow-naped amazon parrot contact calls from three dialects. (**A**) Sonograms and waveforms of contact calls from two birds at three roosts (i–iii) in the (a) northern, (b) southern, and (c) Nicaraguan dialects. (**B**) Multidimensional scaling of spectrogram cross correlations with calls from 16 roosts. Each point represents an individual coded by roost: northern (open symbols), southern (closed symbols), Nicaraguan (heavy-lined open squares), border roosts (cross and plus symbols). Minimum-area polygons enclose birds classified from each of the three dialects. (From Wright 1996. Reprinted with permission from The Royal Society.)

call types (Fig. 4.9). Several birds at roosts near the border between the southern and northern dialects are "bilingual"; that is, they produce some calls similar to each dialect. In addition to dialect variation, birds from roosts within a dialect differ in frequency and temporal characteristics of their shared call type. The amount of differentiation correlates with geographic distance, at least within the southern dialect (Wright 1996). Wright (1996) also reports slight differences among individuals within a roost. Several other calls in the repertoire, including short contact calls, growls, and pair duets, show dialect variation (Wright 1997). One call class, squeals and gurgles, does not show a dialect pattern but is highly variable within individuals. Pairs respond strongly to playback of roost mates' pair duets near their nest site but not to playback of calls from outside the dialect region (Wright and Dorin 2001), indicating that parrots perceive these dialect differences, that they are functionally relevant, and that local birds are attended to and thus more likely to be copied.

Genetic analyses indicate substantial gene flow across dialect boundaries (Wright and Wilkinson 2001). Thus, dialects are not a result of population genetic structure. The parsimonious interpretation is that birds copy the local dialect after dispersal, indicating strong social pressure to conform to the local dialect.

4.1.2. Trochilids—Hummingbird Song Dialects

Several studies have been conducted on dialect variation in hummingbird song using the little hermit, *Phaethornis longuemareus* (Snow 1968; Wiley 1971), Anna's hummingbird, *Calypte anna* (Baptista and Schuchmann 1990), the sparkling violet-ear, *Colibri coruscans*, and the green violet-ear, *Colibri thalassinus* (Gaunt et al. 1994). Unfortunately, no single species has been studied intensively, and the hermits are not closely related to the other hummingbirds (Bleiweiss 1998).

Hummingbirds are surprisingly long-lived birds for their size and metabolic rate, living 6–10 years in captivity (Skutch 1972). Most tropical species are rather sedentary, although species that breed in North America migrate to the tropics to overwinter. We know little of population structure in any of these species where vocal learning has been studied. Males of many tropical genera, including *Phaethornis* and *Colibri*, congregate and sing at traditional sites in dispersed leks. Several North American species perform complex aerial courtship displays and are less likely to lek, although *Calypte anna* males do sing within hearing distance of one another.

Snow (1968) first discovered dialects of little hermits, *P. longuemareus*, in Trinidad, and this discovery was further investigated by Wiley (1971). Males sing songs that are very similar to those of their nearest neighbors, yet several song groups occur on a single lek (Snow 1968; Wiley 1971). Within an individual, songs vary in the number of introductory notes and the structure of the song's terminal part (Wiley 1971). Spectral variation has not been quantitatively studied, but inspection of published sonograms suggests that individuals may vary from one another in frequency and temporal characteristics of all song components, even when the basic structure of notes is quite similar, a possible example of by-product distinctiveness. Songs sung by different males within a song group show striking similarity in the frequency contour from the song's middle portion, and this feature differs from other song groups on a lek (Wiley 1971). Visual comparison of sonograms indicates that song at the lek studied by Snow (1968) differs from that at the lek studied by Wiley (1971). Both Snow and Wiley suggest that vocal learning leads to the song sharing they observed. To date, no experimental tests of the learning hypothesis have been conducted on little hermits.

Dialect variation has also been found in the song of Anna's hummingbird, *C. anna* (Mirsky 1976; Baptista and Schuchmann 1990). The differences between island and mainland populations include the structure of syllables and temporal phrasing of the song (Mirsky 1976). Mainland males show no response to playback of island songs, whereas they respond strongly to playback of mainland songs (Mirsky 1976), indicating that dialects are recognized as distinct by the birds. Visual comparison of sonograms from southern (Mirsky 1976) and northern California (Baptista and Schuchmann 1990) shows differences that are much less marked. The island population is only about 100 birds and was probably founded quite recently; song varies little among individuals. Mirsky (1976) speculates that the oddity of island song results from the initial founding event coupled with a subsequent bottleneck. Repeated bottlenecks may have reduced genetic and cultural variation, resulting in strong cultural drift. This hypothesis would be supported if a song resembling that found in the island song were part of the mainland repertoire.

Gaunt et al. (1994) obtained indirect evidence of vocal learning by comparing the songs of neighbors to more distant individuals in two species, the sparkling violet-ear, *Colibri coruscans*, and green violet-ear, *C. thalassinus*. They find that neighbors' songs are more highly correlated with each other than with those of nonneighbors. The degree of acoustic difference correlates with geographic distance between neighborhoods. Oddly, the within-individual correlation for *C. coruscans* is quite low (0.28), and the authors do not comment on this in their paper. They find occasional strong note similarities between distant birds in *C. thalassinus* but argue that these similarities are not as strong as those between neighbors. The lack of genetic data makes interpretation of these patterns inconclusive.

More direct evidence that song is acquired through learning comes from a single study of song development in isolated Anna hummingbirds (Fig. 4.10; Baptista and Schuchmann 1990). One male reared without exposure to conspecific song developed a song that differed significantly from that of wild adults in frequency range and syllable duration, although it retained several species-specific characteristics. Three males reared with acoustic

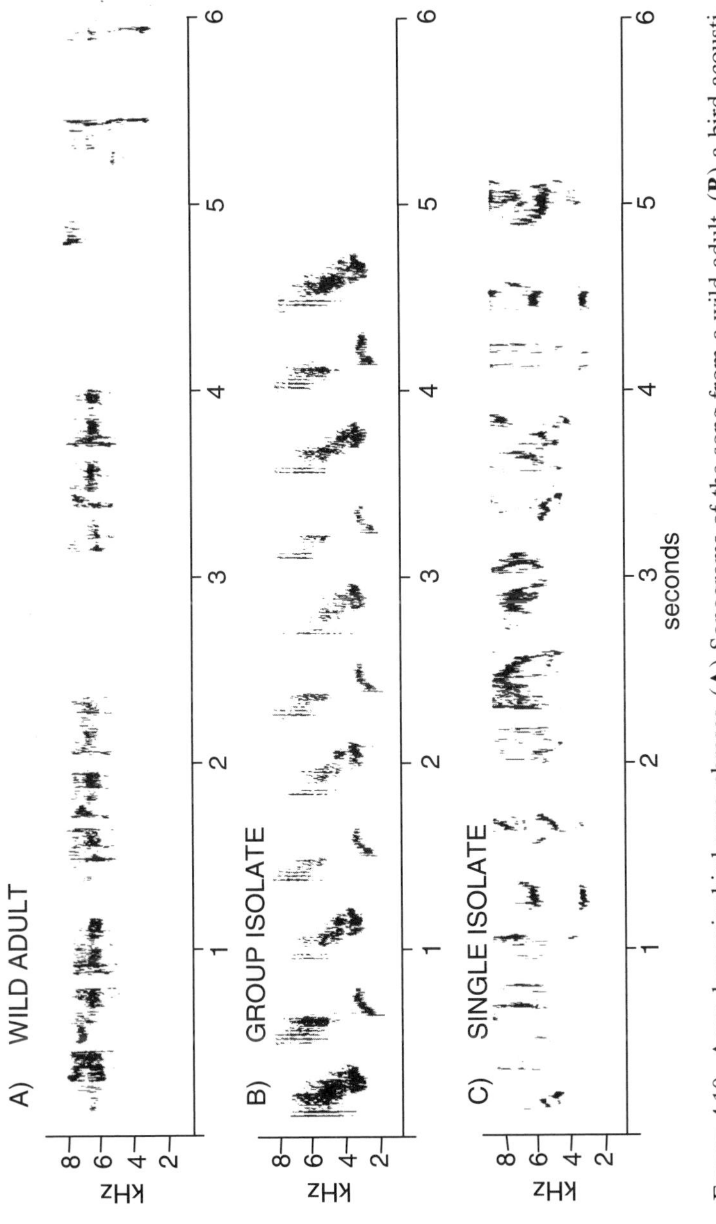

FIGURE 4.10. Anna hummingbird song phrases. (**A**) Sonograms of the song from a wild adult, (**B**) a bird acoustically isolated with two other juveniles, and (**C**) a single isolate. Note the abnormal structure of the song from the isolated birds. (From Baptista and Schuchmann 1990. Reprinted with permission from Blackwell Wissenschafts-Verlag Berlin, GmgH.)

exposure only to each other's song shared many song characteristics, including the structure of individual syllables and temporal pattern of song. The group isolates' songs differed from those of wild males in cadence and minimum frequency. One group isolate was initially raised in the nest and may have heard an adult male singing, so this group's experience differed in two ways from the single isolate, confounding tests of the relative importance of social and acoustic isolation. Both single and group isolates used a chitter that was not heard in the songs of wild males. Baptista and Schuchmann (1990) noted that the single isolate's song resembled the island songs of Mirsky's study and suggested that the initial founder of the island population may have been a young bird who had not completed song learning. Their study, although small, suggested that both learned acquisition and social modification might be involved and mirrors the patterns described in observational studies, lending credence to the vocal learning interpretation. Additional experimental tests are necessary before firm conclusions can be drawn.

4.2. Mammals

4.2.1. Cetaceans—Humpback Song

Humpback whales (*Megaptera novaeangliae*) migrate from summer feeding grounds in northern latitudes of the Pacific and Atlantic oceans to equatorial breeding grounds in each ocean basin. They are also found in the southern hemisphere, where they feed off Antarctica and winter in the South Pacific near Australia, New Zealand, and Tonga. Although highly social, they have few long-term associations except for mother–calf pairs. On the breeding grounds, mature females and their calves are often accompanied by an escort. Escorts are likely to be males, and these escorts compete for access to mature females (Tyack and Whitehead 1983), but females show no fidelity to individual males (Clapham and Palsboll 1997). In some cases, individuals feeding on locally abundant schooling fish will associate repeatedly (Perry et al. 1990), and a few recurring associations among adults are known (Weinrich 1991), but large, loosely structured aggregations are common when humpbacks feed on krill, and individuals often feed and travel alone.

Information on population structure greatly facilitates evaluation of the potential influence of cultural transmission and genetic factors on song sharing. Even though few physical barriers exist, both resighting and genetic data indicate that humpback populations are structured. Numerous genetic analyses provide detailed information on population structure, an enviable position compared with the dearth of genetic data for other species in which dialects have been described. Analyses of both mitochondrial (mtDNA) and nuclear DNA (microsatellite and intron sequence) variation indicate that ocean basins are largely isolated from one another (Valsecchi et al.

1997). Even so, the effective migration rate between ocean basins is estimated to be between 0.5 (Baker et al. 1993) and 10 (Palumbi and Baker 1994) whales per generation.

Each ocean basin contains several populations, and movement of individuals is primarily within ocean basins (Fig. 4.11). In the North Pacific Ocean, one population migrates between Alaska and Hawaii, a second between California and Mexico, and a third winters near Japan and has unknown feeding grounds. Infrequent movement of individuals between these populations occurs (Perry et al. 1990; Calambokidis et al. 1996). Isolation of California–Mexico and Alaska–Hawaii populations is indicated by differences in mtDNA haplotypes (Palumbi and Baker 1994) and nuclear DNA allele frequencies (Baker et al. 1998). However, occasional interbreeding between California and Hawaii whales must have occurred because they share some nuclear DNA alleles (Palumbi and Baker 1994). MtDNA data suggest that other California whales breed with humpbacks that feed in the South Pacific (Baker et al. 1990; Stone et al. 1990; Medrano-Gonzalez et al. 1995).

North Atlantic humpbacks may also segregate into several subpopulations (Fig. 4.11), one migrating between the West Indies and the Western North Atlantic, the other between the Cape Verde Islands and the Eastern North Atlantic (Mattila et al. 1989, 1994; Katona and Beard 1990; Palsboll et al. 1995; Larsen et al. 1996). MtDNA variation indicates that these populations form two distinct matrilineal aggregations (Palsboll et al. 1995; Larsen et al. 1996). However, some gene flow occurs because no differences were found in allele frequencies at six microsatellite loci (Larsen et al. 1996).

Genetic data support the hypothesis that males move more often between populations than females (Palumbi and Baker 1994; Baker et al. 1998). However, one study found that males were more likely to be resighted on the breeding grounds than females (Craig and Herman 1997), so whether migration is sex-biased remains unclear.

Male humpback whales sing long, complex, stereotyped songs (Payne and McVay 1971). Humpback song consists of repetitive elements ranging in frequency from 30 to 4,000 Hz combined into "themes." Themes are combinations of elements given in set patterns. A number of themes are combined in apparently fixed order to produce a single song (Payne and McVay 1971). Variation occurs in the number of repetitions of elements in a theme, the presence of themes in a song, and song duration, but not in the order of themes (Payne and McVay 1971; Payne and Payne 1985). Males sing primarily on the breeding ground, and other adult males move away from singing whales (Tyack 1981) and from song playback (Tyack 1983), suggesting that song serves to space individuals.

There are two lines of observational evidence for social modification and perhaps learned acquisition of humpback songs: shared temporal changes and dialects. Themes are shared by most of the males in a population during

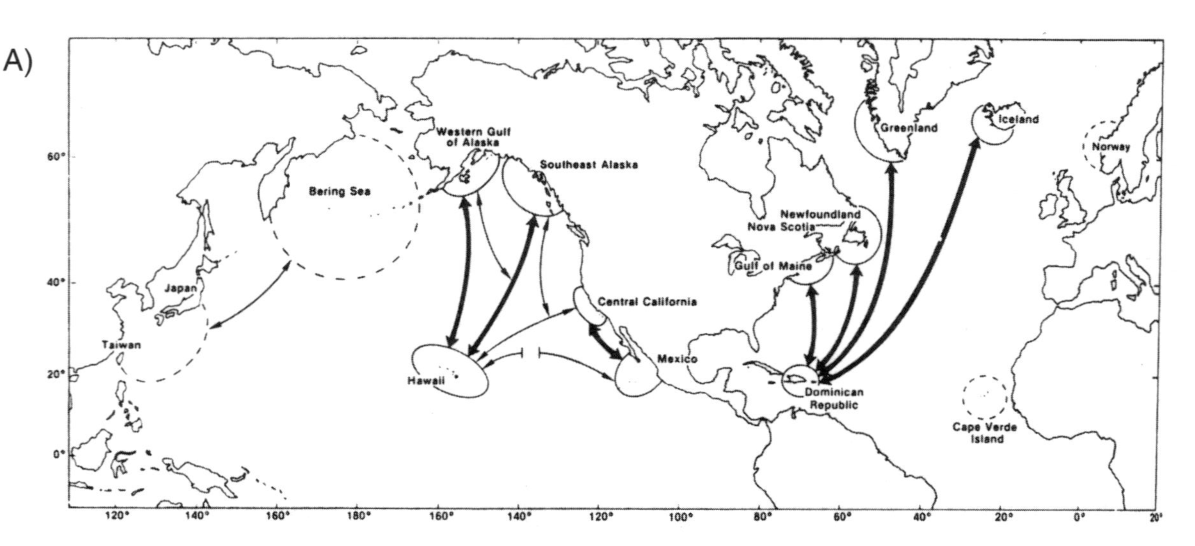

FIGURE 4.11. Humpback whale population structure and song structure. (**A**) Map showing migration patterns in the northern hemisphere based on resightings of marked individuals. Arrows connect regions with strong (heavy arrows) or weak (thin arrows) migratory exchange. (From Baker et al. 1990 Reprinted with permission from Nature. © 1990 Macmillan Magazines Ltd.) (**B**) Spectrogram tracings of humpback whale song themes. One theme is shown for two Atlantic regions in 1979 and two Pacific regions for 1978 and 1979. This theme is given repeatedly while singing. Syllables are described by phonetic terms: GRN (groan), CRY (cry), WO (wavery oo), WVPMN (wavery pulsed moan), SN (snore). (From Winn et al. 1981. Reprinted with permission from Springer-Verlag.)

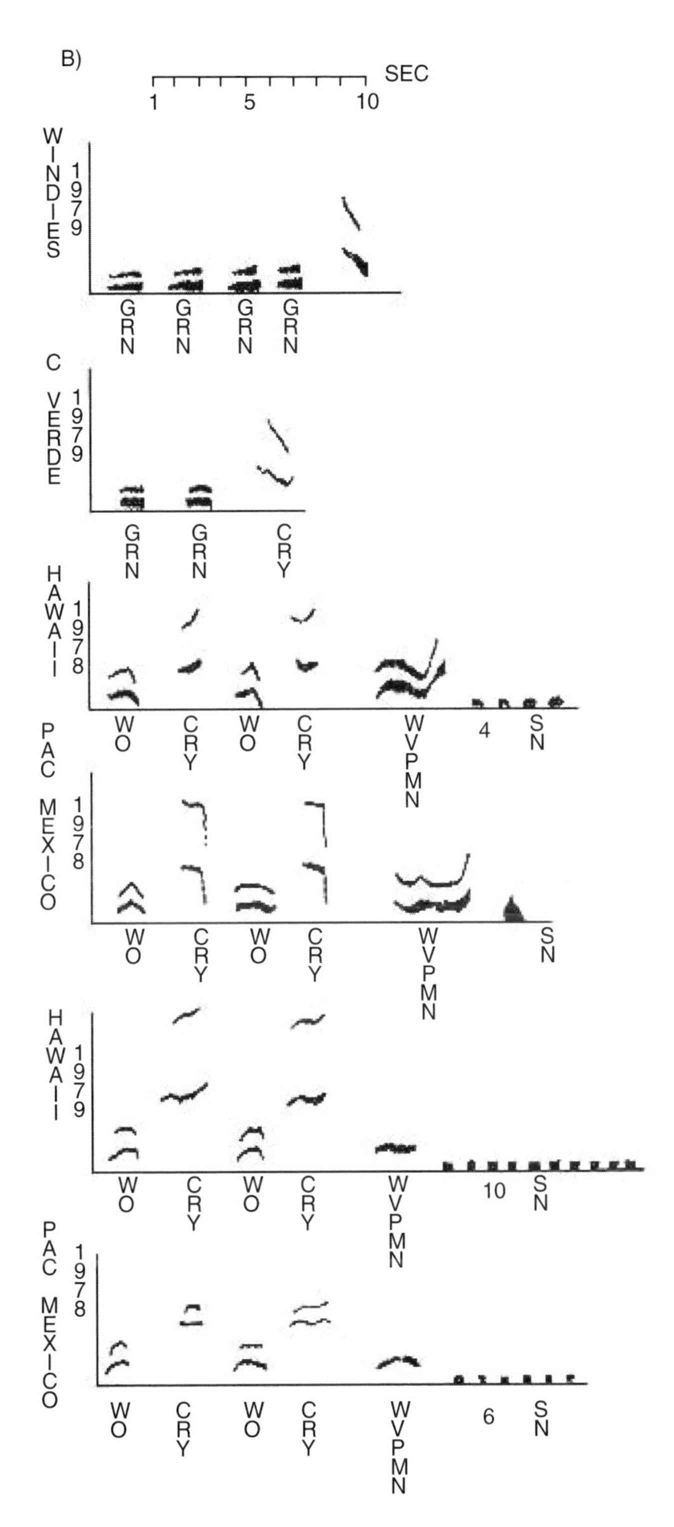

FIGURE 4.11. *Continued*

a single breeding season. However, the song is not static; progressive changes occur (Payne and Guinee 1983; Cato 1991). At the start of a breeding season, the song resembles that from the prior year. As the breeding season progresses, new themes are added and old themes deleted. Thus, the song at the end of a breeding season may bear little resemblance to that at the beginning of that year (Payne and Payne 1985). These changes can be extensive, but it is still sometimes possible to recognize songs from one year as coming from the same population as songs several years later. Observed temporal changes do not result from variation among individuals recorded because individuals recorded over two breeding seasons alter their songs in concert with others in the population (Guinee et al. 1983). The most striking finding with respect to vocal learning is that most individual males in the population make parallel changes (Payne and Payne 1985). The general consensus is that this pattern results from cultural transmission of specific song characteristics, and indeed it is difficult to imagine how such shared changes could occur without copying. The mechanism by which males copy one another and the function of copying is, however, completely unknown.

Additional data supporting the role of social modification come from studies of geographic variation in song. Three ocean basins—North Pacific, North Atlantic, and South Pacific—exhibit distinctly different songs. Recordings within those basins as far as 5,000 km apart show that songs from different populations have very similar structure and the same annual changes (Fig. 4.11; Winn et al. 1981). However, studies of several locations within the North Pacific indicated that whales in Japan, Hawaii, and Mexico share only a portion of their themes (Helweg et al. 1990). Vocal differences correspond roughly to genetic differences in the North Pacific. A similar pattern emerges for humpbacks recorded in the South Pacific in Tonga, New Caledonia, New Zealand, and Australia, although most themes were shared (Helweg et al. 1998), except between the east and west coasts of Australia (Cato 1991).

These patterns suggest a degree of acoustic isolation between populations within ocean basins with some acoustic contact or migratory exchange (Helweg et al. 1990, 1998). The extent of geographic distance, genetic differentiation, and song differentiation appear to be highly correlated. Thus, dialects could arise because of genetic differences, social modification, or both. Currently, we cannot tease apart their separate effects.

Cultural exchange of songs between populations could occur if males sing occasionally on feeding grounds, which they may share with males from other breeding populations (Payne and Guinee 1983). Winn et al. (1981) discuss the possibility that geographic variation arises because populations in one ocean basin are out of synchrony with those in another, even though they may have similar song elements in their repertoires. If this hypothesis is correct, then vocal sharing does not require learned acquisition but is more reminiscent of whistle sharing in dolphins (Smolker and Pepper 1999). This hypothesis is certainly possible; a large number of elements and syntactical arrangements have been described, and current recordings probably do not exhaustively catalog regional repertoires. Asynchrony could result from either genetic or acoustic isolation, and thus this hypothesis does not discriminate between these two mechanisms. To our knowledge, no direct tests of any hypotheses about song sharing or song changes have been conducted. Direct tests are difficult to accomplish but would certainly be worthwhile. Comparing available long-term data sets on a larger scale than has been done in the past to determine to what extent regional repertoires of elements and themes do overlap would be an interesting first step. More direct tests of congruence between genetically defined populations and dialect groups are possible and warranted. The function of dialects and song sharing in humpbacks is completely unknown. Further playback studies to investigate responses of male and female whales to their own and foreign dialects could provide insight into why song sharing is favored and how it occurs.

4.2.2. Primates—Tamarin Long Calls

Dialects have been studied in several species of tamarins in the genus *Saguinus*. Tamarins are distributed throughout the neotropics (Kinzey 1997c) and live in small groups of multiple males and females (Garber et al. 1993; Savage et al. 1996; Kinzey 1997c) that are not close relatives. The dominant female gives birth to twins, which are cared for communally by most adult males in a group (Sussman and Kinzey 1984). Group size correlates with offspring survival (Sussman and Garber 1987) and may reduce care costs for individuals (Price 1992). Groups travel together through their large home ranges (Kinzey 1997c) and feed primarily on widely dispersed insects, fruit, nectar, and exudates from plants (Garber 1993). Home ranges of tamarin groups overlap extensively, and this overlap centers around primary food trees. Groups defend food sources aggressively against other groups to obtain priority access (Garber 1988; Payne and Payne 1997) and experience costs if such defense is unsuccessful (Pruetz and Garber 1991). Group defense involves both vocal display and physical combat (Payne and Payne 1997).

A primary vocalization involved in group defense and group cohesion is the long call. Long calls are 1–2-sec series of frequency-modulated syllables (Fig. 4.12). Calls differ among species and have been used to clarify phylogenetic relationships among tamarin species (Hodun et al. 1981; Kinzey 1997b) and the other members of the callitrichidae, including lion tamarins and marmosets (Snowdon 1993). Long calls also differ between individuals and sexes within a species (Snowdon and Hodun 1985; Maeda and Masataka 1987). Although there is no evidence of call sharing among group members, individuals do respond preferentially to long calls of their own group mates (Snowdon and Hodun 1985), suggesting that they learn to recognize each other's calls. Long calls are often given by animals separated

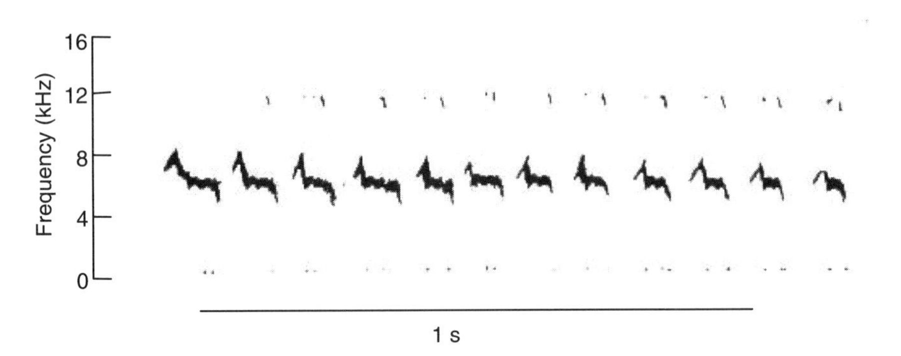

FIGURE 4.12. Sonogram of red-chested moustached tamarin long calls from Cobija region. (From Masataka 1988. Reprinted with permission from Academic Press.

from their group (Snowdon and Hodun 1985; Maeda and Masataka 1987; Masataka 1988) and are thought to facilitate cohesion within the group as it travels through the forest (Waser and Waser 1977). Long calls also function in territory defense during encounters between groups at feeding trees (Waser and Waser 1977; Garber et al. 1993). Females will respond to the calls of an intruder male unless her mate evicts him (Masataka 1988).

Studies of vocal learning in the saddle-back tamarin, *S. fuscicollis* (Cheverud et al. 1993) and the red-chested moustached tamarin, *S. labiatus*, have focused mainly on comparing calls from different populations (Hodun et al. 1981; Maeda and Masataka 1987; Masataka 1988). Contiguous populations of *S. fuscicollis* subspecies (*S. fuscicollis nigrifrons* and *S. fuscicollis illigeri*) differed in one temporal variable and three frequency variables (Hodun et al. 1981). One individual resembled *S. f. nigrifrons* morphologically, but was intermediate between *S. f. nigrifrons* and *S. f. illigeri* in long-call features. Comparison of long calls from captive hybrids between these two subspecies revealed that their long calls resembled *S. f. nigrifrons* with the exception of one frequency variable that most closely resembled the subspecies with which they were housed, *S. f. fuscicollis*. Hodun et al. (1981) suggested that learning must have taken place at least in this one variable. However, inspection of sonograms of these three subspecies' calls indicates that *S. f. fuscicollis* calls were not close matches to either subspecies but were intermediate between *S. f. nigrifrons* and *S. f. illigeri*, which is expected in hybrids if calls are heritable.

Populations of red-moustached tamarins (*S. labiatus*) only 27 km apart differ in frequency characteristics (Maeda and Masataka 1987). The proximity of these populations led the authors to suggest that they are not genetically isolated from one another but that the differences resulted from learning. Masataka (1988) subsequently determined that individuals responded differently to calls from their own and a foreign dialect, suggesting that these differences were perceived by the animals and contributed to call function.

Morphological and genetic data suggest that the subspecies of *S. fuscicollis* are independent and isolated from one another (Cheverud et al. 1993), and even populations of the same species are isolated by geographic barriers (Peres et al. 1996). Combined with the data on call structure of hybrids, this suggests that vocal differences between subspecies (Hodun et al. 1981) are likely to be influenced heavily by genetic differences. It appears that some call features are more strictly under genetic control than others. This finding for *S. fuscicollis* raises the distinct possibility that dialects in *S. labiatus* (Maeda and Masataka 1987) are also influenced by genetic differences among populations. Indeed, under an additive model of genetic variation, intermediacy in acoustic structure is predicted and was found. Further study comparing genetic and acoustic variation for populations within species and between subspecies would help to reveal the extent of genetic control and whether vocal learning is possible.

5. Conclusions

5.1. Comparing Patterns

Call learning in mammals and birds is less well-understood than song learning in birds; nonetheless, progress has been made. Patterns are beginning to emerge in the preponderance of both forms of vocal learning—learned acquisition and social modification—and we have gained insight into what factors favor vocal learning. We discuss these patterns and insights here.

5.1.1. Patterns of Learned Acquisition Compared

Birds from all three families studied—oscines, psittacines, and trocholids—acquire vocalizations through learning, but the evidence for mammals is scant, inconclusive, and limited to cetaceans. Abnormal development of vocalizations in acoustically isolated or deafened birds indicates that chickadees, budgerigars, and Anna's hummingbirds acquire some aspects of calls or song through learning. Other features are apparently innate and still others socially modified. The B and C notes in chickadee calls, but not the A note, require acoustic and social input to develop normally. Tonal features of budgerigar contact calls are acquired through learning, but bandwidth, maximum and minimum frequency, and duration appear to be innate. Some features of budgerigar warble song require acoustic input for normal development. Several frequency and temporal features of Anna hummingbird song are abnormal in isolated birds, although other frequency characteristics develop normally. All members of stripe-backed wren family groups share a sex-specific repertoire. The occasional unrelated male shares its group's repertoire, implicating learned acquisition (Table 4.1). The one phasianid studied, bobwhite quail, shows no evidence of any kind of vocal learning (Table 4.1).

The only mammalian vocalizations that seem to allow for learned acquisition are whistle signatures in bottlenose dolphins and song themes in humpback whales. Learned acquisition is implied, but not demonstrated, by the appearance of shared whistle types in dolphin repertoires and by the addition of new, shared themes to humpback songs. These instances suggest learned acquisition because complete vocalization types are shared and seem to appear relatively intact in individual repertoires after hearing them in socially relevant contexts. Current data are insufficient to rule out social modification as a mechanism for sharing of call types. Individual repertoires of dolphins appear to be large and are incompletely known; thus, whistle sharing could reflect a change in how often an individual produces a particular whistle that it already knows rather than acquisition of a novel type. Experiments should be conducted and published to test directly the importance of learned acquisition to whistle structure in dolphins. Unfortunately, experimental tests in humpbacks would be difficult.

The types of vocalizations that are acquired through learning include both song and contact calls that function as group and individual signatures and show dialect variation in birds and possibly mammals. This suggests that similarity to others is important in all types of vocalizations.

5.1.2. Patterns of Social Modification Compared

Social modification seems to be a widespread mechanism of vocal learning both in birds and in mammals. Again, birds from all three groups show social modification (Table 4.3). Convergence in frequency features of chickadee D notes depends on social modification in adults, although temporal features and note-type proportions do not seem to rely on any form of vocal learning. Juvenile budgerigars incorporate many frequency features and some temporal features of social partners' calls into their contact calls. Vocal plasticity persists into adulthood in this species also because the adult members of social groups give similar contact calls and males sing similar warble song. As with chickadees, some call and song features do not rely on learning. The structure of individual notes and the temporal pattern of song is shared by Anna hummingbird males raised together, suggesting that, in addition to learned acquisition, social modification influences song structure in this species. In addition, individual Australian magpies share songs with their group mates (Table 4.1).

The incidence of social modification in mammals, including bats and cetaceans but not nonhuman primates, is more widespread than for learned acquisition (Tables 4.2 and 4.3). Lesser spear-nosed bat infants appear to modify frequency and temporal features of their isolation calls to increase similarity to their mothers' directive calls. Greater spear-nosed bat females change the fine structure of screech calls, including frequency and

TABLE 4.3. Distribution of social modification among taxa reviewed.

Taxon	Family	Level	Function	Social modification	Species
bird	songbird	individual sig.	contact	unknown	zebra finch
bird	songbird	group sig.	contact	yes	chickadee
bird	songbird	group sig.	territory/ contact	yes	Australian magpie
bird	songbird	group sig.	territory/ contact	unknown	stripe-backed wren
bird	songbird	group sig.	contact	unknown	cacique
bird	parrot	group sig.	contact	yes	budgerigar
bird	quail	group sig.	contact	no	bobwhite
bird	parrot	warble song	song	yes	budgerigar
bird	hummingbird	dialect	song	yes	Anna's hummingbird
bird	hummingbird	dialect	song	unknown	hermit hummingbird
bird	hummingbird	dialect	song	unknown green	violet-ear hummingbird
bird	hummingbird	dialect	song	unknown sparkling	violet-ear hummingbird
bird	parrot	dialect	contact	unknown	yellow-naped amazon
mammal	bat	individual sig.	echo	yes	horseshoe bat
mammal	bat	individual sig.	contact	yes	lesser spear-nosed bat
mammal	bat	individual sig.	contact	no	evening bat
mammal	bat	individual sig.	echo	no	big brown bat
mammal	cetacean	individual sig.	contact	yes	bottlenose dolphin
mammal	primate	individual sig.	alarm	no	vervet monkey
mammal	primate	individual sig.	food	no	rhesus macaque
mammal	primate	individual sig.	play	no	rhesus macaque
mammal	primate	individual sig.	food	no	Japanese macaque
mammal	primate	individual sig.	play	no	Japanese macaque
mammal	bat	group sig.	contact	yes	greater spear-nosed bat
mammal	cetacean	group sig.	contact	unknown	orca
mammal	cetacean	group sig.	contact	unknown	sperm whale
mammal	primate	group sig.	contact	no	marmoset
mammal	primate	group sig.	contact	no	pigtail macaque
mammal	primate	group sig.	territory/ contact	no	tamarin
mammal	cetacean	song	song	yes	humpback whale
mammal	primate	dialect	territory/ contact	no	gibbon
mammal	seal	dialect	song	no	Weddell seal
mammal	seal	dialect	song	no	bearded seal
mammal	seal	dialect	song	no	N. elephant seal
mammal	rodent	dialect	alarm	no	prairie dog

temporal features, to converge on group mates' calls. Juvenile horseshoe bats alter the resting frequency of CF echolocation pulses to be more similar to their mothers' (Table 4.2). Humpback whales match the themes they sing to those that other whales are singing, although the nature of this process and how heavily it relies on learned acquisition or social modification remain to be explained. In all of these cases, social modification increases similarity and may facilitate recognition of individuals and group mates. Bottlenose dolphins may modify whistle signatures to be either more or less similar to certain other individuals, although who is copied is inconsistent. Decreased similarity may facilitate individual recognition, whereas increased similarity may facilitate social cohesion. No uncontested evidence demonstrates that primates socially modify vocalization structure, although they may learn correct usage of vocalizations through social interactions. Social modification appears to be more common in birds than in mammals, although mammals have been less well-studied so this pattern may reflect an experimental bias.

For several mammals, individual signatures are socially modified. These individually distinctive calls function as contact calls (e.g., lesser spear-nosed bats), sonar signals (e.g., moustache bats; Table 4.2), or territorial advertisements (kangaroo rats; Table 4.2). Social modification occurs quite frequently in group signatures for both birds and mammals (Tables 4.1 and 4.2). In addition, dialect variation in song of two bird species (Anna's hummingbird and budgerigar) and one mammal species (humpback whale) appears to arise, at least in part, from social modification.

5.1.3. General Principles that Can Be Derived from These Patterns

Social modification may be more widespread because it is a simpler mechanism than learned acquisition with respect to required motor, perceptual, and neural substrates, or because social modification is less risky in an evolutionary sense. Normal vocal development is more certain than with learned acquisition, but enough flexibility in motor control, perceptual discrimination, and neural processing can be retained to allow vocalizations to be fine-tuned to the social circumstances. The apparent predominance of social modification may also result from our patterns of investigation. Research into call learning in birds and mammals is much less well-developed than song learning in oscine birds, and we have yet to work out the mechanism of call sharing and dialect variation in many cases. Demonstrating learned acquisition is more complex than demonstrating social modification, so we may just have more known instances of the latter. More work to elucidate mechanisms in taxa where vocal learning occurs is warranted. However, this work is not without obstacles and will require creativity in experimental design. Deafening is both unethical and likely to be unproductive in taxa such as bats and odontocetes that rely on hearing for navigation. Social isolation is less extreme but can produce abnormal behavior in taxa that are extremely social, making interpretation of call development patterns difficult. Vocalizations may appear abnormal not because acoustic input and motor flexibility are necessary for normal development but because of pathological behavior or inappropriate social context.

The vocal flexibility required for effective echolocation in odontocetes and chiropterans may have allowed vocal learning of social signals as well, an idea also developed by Tyack (personal communication). Bats, whose echolocation behavior is well-studied, modify many aspects of their vocal output to facilitate navigation and prey capture (Griffin 1958, 1986; Schnitzler and Henson 1980). They alter the structure of sonar signals to provide appropriate information at each stage of searching for and approaching an insect prey, changing duration, extent of frequency modulation, bandwidth, and repetition rate of pulses (e.g., Kalko and Schnitzler 1993; Kalko 1995). Some species also exhibit Doppler-shift compensation, adjusting the frequency of sonar emissions to stabilize echo frequency during flight (Schnitzler 1967). This flexibility allows bats to finely tune their vocal behavior to its specific function, and this modification occurs very rapidly, often in less than a second (for reviews, see Schnitzler and Henson 1980; Griffin 1986). Such vocal flexibility in echolocation pulses is probably under fairly strong selection because of its role in foraging. Once vocal flexibility for echolocation evolved, social calls may have been readily learned, requiring little modification to vocal production or neural processing mechanisms. Indeed, it appears that processing of social calls is done by the same neurons involved in echolocation processing (Ohlemiller et al. 1996; Esser et al. 1997). Thus, adapting the similarity of individual or group signatures to social partners becomes an easy task.

Individual signatures function best when they are maximally different from others, whereas group signatures function best when group mates sound similar yet distinct from other groups. The common effect of vocal learning is to modify the extent of similarity between individuals; however, changes might be in opposite directions, depending on call function.

5.2. Are Models of Song Learning in Birds Relevant to Call Learning?

5.2.1. Sensorimotor Model

The most widely considered model for the process of song learning in oscine birds is the auditory template model (Marler 1976) that builds on the sensorimotor model developed by Konishi (1965). This idea suggests that juvenile songbirds are born with a crude neural template of their species' song. This crude template guides song memorization and attention to certain song types during the early phases of song learning. During the memorization phase, young birds memorize songs and song elements that they hear, refining the neural template to match these memorized songs. In many species, there is a critical period during which songs are memorized. Songs heard after this critical period are not memorized and copied. During the motor phase, young males begin to sing. Song output is guided by auditory feedback, and close matching is achieved by comparing an individual's own song to the neural template. Initially, acoustic output is highly variable subsong, likened to babbling in human infants. Gradually, song comes to resemble the adult songs that young males heard during the memorization phase as they more closely match song output to the internal template. Finally, males crystallize their songs and produce adult-form songs.

5.2.2. The Process of Song Matching: Delayed Critical Period or Selective Attrition?

When exposure to normal calls as adults allows isolates whose calls are abnormal to develop normal calls, a critical period is unlikely or has an extended duration. This pattern cannot result from selective attrition because juveniles neither heard nor produced normal calls early in development. Unfortunately, studies do not regularly report repertoire size during the various stages of normal vocal development, which is essential data for testing selective attrition.

Song matching to neighbors seems to occur when a male first settles on a breeding territory. Controversy centers on whether this matching involves acquisition of those songs at that time, and hence extension of the sensitive phase (Slater 1989), or selective attrition from an overproduced repertoire of songs memorized during an early sensitive phase (Nelson and Marler 1994). While those working on song learning debate and test these two hypotheses, the controversy can ignite some very exciting work on call learning in birds and mammals. For each taxonomic group, we can begin to address several questions: Is there a sensitive phase for call learning? Does it differ depending on call function? Can it be modified or extended by social interaction? Is there a phase of overproduction followed by a loss of call types to achieve matching? Or are novel call types acquired de novo to achieve matching? Broadening our inquiry into vocal learning by including other taxa besides oscines and other vocalizations besides song may give us insight into the process in songbirds by helping to reveal which ecological and behavioral factors shape the vocal learning process. What we currently know about the process of vocal learning in taxa other than oscines is discussed below.

5.2.3. The Importance of Social Interaction

Social interaction is often claimed to override the critical period and be a powerful force shaping songs. The keen interest in how social processes affect vocal learning is highlighted in the title of a recent book on vocal learning: *Social Influences on Vocal Development* (Snowdon and

Hausberger 1997). If our premise here is correct—that call function reveals where selection operates—then social interactions should have a very strong part to play in call learning because calls often function to mediate complex social interactions. When the degree of similarity affects call function, social partners are likely to serve as call tutors, and the absence of social interaction is likely to have a profound influence on call development, especially when learned acquisition occurs.

Primates may not show vocal learning because other cues are used for identification and social facilitation, with vocalizations playing only a supportive role. Alternatively, individuals' distinctive calls are a by-product of morphological and body-size differences, which do not require vocal learning. Long lives and close association allow troop members to recognize group mates' individually distinctive calls. Under both hypotheses, vocal learning simply is not necessary.

5.2.4. Relevance to Call Learning in Birds

We would next like to apply these ideas to the few cases where data on the process of call learning are available for birds and mammals. This allows us to evaluate the extent to which auditory template or sensorimotor models describe call learning in mammals and birds. We also describe the process of call matching to evaluate the universality of critical periods and the extent of social influences on call development. Ideal studies for evaluating these hypotheses include replicated data from deafened, totally isolated, acoustically isolated, socially isolated, and normally reared individuals. Unfortunately, we have these comprehensive data for very few taxa; therefore, our conclusions are tentative. Currently, insufficient data are available for most taxa to test the selective attrition hypothesis for call learning.

5.2.4.1. Sensorimotor Model in Birds

Evidence that sensory input is essential for normal call learning comes from studies on contact calls in budgerigars (Dooling et al. 1987; Brittan-Powell et al. 1997; Heaton and Brauth 1999; Heaton et al. 1999) and chickadees (Ficken et al. 1985; Hughes et al. 1998) and for song learning in Anna's hummingbirds (Baptista and Schuchmann 1990). Early vocalizations are often structurally simple and highly variable, and they become more complex and stereotyped with age (Clemmons and Howitz 1990; Brittan-Powell et al. 1997). Deafened budgerigars developed abnormal contact calls that bore little resemblance to the species' typical call (Dooling et al. 1987; Heaton and Brauth 1999). Calls of adults without auditory feedback deteriorated (Heaton et al. 1999). Juvenile chickadees who did not hear adult calls developed abnormal variable see, gargle (Ficken et al. 1985), and chickadee calls (Hughes et al. 1998), indicating that appropriate acoustic input is essential. Isolated juvenile Anna's hummingbirds sang songs that differed from normal song in several acoustic features (Baptista and Schuchmann 1990), again showing the importance of acoustic input. Too little work has been done to directly test the auditory template model; however, these data suggest that sensorimotor processes are important in budgerigars, chickadees, and hummingbirds. There is no evidence of a delay between memorization and production of calls in budgerigars or chickadees as has been found for song learning in many songbirds.

5.2.4.2. The Process of Call Matching for Birds: Delayed Critical Period or Selective Attrition?

Budgerigars (Farabaugh et al. 1994; Brittan-Powell et al. 1997), chickadees (Mammen and Nowicki 1981; Nowicki 1989), and Australian magpies (Table 4.1) retain the ability to match contact calls of social partners past fledging and into adulthood. This could result from a delayed closure to the critical period for call memorization, the absence of a critical period, or because of selective attrition of previously memorized calls. Taxa for which acoustic isolation experiments have been done supported extended critical periods.

When acoustically isolated budgerigars were subsequently housed with individuals who gave normal calls, they modified their calls to closely resemble species-specific calls (Brittan-Powell et al. 1997). An extended critical period is more likely to explain this result because juveniles neither heard nor produced normal calls early in development. Adult birds also changed calls to match group mates' calls (Farabaugh et al. 1994), again suggesting the absence of a critical period, although in adults this pattern could result from selective attrition. In budgerigars, group signatures clearly rely on social interaction.

Critical periods differ for each note in the chick-a-dee call. A notes develop normally in isolated birds, as do B notes, but Hughes et al. (1998) suggest that selective attrition may be involved in fine-tuning A note structure. A critical period for C notes is implied by the finding that juvenile chickadees exposed to adult chick-a-dee calls only after the age of 38 days developed C notes as abnormal as total isolates (Hughes et al. 1998). Vocal plasticity of D notes is retained into adulthood (Nowicki 1989), and selective attrition may be involved in D note convergence within flocks (Hughes et al. 1998). However, there is no direct test of the selective attrition hypothesis for group-signature development.

Hand-raised juvenile Anna's hummingbirds that were housed together from 64 days to adulthood sang remarkably similar, although abnormal, song (Baptista and Schuchmann 1990), suggesting that vocal plasticity was retained to 64 days or beyond. A single isolate showed no evidence of learning from an adult with which it was housed after one year, suggesting that this species does have a critical period that ended before this time. The data from this one study are somewhat contradictory with regard to critical period, and strong conclusions are not possible with such small sample sizes.

5.2.4.3. The Importance of Social Interaction for Birds

Social interaction strongly influences call learning in budgerigar and chickadee group signatures, and may influence Anna's hummingbird song and stripe-backed wren and cacique group signatures. Budgerigars preferentially converge on contact calls of group mates over other birds that they can hear but with which they have no social contact (Farabaugh et al. 1994). Young birds copy adults with which they have social bonds even when those adults give abnormal or heterospecific calls (Brittan-Powell et al. 1997). Social interaction is clearly important to convergence in chickadee calls. Isolated juveniles housed together shared the same abnormal song (Hughes et al. 1998), and adults converged rapidly in D note structure (Nowicki 1989). All male stripe-backed wrens in a group share a repertoire, and all females share a different repertoire (Table 4.1). Although male group mates are close relatives and could share songs because of this, female group mates are not closely related; thus, at least female repertoire sharing results from social contact. The contrasting support for a critical period in Anna's hummingbird group and single isolates could be explained by social factors. The group isolates had developed social bonds and thus were effective tutors for each other, whereas the single isolate may not have developed a social bond with the adult male, making this adult an ineffective tutor. Data to evaluate this hypothesis are not available in the original paper (Baptista and Schuchmann 1990). In many birds, social partners are copied, resulting in acoustic convergence among social groups. In the three best-studied birds, sensorimotor input and social interaction are necessary for at least some of call development, and call learning persists well past the nestling stage.

5.2.5. Relevance to Call Learning in Mammals

5.2.5.1. Sensorimotor Model in Mammals

Very few studies on the effects of deafening or acoustic isolation have been done in mammals other than primates, making it difficult to evaluate the importance of sensory input and the sensorimotor or auditory template models. The data we do have are a few observational studies of normal vocal development. Studies of sonar signal development in infant bats generally show that infants begin to produce sonar signals within a few days of birth, coinciding with the onset of hearing during the second postnatal week in many species. Infant sonar signals tend to be of lower frequency, longer duration, show less steep frequency modulation, and be more variable than adult signals for both constant frequency (CF/FM) bats (Brown et al. 1983; Habersetzer and Marimuthu 1986) and FM bats (Moss 1988; Moss et al. 1997). An increase in signal frequency corresponds with an increase in auditory responses to high-frequency sounds (Konstantinov 1973; Rübsamen et al. 1989). The auditory and vocal-production systems develop in concert.

Deafened young horseshoe bats shift the frequency of their CF sonar signals (Rübsamen 1987), indicating that auditory feedback does influence vocal output, resulting in precise matching of cochlear tuning and vocalization frequency in normal individuals. However, vocal output has no effect on cochlear tuning or responsiveness of the auditory system, so this influence is unidirectional (Rübsamen 1987) and results in fine-tuning the frequency of existing vocalizations rather than substantially reorganizing signal structure.

Several researchers hypothesize that sonar signals develop from communication signals (Moss 1988). Isolation calls in infants who had no acoustic input are normal, infants give these calls within a few hours of birth, and species that cannot hear at birth produce isolation calls. Acoustic input is clearly not essential for normal development (Ehret 1980). This appears to be true also for sonar signals (Gould 1975). In none of these instances is there direct evidence testing the role of vocal learning. Isolation calls do show age-related changes (Jones et al. 1991; Scherrer and Wilkinson 1993). Most changes are consistent with maturation of laryngeal and respiratory function and of peripheral and central nervous control, allowing greater control of vocal production (Could 1975). Functional considerations make a reliance on acoustic input for call acquisition and normal development unlikely. A bat with abnormal sonar signals would be severely handicapped in foraging. An infant unable to produce normal isolation calls risks permanent separation from its mother, on whom it is completely dependent early in life. Vocal learning, when it occurs, is probably restricted to social modification to increase individual distinctiveness (Masters et al. 1995) or increase similarity to the infant's mother (Jones and Ransome 1993; Esser 1994) and to tune vocal output to the best frequency of the auditory system (Rübsamen 1987). Thus, individual signatures and echolocation calls are more likely to be socially modified than acquired through learning.

Very young infant bottlenose dolphins give whistles, suggesting that acoustic input is unnecessary for normal call production; however, direct tests have not been done. As with infant bats, a reliance on acoustic input for normal development is risky, so social modification to adjust the degree of similarity with social partners is likely to be the predominant form of vocal learning. An interesting question is whether the preponderance of social modification over learned acquisition in bats and cetaceans is due to similar call function or because both echolocate.

5.2.5.2. The Process of Call Matching for Mammals: Delayed Critical Period or Selective Attrition?

In both bats and dolphins, the critical period seems to be either nonexistent or labile to accommodate changing group composition and social relationships. However, the mechanism of call matching is unclear. In bats, convergence between mothers and offspring or among adult group mates results from subtle acoustic changes to a single call type (e.g., sonar signal, isolation call, or contact call). This convergence may arise through mutual imitation of acoustic characteristics or selective attrition from a more variable acoustic space. In dolphins, several studies have inferred mimicry in both juveniles and adults, suggesting open-ended learned acquisition. This is certainly possible, yet the size and turnover of juvenile repertoires suggest that individuals may choose matching whistles from a large repertoire. Experimental tests of these two mechanisms are lacking and would be fruitful.

5.2.5.3. The Importance of Social Interaction for Mammals

Experimental evidence of the relative importance of social interaction is also lacking. Much of the data on vocal learning in mammals are observations and experiments showing convergence among social partners. The apparent advantage of vocal learning is to modify the degree of acoustic similarity to enhance social function. All of this points to an important role for social factors. Individuals who serve as models for imitation are those who are the focus of other social interactions.

Recent research is just beginning to outline patterns for the importance of sensorimotor input, call matching, and social interaction in mammalian call learning. As yet, we have little data on the actual processes that underlie these patterns and cannot yet evaluate how closely call learning in mammals matches song learning in birds. Research is likely to progress most rapidly with small-bodied mammals, such as bats, because of the greater ease with which they can be housed. We hope that methods will be developed to study processes in cetaceans as well. Interest in comparing the process of call learning in both birds and mammals to song learning is high and should lead to further work on the underlying processes. This area promises rich rewards.

5.3. Future Directions

5.3.1. Studying Vocal Learning in a Phylogenetic Context

We are close to the point where we can consider comparative tests for hypotheses about the mechanisms, function, and evolution of vocal learning in a phylogenetic framework. Many fundamental questions regarding vocal learning would be possible within this framework, such as: How often has vocal learning evolved? Do learned acquisition and social modification represent distinct processes, or are they different outcomes of the same process? A comparative approach opens up the possibility to test specific hypotheses about the motor, sensory, and physiological underpinnings of learned vocalizations. Which mechanisms, if any, are common to the taxa that learn vocalizations? A comparative approach also allows us to test hypotheses of the ecological and behavioral factors that favor vocal learning. Are certain social structures conducive to vocal learning? Are individual signatures and group signatures influenced equally by vocal learning? Is there support for the hypothesis that echolocation predisposes species to learn social calls? Although we are close to the point where we can ask these questions, we need to broaden even further the taxonomic groups we choose to study. When designing new studies, careful consideration of the phylogenetic relationship of a taxon to those for which we already have data on vocal learning is warranted.

5.3.2. Cultural Evolution and Vocal Learning

The study of learned vocalizations is an especially pertinent type of social learning that can inform both theoretical and empirical work on gene–culture evolution. Nonvocal traits that are transmitted culturally can affect the course of evolution for other traits (Feldman and Laland 1996). In the nonhuman literature, examples have focused on cultural transmission of foraging strategies, probably because foraging was thought to be under strong selection to be optimal. This focus has yielded important advances in theoretical work and some very interesting empirical results. Yet, the ability to modify the evolution of other traits might be especially true for behavioral traits that affect social organization or mating. Social interaction is nearly always involved in cultural transmission, and traits are often transmitted between the members of social groups. Consequently, cultural traits that themselves affect social organization can powerfully influence both their own transmission and other aspects of a species' biology. Culturally transmitted traits that affect the likelihood of mating almost certainly alter the fate of traits transmitted genetically. Many of the vocalizations reviewed here are integral to the smooth functioning of social groups. In at least one case, learned vocalizations actually facilitate another form of social learning—social foraging (greater spear-nosed bats—Boughman 1998; Wilkinson and Boughman 1998).

Certainly, there is a rich history of this kind of work in the birdsong literature, including applications of the "meme" concept (Payne et al. 1988; Payne and Westneat 1988; Trainer 1989; Ficken and Popp 1995). Much of this work has focused on the fate of song variants, and fewer studies have explored genetic consequences or tested gene–culture models (Gibbs 1990; Grant and Grant 1996). We urge increased focus on the processes and consequences of culturally transmitted vocalizations outside of oscine song. Both a fine-scale focus on individual species and large-scale consideration of mechanisms and patterns across species are likely to yield plentiful insights into how genes and culture coevolve. This work will require simultaneous study of vocal and genetic variation, both time-consuming occupations but likely to be well worth the effort.

6. Summary

We have reviewed data on call learning in birds and mammals for calls that function as individual or group signatures and dialects and focused on call learning rather than song learning. We distinguished two forms of vocal learning—learned acquisition and social modification. We began by predicting when vocal learning was expected and detailing the evidence necessary to demonstrate that vocalizations were learned. Studies should exclude or control factors that can produce patterns similar to those induced by learning, including maturation, body size, genetics, population structure, and local ecology. We then presented case studies for three functional levels: (1) individual signatures: lesser spear-nosed bats, bottlenose dolphins, and Japanese and rhesus macaques; (2) group signatures: chickadees, budgerigars, greater spear-nosed bats, killer whales, and pygmy marmosets; and (3) dialects: Amazon parrots, hummingbirds, humpback whales, and tamarins.

Learned acquisition of calls is rare in mammals although fairly common in birds. Social modification is more common in both mammals and birds. Social modification of group signatures is particularly prevalent. Too little data are available to carefully test the validity of song-learning theories for call learning. This area should receive further attention to evaluate the necessity of sensorimotor input, the existence and duration of critical periods, and the possibility of selective attrition. Social interaction is clearly necessary for most species. The study of call learning in mammals and birds has much promise. Not only can it provide critical insight into the neural, perceptual, and motor mechanisms that underlie vocal learning across taxa, but also into the ecological and behavioral factors that favor its evolution. These questions will benefit from putting vocal-learning studies squarely in a phylogenetic framework. Vocal learning also has potential to give us insight into cultural evolution and help us understand how the dual processes of genetic and cultural change influence biological evolution and diversity.

Acknowledgments. We thank Kari and Kisi Bohn for help in preparing the figures. Thanks also to Kisi Bohn and the editors, A. Megela-Simmons and R. Fay, for valuable comments on an earlier version of this chapter. This work was supported by NSF and NIH grants to CFM. JWB was supported by an NIH training grant in The Comparative and Evolutionary Biology of Hearing, an NSF–NATO postdoctoral fellowship, and an NSF International Research Fellowship.

References

Andrew RJ (1962) Evolution of intelligence and vocal mimicking. Science 137: 585–589.

Bailey ED, Baker JA (1982) Recognition characteristics in covey dialects of bobwhite quail. Condor 84:317–320.

Bailey K (1978) The structure and variation in the separation call of the bobwhite quail (*Colinus virginianus*). Anim Behav 26:296–303.

Bain DE (1988) An evaluation of evolutionary processes: Studies of natural selection, and cultural evolution in killer whales (*Orcinus orca*). Ph.D. Thesis, University of California, Santa Cruz, Santa Cruz, CA.

Baker CS, Medrano-Gonzalez L, Calambokidis J, Perry A, Pichler F, Rosenbaum H, Straley JM, Urban-Ramirez J, Yamaguchi M, von Ziegesar O (1998) Population structure of nuclear and mitochondrial DNA variation among humpback whales in the North Pacific. Mol Ecol 7:695–707.

Baker CS, Palumbi SR, Lambertsen RH, Weinrich MT, Calambokidis J, O'Brien SJ (1990) Influence of seasonal migration on geographic distribution of mitochondrial DNA haplotypes in humpback whales. Nature 344:238–240.

Baker CS, Perry A, Bannister JL, Weinrich MT, Abernethy RB, Calambokidis J, Lien J, Lambertsen RH, Urban Ramirez J, Vasquez O, Clapham PJ, Alling A, O'Brien SJ, Palumbi SR (1993) Abundant mitochondrial DNA variation and worldwide population structure in humpback whales. Proc Natl Acad Sci USA 90:8239–8243.

Baker JA, Bailey ED (1987) Sources of phenotypic variation in the separation call of northern bobwhite (*Colinus virginianus*). Can J Zool 65:1010–1015.

Baker MC, Cunningham MA (1985) The biology of bird-song dialects. Behav Brain Sci 8:85–133.

Baptista LF, Schuchmann K-L (1990) Song learning in the Anna hummingbird (*Calypte anna*). Ethology 84:15–26.

Barclay RMR, Fullard JH, Jacobs DS (1999) Variation in the echolocation calls of the hoary bat (*Lasiurus cinereus*): Influence of body size, habitat structure, and geographic location. Can J Zool 77:530–534.

Bartlett P, Slater PJB (1999) The effect of new recruits on the flock specific call of budgerigars (*Melopsittacus undulatus*). Ethol Ecol Evol 11:139–147.

Bigg MA, Ellis GM, Ford JKB, Balcomb KC (1987) Killer Whales. A Study of their Identification, Genealogy, and Natural History in British Columbia and Washington State. Nanaimo, BC, Canada: Phantom Press.

Bigg MA, Olesiuk PF, Ellis GM, Ford JKB, Balcomb KC (1990) Social organization and genealogy of resident killer whales (*Orcinus orca*) in the coastal waters of British Columbia and Washington State. Individual recognition of cetaceans. Rep Int Whaling Commis 12:383–406.

Bleiweiss R (1998) Tempo and mode of hummingbird evolution. Biol J Linn Soc 65:63–76.

Boinski S, Mitchell CL (1997) Chuck vocalizations of wild female squirrel monkeys (*Saimiri sciureus*) contain information on caller identity and foraging activity. Int J Primatol 18:975–993.

Boughman JW (1997) Greater spear-nosed bats give group distinctive calls. Behav Ecol Sociobiol 40:61–70.

Boughman JW (1998) Vocal learning by greater spear-nosed bats. Proc R Soc Lond B Biol Sci 265:227–233.

Boughman JW, Wilkinson GS (1998) Greater spear-nosed bats discriminate group mates by vocalizations. Anim Behav 55:1717–1732.

Brittan-Powell EF, Dooling RJ, Farabaugh SM (1997) Vocal development in budgerigars (*Melopsittacus undulatus*): Contact calls. J Comp Psychol 111:226–241.
Brockelman WY, Schilling D (1984) Inheritance of stereotyped gibbon calls. Nature 312:634–636.
Brockelman WY, Srikosamatara S (1980) Maintenance and evolution of social structure in gibbons. In: Preuschoft H, Chivers DJ, Brockelman WY, Creel N (eds) The Lesser Apes: Evolutionary and Behavioural Biology. Edinburgh, UK: Edinburgh University Press, pp. 298–323.
Brockway BF (1964) Ethological studies of the budgerigar (*Melopsittacus undulatus*): Reproductive behavior. Behaviour 23:294–324.
Brown ED, Farabaugh SM (1991) Song sharing in a group-living songbird, the Australian magpie, *Gymnorhina tibicen*. III. Sex specificity and individual specificity of vocal parts in communal chorus and duet songs. Behaviour 118:244–274.
Brown PE, Brown TW, Grinnell AD (1983) Echolocation, development, and vocal communication in the lesser bulldog bat, *Noctilio leporinus.* Behav Ecol Sociobiol 13:287–298.
Brown SD, Dooling RJ, O'Grady K (1988) Perceptual organization of acoustic stimuli by budgerigars (*Melopsittacus undulatus*) III. Contact calls. J Comp Psychol 102:236–247.
Buck JR, Tyack PL (1993) A quantitative measure of similarity for *Tursiops truncatus* signature whistles. J Acoust Soc Am 94:2497–2506.
Calambokidis J, Steiger GH, Evenson JR, Flynn KR, Balcomb KC, Claridge DE, Bloedel P, Straley JM, Baker CS, von Ziegesar O, Dahlheim ME, Waite JM, Darling JD, Ellis GM, Green GA (1996) Interchange and isolation of humpback whales off California and other North Pacific feeding grounds. Mar Mamm Sci 12:215–226.
Caldwell MC, Caldwell DK (1965) Individualized whistle contours in bottlenose dolphins (*Tursiops truncatus*). Nature 207:434–435.
Caldwell MC, Caldwell DK (1979) The whistle of the Atlantic bottlenose dolphin (*Tursiops truncatus*): Ontogeny. In: Winn HE, Olla BL (eds) Behavior of Marine Animals: Current Perspectives in Research, vol 3: Cetaceans. New York: Plenum Press, pp. 369–401.
Caldwell MC, Caldwell DK, Tyack PL (1990) A review of the signature whistle hypothesis for the Atlantic bottlenose dolphin, *Tursiops truncatus.* In: Leatherwood S, Reeves R (eds) The Bottlenose Dolphin: Recent Progress in Research. San Diego: Academic Press, pp. 199–234.
Catchpole CK, Slater PJB (1995) Birdsong: Biological Themes and Variations. Cambridge, UK: Cambridge University Press.
Cato DH (1991) Songs of the humpback whales: The Australian perspective. Mem Queensl Mus 30:277–290.
Cheney DL (1987) Interactions and relationships between groups. In: Smuts BB, Cheney DL, Seyfarth RM, Wrangham RW, Struhsaker TT (eds) Primate Societies. Chicago: University of Chicago Press, pp. 267–281.
Cheverud JM, Jacobs SC, Moore AJ (1993) Genetic differences among subspecies of the saddle-back tamarin (*Saguinus fuscicollis*): Evidence from hybrids. Am J Primatol 31:23–39.
Chivers DJ, Raemaekers JJ (1980) Long term changes in behaviour. In: Chivers DJ (ed) Malayan Forest Primates: Ten Years' Study in Tropical Rain Forest. New York: Plenum, pp. 209–260.

Christal J, Whitehead H, Lettevall E (1998) Sperm whale social units: Variation and change. Can J Zool 76:1431–1440.
Clapham PJ, Palsboll PJ (1997) Molecular analysis of paternity shows promiscuous mating in female humpback whales (*Megaptera novaeangliae*, Borowski). Proc R Soc Lond B Biol Sci 264:95–98.
Clark A, Wrangham RW (1993) Acoustic analysis of wild chimpanzee pant hoots: Do Kibale Forest chimpanzees have an acoustically distinct food arrival pant hoot? Am J Primatol 31:99–109.
Cleator HJ, Stirling I, Smith TG (1989) Underwater vocalizations of the bearded seal (*Erignathus barbatus*). Can J Zool 67:1900–1910.
Clemmons J, Howitz JL (1990) Development of early vocalizations and the chick-a-dee call in the black-capped chickadee, *Parus atricapillus*. Ethology 86:203–223.
Conner DA (1985) Dialects versus geographic variation in mammalian vocalizations. Behav Brain Sci 8:297–298.
Connor RC, Smolker RA (1995) Seasonal changes in the stability of male–male bonds in Indian Ocean bottlenose dolphins (*Tursiops* sp.). Aquat Mamm 21: 213–216.
Connor RC, Smolker RA (1996) Patterns of female attractiveness in Indian Ocean bottlenose dolphins. Behaviour 133:37–69.
Connor RC, Smolker RA, Richards AF (1992a) Aggressive herding of females by coalitions of male bottlenose dolphins (*Tursiops* sp.). In: Harcourt AH, de Waal FBM (eds) Coalitions and Alliances in Humans and other Animals. Oxford, UK: Oxford University Press, pp. 415–443.
Connor RC, Smolker RA, Richards AF (1992b) Two levels of alliance formation among male bottlenose dolphins (*Tursiops* sp.). Proc Natl Acad Sci USA 89: 987–990.
Craig AS, Herman LM (1997) Sex differences in site fidelity and migration of humpback whales (*Megaptera novaeangliae*) to the Hawaiian Islands. Can J Zool 75:1923–1933.
Davidson S (1999) The vocal repertoire of male greater white-lined bats: Context, variation, and relationship to females. M.S. Thesis, University of Maryland, College Park, MD.
Deecke VB (1998) Stability and change of killer whale (*Orcinus orca*) dialects. M.Sc. Thesis, University of British Columbia, Vancouver, BC, Canada.
Deputte BL (1982) Duetting in male and female songs of the white-cheeked gibbon (*Hylobates concolor leucogenys*). In: Snowdon CT, Brown CH, Petersen MR (eds) Primate Communication. Cambridge, UK: Cambridge University Press, pp. 67–93.
Ding W, Wursig B, Evans WE (1995) Whistles of bottlenose dolphins: Comparisons among populations. Aquat Mamm 21:65–77.
Dooling RJ, Gephart BF, Price PH, McHale C, Brauth SE (1987) Effects of deafening on the contact call of the budgerigar, *Melopsittacus undulatus*. Anim Behav 35:1264–1266.
Ehret G (1980) Development of sound communication in mammals. Adv Study Behav 11:179–225.
Elowson AM, Snowdon CT (1994) Pygmy marmosets, *Cebuella pygmaea*, modify vocal structure in response to changed social environment. Anim Behav 47: 1267–1277.
Elowson AM, Snowdon CT, Sweet CJ (1992) Ontogeny of trill and J-call vocalizations in the pygmy marmoset, *Cebuella pygmaea*. Anim Behav 43:703–715.

Elowson AM, Snowdon CT, Lazaro-Perea C (1998a) 'Babbling' and social context in infant monkeys: Parallels to human infants. Trends Cogn Sci 2:31–37.
Elowson AM, Snowdon CT, Lazaro-Perea C (1998b) Infant 'babbling' in a non-human primate: Complex vocal sequences with repeated call types. Behaviour 135:643–664.
Esser K-H (1994) Audio-vocal learning in a non-human mammal: The lesser spear-nosed bat *Phyllostomus discolor*. Neuroreport 5:1718–1720.
Esser K-H, Daucher A (1996) Hearing in the FM-bat *Phyllostomus discolor*. A behavioral audiogram. J Comp Physiol A 178:779–785.
Esser K-H, Kiefer R (1996) Detection of frequency modulation in the FM-bat *Phyllostomus discolor*. J Comp Physiol A 178:787–796.
Esser K-H, Lud B (1997) Discrimination of sinusoidally frequency-modulated sound signals mimicking species-specific communication calls in the FM bat *Phyllostomus discolor*. J Comp Physiol A 180:513–522.
Esser K-H, Schmidt U (1989) Mother–infant communication in the lesser spear-nosed bat *Phyllostomus discolor* (Chiroptera, Phyllostomidae)—evidence for acoustic learning. Ethology 82:156–168.
Esser K-H, Schmidt U (1990) Behavioral auditory thresholds in neonate lesser spear-nosed bats, *Phyllostomus discolor*. Naturwissenschaften 77:292–294.
Esser K-H, Schubert J (1998) Vocal dialects in the lesser spear-nosed bat *Phyllostomus discolor*. Naturwissenschaften 85:347–349.
Esser K-H, Condon CJ, Suga N, Kanwal JS (1997) Syntax processing by auditory cortical neurons in the FM-FM area of the mustached bat. Proc Natl Acad Sci USA 94:14019–14024.
Evans WE, Dreher JJ (1962) Observations on scouting behavior and associated sound production by the Pacific bottlenosed porpoise, (*Tursiops gilli* Dall). Bull Calif Acad Sci 61:217–226.
Farabaugh SM, Dooling RJ (1996) Acoustic communication in parrots: Laboratory and field studies of budgerigars, *Melopsittacus undulatus*. In: Kroodsma DE, Miller EH (eds) Ecology and Evolution of Acoustic Communication in Birds. Ithaca, NY: Comstock Publishers, pp. 97–117.
Farabaugh SM, Brown ED, Veltman CJ (1988) Song sharing in a group-living songbird, the Australian magpie. II. Vocal sharing between territorial neighbors, within and between geographic regions, and between sexes. Behaviour 104: 105–125.
Farabaugh SM, Brown ED, Dooling RJ (1992a) Analysis of warble song of the budgerigar *Melopsittacus undulatus*. Bioacoustics 4:111–130.
Farabaugh SM, Brown ED, Hughes JM (1992b) Cooperative territorial defense in the Australian magpie, *Gymnorhina tibicen* (Passeriformes, Cracticidae), a group-living songbird. Ethology 92:283–292.
Farabaugh SM, Linzenbold A, Dooling RJ (1994) Vocal plasticity in budgerigars (*Melopsittacus undulatus*): Evidence for social factors in the learning of contact calls. J Comp Psychol 108:81–92.
Feekes F (1977) Colony-specific song in *Cacicus cela* (Icteridae, Aves): The password hypothesis. Ardea 3:197–202.
Feekes F (1982) Sound mimesis within colonies of *Cacicus c. cela*. A colonial password? Z Tierpsychol 58:119–152.
Feldman MW, Laland KN (1996) Gene–culture coevolutionary theory. Trends Ecol Evol 11:453–457.

Ficken MS, Popp JW (1995) Long-term persistence of a culturally transmitted vocalization of the black-capped chickadee. Anim Behav 50:683–693.
Ficken MS, Weise CM (1984) A complex call of the black-capped chickadee (*Parus atricapillus*). I. Microgeographic variation. Auk 101:349–360.
Ficken MS, Ficken RW, Witkin SR (1978) Vocal repertoire of the black-capped chickadee. Auk 95:34–48.
Ficken MS, Witkin SR, Weise CM (1981) Associations among members of a Black-capped Chickadee flock. Behav Ecol Sociobiol 8:245–249.
Ficken MS, Ficken RW, Apel KM (1985) Dialects in a call associated with pair interaction in the black-capped chickadee. Auk 102:145–151.
Ford JKB (1989) Acoustic behaviour of resident killer whales (*Orcinus orca*) off Vancouver Island, British Columbia. Can J Zool 67:727–745.
Ford JKB (1991) Vocal traditions among resident killer whales (*Orcinus orca*) in coastal water of British Columbia. Can J Zool 69:1454–1483.
Ford JKB, Fisher HD (1982) Killer whale (*Orcinus orca*) dialects as an indicator of stocks in British Columbia. Rep Int Whaling Comm 32:671–679.
Ford JKB, Fisher HD (1983) Group-specific dialects of killer whales (*Orcinus orca*) in British Columbia. In: Payne RS (ed) Communication and Behavior of Whales. Boulder, CO: Westview Press, pp. 129–161.
Garber PA (1988) Diet, foraging patterns, and resource defense in a mixed species troop of *Saguinus mystax* and *Saguinus fuscicollis* in Amazonian Peru. Behaviour 105:18–34.
Garber PA (1993) Feeding ecology and behavior of the genus *Saguinus*. In: Rylands AB (ed) Marmosets and Tamarins: Systematics, Behaviour, and Ecology. Oxford, UK: Oxford University Press, pp. 273–295.
Garber PA, Pruetz JD, Isaacson J (1993) Patterns of range use, range defense, and intergroup spacing in moustached tamarin monkeys (*Saguinus mystax*). Primates 34:11–25.
Gaunt SLL, Baptista LF, Sanchez JE, Hernandez D (1994) Song learning as evidenced from song sharing in two hummingbird species (*Colibri coruscans* and *C. thalassinus*). Auk 111:87–103.
Gibbs HL (1990) Cultural evolution of male song types in Darwin's medium ground finches, *Geospiza fortis*. Anim Behav 39:253–263.
Goldstein RB (1978) Geographic variation in the 'hoy' call of the bobwhite. Auk 95:85–94.
Gould E (1975) Experimental studies of the ontogeny of ultrasonic vocalizations in bats. Psychobiology 8:333–346.
Gould E (1983) Mechanisms of mammalian auditory communication. In: Eisenberg JF (eds) Advances in the Study of Mammalian Behavior. Stillwater, American Society of Mammalogists, pp. 265–342.
Gouzoules H, Gouzoules S (1989a) Design features and developmental modification of pigtail macaque, *Macaca nemestrina*, agonistic screams. Anim Behav 37:383–401.
Gouzoules H, Gouzoules S (1989b) Sex differences in the acquisition of communicative competence by pigtail macaques (*Macaca nemestrina*). Am J Primatol 19:163–174.
Gouzoules H, Gouzoules S (1990a) Body size effects on the acoustic structure of pigtail macaques (*Macaca nemestrina*) screams. Ethology 85:324–334.
Gouzoules H, Gouzoules S (1990b) Matrilineal signatures in the recruitment screams of pigtail macaques, *Macaca nemestrina*, Behaviour 115:327–347.

Gouzoules H, Gouzoules S (1995) Recruitment screams of pigtail monkeys (*Macaca nemestrina*): Ontogenetic perspectives. Anim Behav 132:431–450.
Grafen A (1990) Do animals really recognize kin? Anim Behav 39:42–54.
Grant BR, Grant PR (1996) Cultural inheritance of song and its role in the evolution of Darwin's finches. Evolution 50:2471–2487.
Graycar P (1976) Whistle dialects of the Atlantic bottlenose dolphin. Thesis, University of Florida, Gainesville, FL.
Green K, Burton HR (1988) Annual and diurnal variations in the underwater vocalizations of Weddell seals. Polar Biol 8:161–164.
Green SM (1975) Dialects in Japanese monkeys: Vocal learning and cultural transmission of locale-specific vocal behavior? Z Tierpsychol 38:304–314.
Griffin DR (1958) Listening in the Dark. New Haven, CT: Yale University Press.
Griffin DR (1986) Listening in the Dark. 2nd ed. Ithaca, NY: Cornell University Press.
Guinee LN, Chu K, Dorsey EM (1983) Changes over time in the songs of known individual humpback whales (*Megaptera novaeangliae*). In: Payne RS (ed) Communication and Behavior of Whales. Boulder, CO: Westview Press, pp. 59–80.
Habersetzer J, Marrmuthu G (1986) Ontogeny of sounds in the echolocating bat *Hipposideros speoris*. J Comp Psychol 158:247–257.
Hafen T, Neveu H, Rumpler Y, Wilden I, Zimmermann E (1998) Acoustically dimorphic advertisement calls separate morphologically and genetically homogeneous populations of the grey mouse lemur (*Microcebus murinus*). Folia Primatol 69:342–356.
Hailman JP, Griswold CK (1996) Syntax of black-capped chickadee (*Parus atricapillus*) gargles sorts many types into few groups: Implications for geographic variation, dialect drift, and vocal learning. Bird Behav 11:39–57.
Handford P (1988) Trill rate dialects in the rufous-collared sparrow, *Zonotrichia capensis*, in northwestern Argentina. Can J Zool 66:2658–2670.
Hansen EW (1976) Selective responding by recently separated juvenile rhesus monkeys to the calls of their mothers. Dev Psychobiol 9:83–88.
Hansen P (1979) Vocal learning: Its role in adapting sound structures to long-distance propagation and a hypothesis on its evolution. Anim Behav 27:1270–1271.
Hauser MD (1993) The evolution of nonhuman primate vocalizations: Effects of phylogeny, body weight, and social context. Am Nat 142:528–542.
Hauser MD (1996) Vocal communication in macaques: Causes of variation. In: Fa JE, Lindburg DG (eds) Evolution and Ecology of Macaque Societies. Cambridge, MA: Cambridge University Press, pp. 551–577.
Hayes SL, Snowdon CT (1990) Responses to predators in cotton-top tamarins. Am J Primatol 20:283–291.
Heaton JT, Brauth SE (1999) Effects of deafening on the development of nestling and juvenile vocalizations in budgerigars (*Melopsittacus undulatus*). J Comp Psychol 113:314–320.
Heaton JT, Dooling RJ, Farabaugh SM (1999) Effects of deafening on the calls and warble song of adult budgerigars (*Melopsittacus undulatus*). J Acoust Soc Am 105:2010–2019.
Helweg DA, Herman LM, Yamamoto S, Forestell PH (1990) Comparison of songs of humpback whales, *Megaptera novaeangliae*, recorded in Japan, Hawaii, and Mexico during the winter of 1989. Sci Rep Cetacean Res 1:1–20.
Helweg DA, Cato DH, Jenkins PF, Garrigue C, McCauley RD (1998) Geographic variation in South Pacific humpback whale songs. Behaviour 135:1–27.

Hile AG, Plummer TK, Striedter GF (2000) Male vocal imitation produces call convergence during pair bonding in budgerigars, *Melopsittacus undulatus*. Anim Behav 59:1209–1218.
Hodun A, Snowdon CT, Soini P (1981) Subspecific variation in the long calls of the tamarin *Saguinus fuscicollis*. Z Tierpsychol 57:97–110.
Hoelzel AR (1993) Foraging behaviour and social group dynamics in Puget Sound killer whales. Anim Behav 45:581–591.
Hoelzel AR, Dahlheim ME, Stern SJ (1998) Low genetic variation among killer whales (*Orcinus orca*) in the eastern north Pacific and genetic differentiation between foraging specialists. J Hered 89:121–128.
Hoese HD (1971) Dolphin feeding out of water in a salt marsh. J Mammal 52:222–223.
Hughes M, Nowicki S, Lohr B (1998) Call learning in black-capped chickadees (*Parus atricapillus*): The role of experience in the development of "chick-a-dee" calls. Ethology 104:232–249.
Janik VM (1999) Origins and implications of vocal learning in bottlenose dolphins. In: Box HO, Gibson K (eds) Mammalian Social Learning: Comparative and Ecological Perspectives. Cambridge: Cambridge University Press, pp. 308–326.
Janik VM, Slater PJB (1997) Vocal learning in mammals. Adv Study Behav 26: 59–99.
Janik VM, Slater PJB (1998) Context-specific use suggests that bottlenose dolphin signature whistles are cohesion calls. Anim Behav 56:829–838.
Janik VM, Slater PJB (2000) The different roles of social learning in vocal communication. Anim Behav 60:1–11.
Jones G, Hughes PM, Rayner JMV (1991) The development of vocalizations in *Pipistrellus pipistrellus* (Chiroptera: Vespertilionidae) during post-natal growth and the maintenance of individual vocal signatures. J Zool 225:71–84.
Jones G, Ransome R (1993) Echolocation calls of bats are influenced by maternal effects and change over a lifetime. Proc R Soc Lond B Biol Sci 252:125–128.
Juniper T, Parr M (1998) Parrots: A Guide to Parrots of the World. New Haven, CT: Yale University Press.
Kalko EKV (1995) Insect pursuit, prey capture and echolocation in pipistrelle bats (Microchiroptera). Anim Behav 50:861–880.
Kalko EKV, Schnitzler H-U (1993) Plasticity in echolocation signals of European pipistrelle bats in search flight: Implications for habitat use and prey detection. Behav Ecol Sociobiol 33:415–428.
Katona SK, Beard JA (1990) Population size, migrations and feeding aggregations of the humpback whale (*Megaptera novaeangliae*) in the Western North Atlantic Ocean. Rep Int Whaling Comm 12:295–305.
Kinzey WG (1997a) Cebuella. In: Kinzey WG (ed) New World Primates: Ecology, Evolution, and Behavior. New York: Aldine de Gruyter, pp. 240–247.
Kinzey WG (1997b) New World Primates: Ecology, Evolution, and Behavior. New York: Aldine de Gruyter.
Kinzey WG (1997c) Saguinus. In: Kinzey WG (ed) New World Primates: Ecology, Evolution, and Behavior. New York: Aldine de Gruyter, pp. 289–296.
Konishi M (1965) The role of auditory feedback in the control of vocalization in the white-crowned sparrow. Z Tierpsychol 22:770–778.
Konstantinov AI (1973) Development of echolocation in bats in postnatal ontogenesis. Period Biol 75:13–19.

Krebs JR, Kroodsma DE (1980) Repertoires and geographical variation in bird song. Adv Study Behav 11:143–177.
Kroodsma DE, Byers BE (1991) The function(s) of bird song. Am Zool 31:318–328.
Kroodsma DE, Miller EH (1996) Ecology and Evolution of Acoustic Communication in Birds. Ithaca, NY: Comstock Publishing.
Larsen AH, Sigurjonsson J, Oien N, Vikingsson G, Palsboll PJ (1996) Population genetic analysis of nuclear and mitochondrial loci in skin biopsies collected from central northeastern North Atlantic humpback whales (*Megaptera novaeangliae*): Population identity and migratory destinations. Proc R Soc Lond B Biol Sci 263:1611–1618.
Le Boeuf BJ, Peterson RD (1969) Dialects in elephant seals. Science 166:1654–1656.
Le Boeuf BJ, Petrinovich LF (1974) Dialects in northern elephant seals, *Mirounga angustirostris*: Origin and reliability. Anim Behav 22:656–663.
Lyrholm T, Gyllensten U (1998) Global matrilineal population structure in sperm whales as indicated by mitochondrial DNA sequences. Proc R Soc Lond B Biol Sci 265:1679–1684.
Maeda T, Masataka N (1987) Locale-specific vocal behaviour of the tamarin (*Saguinus l. labiatus*). Ethology 75:25–30.
Mammen DL, Nowicki S (1981) Individual differences and within flock convergence in chickadee calls. Behav Ecol Sociobiol 9:179–186.
Marler P (1976) Sensory templates in species-specific behavior. In: Fentress JC (ed) Simpler Networks and Behavior. Sunderland, MA: Sinauer, pp. 315–329.
Marler P, Hobbett L (1975) Individuality in a long-range vocalization of wild chimpanzees. Z Tierpsychol 38:97–109.
Marshall AJ, Wrangham RW, Arcadi AC (1999) Does learning affect the structure of vocalizations in chimpanzees? Anim Behav 58:825–830.
Marshall JT Jr, Marshall ER (1976) Gibbons and their territorial songs. Science 193:235–237.
Masataka N (1985) Development of vocal recognition of mothers in infant Japanese macaques. Dev Psychobiol 18:107–114.
Masataka N (1988) The response of red-chested moustached tamarins to long calls from their natal and alien populations. Anim Behav 36:55–61.
Masataka N, Fujita K (1989) Vocal learning of Japanese and rhesus macaques. Behaviour 109:191–199.
Masters WM, Raver KAS, Kazial KA (1995) Sonar signals of big brown bats, *Eptesicus fuscus*, contain information about individual identity, age and family affiliation. Anim Behav 50:1243–1260.
Mattila DK, Clapham PJ, Katona SK, Stone GS (1989) Population composition of humpback whales, *Megaptera novaeangliae*, on Silver Bank, 1984. Can J Zool 67:281–285.
Mattila DK, Clapham PJ, Vasquez O, Bowman RS (1994) Occurrence, population composition, and habitat use of humpback whales in Samana Bay, Dominican Republic. Can J Zool 72:1898–1907.
McBride AF, Kritzler H (1951) Observations on the pregnancy, parturition, and postnatal behavior in the bottlenose dolphin. J Mammal 32:251–266.
McCowan B, Reiss D (1995a) Quantitative comparison of whistle repertoires from captive adult bottlenose dolphins (Delphinidae, *Tursiops truncatus*): A re-evaluation of the signature whistle hypothesis. Ethology 100:194–209.

McCowan B, Reiss D (1995b) Whistle contour development in captive-born infant bottlenose dolphins (*Tursiops truncatus*): Role of learning. J Comp Psychol 109: 242–260.
McCowan B, Reiss D (2001) The fallacy of "signature whistles" is bottlenose dolphins: A comparative perspective of "signature information" in animal vocalizations. Anim Behav 62:1151–1162.
McCowan B, Reiss D, Gubbins C (1998) Social familiarity influences whistle acoustic structure in adult female bottlenose dolphins (*Tursiops truncatus*). Aquat Mamm 24:27–40.
McCracken GF (1987) Genetic structure of bat social groups. In: Fenton MB, Racey P, Rayner JMV (eds) Recent Advances in the Study of Bats. Cambridge, UK: Cambridge University Press, pp. 281–298.
McCracken GF, Bradbury JW (1977) Paternity and genetic heterogeneity in the polygynous bat, *Phyllostomus hastatus*. Science 198:303–306.
McCracken GF, Bradbury JW (1981) Social organization and kinship in the polygynous bat *Phyllostomus hastatus*. Behav Ecol Sociobiol 8:11–34.
Medrano-Gonzalez L, Aguayo-Lobo A, Urban-Ramirez J, Baker CS (1995) Diversity and distribution of mitochondrial DNA lineages among humpback whales, *Megaptera novaeangliae*, in Mexican Pacific Ocean. Can J Zool 73:1735–1743.
Medvin MB, Stoddard PK, Beecher MD (1992) Signals for parent–offspring recognition—strong sib–sib call similarity in cliff swallows but not barn swallows. Ethology 90:17–28.
Mirsky EN (1976) Song divergence in hummingbird and junco populations on Guadalupe Island. Condor 78:230–235.
Mitani JC, Brandt KL (1994) Social factors influence the acoustic variability in the long-distance calls of male chimpanzees. Ethology 96:233–252.
Mitani JC, Gros-Louis J (1995) Species and sex differences in the screams of chimpanzees and bonobos. Int J Primatol 16:393–411.
Mitani JC, Nishida T (1993) Contexts and social correlates of long-distance calling by male chimpanzees. Anim Behav 45:735–746.
Mitani JC, Hasegawa T, Gros-Louis J, Marler P, Byrne R (1992) Dialects in wild chimpanzees? Am J Primatol 27:233–243.
Morrice MG, Burton HR, Green K (1994) Microgeographic variation and songs in the underwater vocalisation repertoire of the Weddell seal (*Leptonychotes weddellii*) from the Vestfold Hills, Antarctica. Polar Biol 14:441–446.
Morton ES (1975) Ecological sources of selection on avian sounds. Am Nat 108:17–34.
Morton ES (1982) Grading, discreteness, redundancy, and motivation-structural rules. In: Kroodsma DE, Miller EH, Ouellet H (eds) Acoustic Communication in Birds. New York: Academic Press, pp. 183–213.
Moss CF (1988) Ontogeny of vocal signals in the big brown bat, *Eptesicus fuscus*. In: Nachtigall PE, Moore PWB (eds) Animal Sonar. New York: Plenum Publishing Corp., pp. 115–120.
Moss CF, Redish D, Gounden C, Kunz TH (1997) Ontogeny of vocal signals in the little brown bat, *Myotis lucifugus*. Anim Behav 54:131–141.
Nelson DA, Marler P (1994) Selection-based learning in bird song development. Proc Natl Acad Sci USA 91:10498–10501.
Nelson DA, Marler P, Morton ML (1996) Overproduction in song development: An evolutionary correlate with migration. Anim Behav 51:1127–1140.

Nowicki S (1983) Flock-specific recognition of chickadee calls. Behav Ecol Sociobiol 12:317–320.
Nowicki S (1989) Vocal plasticity in captive black-capped chickadees: The acoustic basis and rate of call convergence. Anim Behav 37:64–73.
Ohlemiller KK, Kanwal JS, Suga N (1996) Facilitative responses to species-specific calls in cortical FM-FM neurons of the mustached bat. Neuroethology 7:1749–1755.
Olesiuk PK, Bigg MA, Ellis GM (1990) Life history and population dynamics of resident killer whales (*Orcinus orca*) in the coastal waters of British Columbia and Washington State. Rep Int Whaling Comm 12:209–244.
Olivier TJ, Ober C, Buettner-Janusch J, Sade DS (1981) Genetic differentiation among matrilines in social groups of rhesus monkeys. Behav Ecol Sociobiol 8:279–285.
Owren MJ, Dieter JA, Seyfarth RM, Cheney DL (1992) "Food" calls produced by adult female rhesus (*Macaca mulatta*) and Japanese (*M. fuscata*) macaques, their normally-raised offspring, and offspring cross-fostered between species. Behaviour 120:218–231.
Owren MJ, Dieter JA, Seyfarth RM, Cheney DL (1993) Vocalizations of rhesus (*Macaca mulatta*) and Japanese (*M. fuscata*) macaques cross-fostered between species show evidence of only limited modification. Dev Psychobiol 26:389–406.
Palsboll PJ, Clapham PJ, Mattila DK, Larsen F, Sears R, Siegismund HR, Sigurjonsson J, Vasquez O, Arctander P (1995) Distribution of mtDNA haplotypes in North Atlantic humpback whales: The influence of behaviour on population structure. Mar Ecol 116:1–10.
Palumbi SR, Baker CS (1994) Contrasting population structure from nuclear intron sequences and mtDNA of humpback whales. Mol Biol Evol 11:426–435.
Payne K, Payne RS (1985) Large scale changes over 19 years in songs of humpback whales in Bermuda. Z Tierpsychol 68:89–114.
Payne RB (1981) Population structure and social behavior: Models for testing the ecological significance of song dialects in birds. In: Alexander RD, Tinkle DW (eds) Natural Selection and Social Behavior: Recent Research and New Theory. New York: Chiron Press, pp. 108–120.
Payne RB, Payne LL (1997) Field observations, experimental design, and the time and place of learning bird songs. In: Snowdon CT, Hausberger M (eds) Social Influences on Vocal Development. Cambridge, UK: Cambridge University Press, pp. 57–84.
Payne RB, Westneat DF (1988) A genetic and behavioral analysis of mate choice and song neighborhoods in indigo buntings. Evolution 42:935–947.
Payne RB, Payne LL, Doehlert SM (1988) Biological and cultural success of song memes in indigo buntings. Ecology 69:104–117.
Payne RS, Guinee LN (1983) Humpback whale (*Megaptera novaeangliae*) songs as an indicator of "stocks". In: Payne RS (ed) Communication and Behavior of Whales. Boulder, CO: Westview Press, pp. 333–358.
Payne RS, McVay S (1971) Songs of humpback whales. Science 173:585–597.
Pearl DL, Fenton MB (1996) Can echolocation calls provide information about group identity in the little brown bat (*Myotis lucifugus*)? Can J Zool 74: 2184–2192.
Peres CA, Patton JL, daSilva MNF (1996) Riverine barriers and gene flow in Amazonian saddle-back tamarins. Folia Primatol 67:113–124.

Perry A, Baker CS, Herman LM (1990) Population characteristics of individually identified humpback whales in the Central and Eastern North Pacific: A summary and critique. Rep Int Whaling Comm 12:307–317.
Piper WH, Parker PG, Rabenold KN (1995) Facultative dispersal by juvenile males in the cooperative stripe-backed wren. Behav Ecol 6:337–342.
Pola YV, Snowdon CT (1975) The vocalizations of the pygmy marmoset (*Cebuella pygmaea*). Anim Behav 23:826–842.
Porter TA, Wilkinson GS (in press) Birth synchrony in greater spear-nosed bats *Phyllostomus hastatus*. J Zool 253:383–390.
Price EC (1992) The benefits of helpers: Effects of group and litter size on infant care in tamarins (*Saguinus oedipus*). Am J Primatol 26:179–190.
Price JJ (1998a) Acoustic communication in a cooperative songbird: Transmission, recognition and use of shared call repertoires. Ph.D. Thesis, University of North Carolina, Chapel Hill, NC.
Price JJ (1998b) Family- and sex-specific vocal traditions in a cooperatively breeding songbird. Proc R Soc Lond B Biol Sci 265:497–502.
Price JJ (1999) Recognition of family-specific calls in stripe-backed wrens. Anim Behav 57:483–492.
Pruetz JD, Garber PA (1991) Patterns of resource utilization, home range overlap, and intergroup encounters in moustached tamarin monkeys. Am J Phys Anthropol 12:146.
Rabenold KN (1984) Cooperative enhancement of reproductive success in tropical wren societies. Ecology 65:871–885.
Raemaekers JJ, Raemaekers PM, Haimoff EH (1984) Loud calls of the gibbon (*Hylobates lar*): Repertoire, organization and context. Behaviour 91:146–189.
Randall JA (1989a) Individual footdrumming signatures in banner-tailed kangaroo rats *Dipodomys spectabilis*. Anim Behav 38:620–630.
Randall JA (1989b) Neighbor recognition in a solitary desert rodent (*Dipodomys merriami*) Ethology 81:123–133.
Randall JA (1994) Discrimination of footdrumming signatures by kangaroo rats, *Dipodomys spectabilis*. Anim Behav 47:45–54.
Randall JA (1995) Modification of footdrumming signatures by kangaroo rats: Changing territories and gaming new neighbors. Anim Behav 49:1227–1237.
Reiss D, McCowan B (1993) Spontaneous vocal mimicry and production by bottlenose dolphins *Tursiops truncatus*: Evidence for vocal learning. J Comp Psychol 107:301–312.
Rendall D, Rodman PS, Emond RE (1996) Vocal recognition of individuals and kin in free-ranging rhesus monkeys. Anim Behav 51:1007–1015.
Richard KR, Dillon MC, Whitehead H, Wright JM (1996) Patterns of kinship in groups of free-living sperm whales (*Physeter macrocephalus*) revealed by multiple molecular genetic analyses. Proc Natl Acad Sci USA 93:8792–8795.
Richards DG, Wolz JP, Herman LM (1984) Vocal mimicry of computer-generated sounds and vocal labeling of objects by a bottlenose dolphin, *Tursiops truncatus*. J Comp Psychol 98:10–28.
Rothe H, Koenig A, Darms K (1993) Infant survival and number of helpers in captive groups of common marmosets (*Callithrix jacchus*). Am J Primatol 30:131–137.
Rothstein SI, Fleischer RC (1987) Vocal dialects and their possible relation to honest status signalling in the brown-headed cowbird. Condor 89:1–23.

Rübsamen R (1987) Ontogenesis of the echolocation system in the rufous horseshoe bat, *Rhinolophus rouxii*: Audition and vocalization in early postnatal development. J Comp Physiol A 161:899–913.

Rübsamen R, Gerhardt HC, Neuweiler G, Marimutha G (1989) Ontogenesis of tonotopy in inferior colliculus of a hipposiderid bat reveals postnatal shift in frequency-place code. J Comp Physiol A 165:755–769.

Rydell J (1993) Variation in the sonar of an aerial-hawking bat *Eptesicus nilssonii*. Ethology 93:275–284.

Savage A, Giraldo LH, Soto LH, Snowdon CT (1996) Demography, group composition, and dispersal in wild cotton-top tamarin (*Saguinus oedipus*) groups. Am J Primatol 38:85-100.

Sayigh LS, Tyack PL, Wells RS, Scott MD (1990) Signature whistles of free-ranging bottlenose dolphins *Tursiops truncatus*: Stability and mother–offspring comparisons. Behav Ecol Sociobiol 26:247–260.

Sayigh LS, Tyack PL, Wells RS, Scott MD, Irvine AB (1995) Sex differences in signature whistle production of free-ranging bottlensoe dolphins, *Tursiops truncatus*. Behav Ecol Sociobiol 36:171–177.

Scherrer JA, Wilkinson GS (1993) Evening bat isolation calls provide evidence for heritable signatures. Anim Behav 46:847–860.

Schnitzler H-U (1967) Compensation of Doppler effects in horseshoe bats. Naturwissenschaften 54:523.

Schnitzler H-U, Henson OW Jr (1980) Performance of airborne animal sonar systems: I. Microchiroptra. In: Busnel RG, Fish JF (eds) Animal Sonar Systems. New York: Plenum Press, pp. 109–181.

Schubert J, Esser K-H (1997) Responses of juvenile lesser spear-nosed bats to playback of natural and digitally modified maternal directive calls: Ontogeny of individual recognition. In: Elsner N, Wassle H (eds) Gottingen Neurobiology Report. New York: Georg Thieme Verlag Stuttgart, p. 375.

Seyfarth RM, Cheney DL (1980) The ontogeny of vervet monkey alarm-calling behavior: A preliminary report. Z Tierpsychol 54:37–56.

Seyfarth RM, Cheney DL (1986) Vocal development in vervet monkeys. Anim Behav 34:1640–1658.

Seyfarth RM, Cheney DL (1997) Some general features of vocal development in nonhuman primates. In: Snowdon CT, Hausberger M (eds) Social Influences on Vocal Development. Cambridge, UK: Cambridge University Press, pp. 249–273.

Shipley C, Hines M, Buchwald JS (1981) Individual differences in threat calls of northern elephant seals. Anim Behav 29:12–19.

Shipley C, Hines M, Buchwald JS (1986) Vocalizations of northern elephant seal bulls: Development of adult call characteristics during puberty. J Mammal 67: 526–536.

Skutch AF (1972) Studies of Tropical American Birds. Cambridge, MA: Nuttall Ornithological Club.

Slater PJB (1989) Bird song learning: Causes and consequences. Ethol Ecol Evol 1:19–46.

Slobidchikoff CN, Coast R (1980) Dialects in the alarm calls of prairie dogs. Behav Ecol Sociobiol 7:49–53.

Smolker RA (1993) Acoustic communication in bottlenose dolphins. Ph.D. Thesis, University of Michigan, Ann Arbor, MI.

Smolker RA, Pepper JW (1999) Whistle convergence among allied male bottlenose dolphins (Delphinidae, *Tursiops* spp.). Ethology 105:595–617.

Smolker RA, Richards AF, Connor RC, Pepper JW (1992) Sex differences in patterns of association among Indian Ocean bottlenose dolphins. Behaviour 123: 38–69.

Smolker RA, Mann J, Smuts BB (1993) Use of signature whistles during separation and reunions by wild bottlenose dolphin mothers and infants. Behav Ecol Sociobiol 33:393–402.

Snow DW (1968) The singing assemblies of little hermits. Living Bird 7:47–55.

Snowdon CT (1993) A vocal taxonomy of the callitrichids. In: Rylands AB (ed) Marmosets and Tamarins: Systematics, Behaviour, and Ecology. Oxford, UK: Oxford University Press, pp. 78–94.

Snowdon CT (1997) Is speech special? Lessons from new world primates. In: Kinzey WG (ed) Primates: Ecology, Evolution, and Behavior. New York: de Gruyter, pp. 75–93.

Snowdon CT, Cleveland J (1980) Individual recognition of contact calls by pygmy marmosets. Anim Behav 28:717–727.

Snowdon CT, Elowson AM (1999) Pygmy marmosets modify call structure when paired. Ethology 105:893–908.

Snowdon CT, Hausberger M (1997) Social Influences on Vocal Development. Cambridge, UK: Cambridge University Press.

Snowdon CT, Hodun A (1981) Acoustic adaptations in pygmy marmoset contact calls: Locational cues vary with distances between conspecifics. Behav Ecol Sociobiol 9:295–300.

Snowdon CT, Hodun A (1985) Troop-specific responses to long calls of isolated tamarins (*Saguinus mystax*). Am J Primatol 8:205–213.

Snowdon CT, Coe CL, Hodun A (1985) Population recognition of infant isolation peeps in the squirrel monkey. Anim Behav 33:1145–1151.

Soini P (1993) The ecology of the pygmy marmoset, *Cebuella pygmaea*: Some comparisons with two sympatric tamarins. In: Rylands AB (ed) Marmosets and Tamarins: Systematics, Behaviour, and Ecology. Oxford, UK: Oxford University Press, pp. 257–261.

Somers P (1973) Dialects in southern rocky mountain pikas, *Ochotona princeps* (Lagomorpha). Anim Behav 21:124–137.

Steiner WW (1981) Species-specific differences in pure tonal whistle vocalizations of five western north Atlantic dolphin species. Behav Ecol Sociobiol 9:241–246.

Stone GS, Florez-Gonzalez L, Katona SK (1990) Whale migration record. Nature 346:705.

Strager H (1995) Pod-specific call repertoires and compound calls of killer whales, *Orcinus orca* Linneaus, 1758, in the waters of northern Norway. Can J Zool 73:1037–1047.

Sussman RW, Garber PA (1987) A new interpretation of the social organization and mating system of the Callitrichidae. Int J Primatol 8:73–92.

Sussman RW, Kinzey WG (1984) The ecological role of the Callitrichidae. Am J Phys Anthropol 64:419–449.

Symmes D, Newman JD, Talmage-Riggs G, Lieblich AK (1979) Individuality and stability of isolation peeps in squirrel monkeys. Anim Behav 27:1142–1152.

Thomas JA, Stirling I (1983) Geographic variation in Weddell seal (*Leptonychotes weddelli*) vocalizations between Palmer Peninsula and McMurdo Sound, Antarctica. Can J Zool 61:2203–2210.

Thomas JA, Puddicombe RA, George M, Lewis D (1988) Variations in the underwater vocalizations of Weddell seals (*Leptonychotes weddelli*) at the Vestfold Hills as a measure of breeding population discreteness. Hydrobiologia 165:279–284.

Trainer JM (1987) Behavioral association of song types during aggressive interactions among male yellow-rumped caciques. Condor 89:731–738.

Trainer JM (1988) Singing organization during aggressive interactions among male yellow-rumped caciques. Condor 90:681–688.

Trainer JM (1989) Cultural evolution of song dialects of yellow-rumped caciques in Panama. Ethology 80:190–204.

Tyack PL (1981) Interactions between singing Hawaiian humpback whales and conspecifics nearby. Behav Ecol Sociobiol 8:105–116.

Tyack PL (1983) Differential response of humpback whales, *Megaptera novaeangliae*, to playback of song or social sounds. Behav Ecol Sociobiol 13:49–55.

Tyack PL (1986) Whistle repertoires of two bottlenose dolphins, *Tursiops truncatus*: Mimicry of signature whistles? Behav Ecol Sociobiol 18:251–257.

Tyack PL (1997) Development and social functions of signature whistles in bottlenose dolphins *Tursiops truncatus*. Bioacoustics 8:21–46.

Tyack PL, Sayigh LS (1997) Vocal learning in cetaceans. In: Snowdon CT, Hausberger M (eds) Social Influences on Vocal Development. Cambridge, UK: Cambridge University Press, pp. 208–233.

Tyack PL, Whitehead H (1983) Male competition in large groups of wintering humpback whales. Behaviour 83:132–154.

Valsecchi E, Palsboll PJ, Hale PT, Glockner-Ferrari D, Ferrari M, Clapham PJ, Larsen F, Mattila DK, Sears R, Sigurjonsson J, Brown MR, Corkeron PJ, Amos B (1997) Microsatellite genetic distances between oceanic populations of the humpback whale (*Megaptera novaeangliae*). Mol Biol Evol 14:355–362.

Wanker R, Apcin J, Jennerjahn B, Waibel B (1998) Discrimination of different social companions in spectacled parrotlets (*Forpus conspicillatus*): Evidence for individual vocal recognition. Behav Ecol Sociobiol 43:197–202.

Waser PM, Waser MS (1977) Experimental studies of primate vocalizations: Specializations for long distance propagation. Z Tierpsychol 43:239–263.

Weary DM, Krebs JR (1992) Great tits classify songs by individual voice characteristics. Anim Behav 43:283–287.

Weilgart L, Whitehead H (1993) Coda communication by sperm whales (*Physeter macrocephalus*) off the Galapagos Islands. Can J Zool 71:744–752.

Weilgart L, Whitehead H (1997) Group-specific dialects and geographical variation in coda repertoire in South Pacific sperm whales. Behav Ecol Sociobiol 40:277–285.

Weinrich MT (1991) Stable social associations among humpback whales, *Megaptera novaeangliae*, in the southern Gulf of Maine. Can J Zool 69:3012–3019.

Weise CM, Meyer JM (1979) Juvenile dispersal and development of site-fidelity in the black-capped chickadee. Auk 96:40–55.

Wells RS (1991) The role of long-term study in understanding the social structure of a bottlenose dolphin community. In: Pryor K, Norris KS (eds) Dolphin Societies. Berkeley, CA: University of California Press, pp. 199–235.

Wells RS, Scott MD, Irvine AB (1987) The social structure of free-ranging bottlenose dolphins. In: Genoways HH (ed) Current Mammalogy. New York: Plenum Press, pp. 247–305.

West MJ, King AP (1988) Female visual displays affect the development of male song in the cowbird. Nature 334:244–246.

Whitehead H, Waters S, Lyrholm T (1991) Social organization of female sperm whales and their offspring: Constant companions and casual acquaintances. Behav Ecol Sociobiol 29:385–389.

Wiens JA (1982) Song pattern variation in the sage sparrow (*Amphispiza belli*): Dialects or epiphenomena. Auk 99:208–229.

Wiley RH (1971) Song groups in a singing assembly of little hermits. Condor 73:28–35.

Wiley RH, Richards DG (1982) Adaptations for acoustic communication in birds: Sound transmission and signal detection. In: Kroodsma DE, Miller EH, Ouellet H (eds) Acoustic Communication in Birds. New York: Academic Press, pp. 132–182.

Wilkinson GS, Boughman JW (1998) Social calls coordinate foraging in greater spear-nosed bats. Anim Behav 55:337–350.

Wilkinson GS, Boughman JW (1999) Social influences on foraging in bats. In: Box HO, Gibson K (eds) Mammalian Social Learning: Comparative and Ecological Perspectives. Cambridge, UK: Cambridge University Press, pp. 188–204.

Winn HE, Thompson TJ, Cummings WC, Hain J, Hudnall J, Hays H, Steiner WW (1981) Song of the humpback whale—population comparisons. Behav Ecol Sociobiol 8:41–46.

Winter P, Handley P, Ploog D, Schott D (1973) Ontogeny of squirrel monkey calls under normal conditions and under acoustic isolation. Behaviour 47:231–239.

Wright TF (1996) Regional dialects in the contact call of a parrot. Proc R Soc Lond B Biol Sci 263:867–872.

Wright TF (1997) Vocal communication in the yellow-naped amazon (*Amazona auropalliata*). Ph.D. Thesis, University of California, San Diego, CA.

Wright TF, Dorin M (2001) Pair duets in the yellow-naped Amazon (Psittaciformes: *Amazona auropalliata*): Responses to playbacks of different dialects. Ethology 107:111–124.

Wright TF, Wilkinson GS (2001) Population genetic structure and vocal dialects in an amazon parrot. Proc R Soc Lond B Biol Sci 268:609–616.

Wyndham E (1980a) Diurnal cycle, behavior and social organization in the budgerigar (*Melopsittacus undulatus*). Emu 80:25–33.

Wyndham E (1980b) Environment and food of the budgerigar (*Melopsittacus undulatus*). Aust J Ecol 5:47–61.

Zann R (1984) Structural variation in the zebra finch distance call. Z Tierpsychol 66:328–345.

Zann R (1985) Ontogeny of the zebra finch distance call: I. Effects of cross fostering to Bengalese finches. Z Tierpsychol 68:1–23.

Zann R (1990) Song and call learning in wild zebra finches in south-east Australia. Anim Behav 40:811–828.

Zimmermann E (1995) Loud calls in nocturnal prosimians: Structure, evolution and ontogeny. In: Zimmermann E, Newman JD, Jurgens U (eds) Current Topics in Primate Vocal Communication. New York, Plenum Press. pp. 47–72.

Zimmermann E, Lerch C (1993) The complex acoustic design of an advertisement call in male mouse lemurs (*Microcebus murinus*, Prosimii, Primates) and sources of its variation. Ethology 93:211–224.

5
Selection on Long-Distance Acoustic Signals

Michael J. Ryan and Nicole M. Kime

1. Introduction

The purpose of this chapter is to discuss the evolution of long-distance acoustic signals through the atmosphere (Bass and Clark, Chapter 2, address problems in underwater communication). We are especially interested in signals that are used in mate recognition. In most cases, these signals are produced by males to attract potential mates and repel male competitors, and they are evaluated by females when they make a mating decision. Although there are cases of females attracting males and males choosing females, we will tend to concentrate on the more typical case.

1.1. Long-Distance Signals, Selection, and Evolution

Sounds that have evolved to attract mates over relatively long distances are common throughout diverse taxa. These signals are often subject to strong natural selection imposed by the environment and by unintended receivers, such as predators and parasitoids, as well as sexual selection imposed by intended receivers, such as potential mates and competitors for those mates. The importance of these signals in reproduction implicates them in two processes central to evolutionary biology—speciation and sexual selection. These are the processes primarily responsible for increasing biodiversity through multiplying and diversifying species.

Long-distance communication signals occur in many taxa, such as mammals and fish, but are especially prevalent and well-studied in insects, frogs, and birds. Within each of these taxa, long-distance signals are better-studied than acoustic signals used in other contexts (see taxon-specific reviews cited below). There are several taxon-specific reviews of various aspects of long-distance communication (e.g., insects: Ewing 1989; Michelson 1998; Römer 1998; frogs: Fritzsch et al. 1988; Gerhardt 1994; Zelick et al. 1998; Ryan 2001; birds: Kroodsma and Miller 1982, 1996; Catchpole and Slater 1995) as well as reviews in more general texts on animal communication (Hauser 1996; Bradbury and Vehrencamp 1998; Owens and Morton

1998). In this chapter, we offer a general survey of various factors that influence the evolution of long-distance signals.

The focus by researchers on long-distance signals is due in part to the conspicuousness of these signals. Except for those secluded in the most urban and sterile environments, most of us have probably heard a dawn chorus of birds, the evening chirping of crickets, and the nocturnal serenading of frogs. In all of these cases, the sounds we hear are not merely random fluctuations in ambient pressure derived from some inconsequential movement of an animal's body parts, as if one happened to step in a puddle and caused a complicated and intricate but rather meaningless pattern of waves and troughs on the water's surface. These sounds are acoustic signals. Thus, by definition, they have evolved under selection to serve a communication purpose.

The purpose of long-distance signals, defined by Littlejohn (2001) as signals that function over a distance of more than several body lengths from the receiver, is to advertise the presence of the sender to a receiver. In many of the cases we consider, the sender is a male who is advertising his presence in a context linked in some way to reproduction, usually to receivers that are potential mates or potential competitors. The evolution of long-distance acoustic signals involves the modification of the animal's morphology and physiology to couple a mechanical displacement to pressure fluctuations in the external environment; the production of these signals sometimes stresses the physical and energetic limitations imposed on bioacoustic production. Furthermore, not all pressure fluctuations will be favored by selection. They must have the temporal and spectral properties that allow them to transmit over a functional distance; that is, to encounter the intended receiver. Nor is efficient transmission through the environment the sole criterion of selection. The signal must interact effectively with the receiver. At a minimum, it must be detected. Thus, the sounds are also constrained to function within temporal and spectral limitations relative to the intended receiver. But this pattern of pressure fluctuations emanating from the sender must have meaning; thus, it is also constrained by the higher-order neural processing and cognitive abilities of the intended receiver, and to complicate matters, evolution of these long-distance communication systems is not one-sided. Whether viewed as an intricate evolutionary dance or an arms race, the properties of the signal and receiver have the potential to influence each other's evolution—thus, the signals and receivers evolve but the communication system coevolves.

Only when all of these criteria are met do we have communication, and when we do, it is an amazing phenomenon. As Pinker (1994) eloquently stated for human language: "Simply by making noises with our mouths, we can reliably cause precise new combinations of ideas to arise in each other's minds. The ability comes so naturally that we are apt to forget what a miracle it is" (p. 15).

2. Morphological and Energetic Constraints

A number of different selective forces can act on any one morphological or behavioral trait, including long-distance communication signals (e.g., Wilczynski and Ryan 1999). These forces may act in unison or in opposition to one another. In addition, a trait may be restricted in its ability to evolve in response to selection either because the necessary genetic variation does not exist or because of the constraints of physical laws. Consequently, the traits that we see are not always at a selective optimum with respect to one fitness component but instead reflect a compromise between a number of different selective forces and constraints.

The physical structures used for sound production play a large role in determining both the temporal and spectral characteristics of acoustic signals. Although these structures can certainly evolve in response to the various selective forces acting on long-distance communication signals, phylogenetic or physical constraints on their morphology can also impose limitations on the form of signals. In this section, we discuss how two such limiting factors, the low energetic efficiency of sound production and the body size of the sender, constrain the signals used for long-distance communication.

2.1. *The Energetics of Signal Production*

2.1.1. Energetics and Efficiency

One important element of natural selection is the energetic cost of an otherwise advantageous trait. The cost associated with the energy required to perform a display or behavior can outweigh the benefit of the trait, especially when it depletes resources necessary for basic maintenance or other activities. The energetic cost of acoustic signaling can be estimated in a number of ways, the best of which is probably the rate of oxygen consumption ($\dot{V}_{O_2}$) during a bout of calling. This measure is most easily procured for animals such as insects and frogs, which will call in a respirometer (MacNally and Young 1981; Ryan 1988). Less reliable methods must usually be applied to studies of birds and mammals (e.g., Brackenbury 1979; but see Eberhard 1994); for this reason, the most well-known studies of calling energetics have focused on anurans and insects.

These studies have repeatedly shown that acoustic signals are extremely costly to produce. The rate of oxygen consumption during calling can be 5–30 times that during rest (Stevens and Josephson 1977; MacNally and Young 1981; Prestwich and Walker 1981; Bucher et al. 1982; Taigen and Wells 1985; Taigen et al. 1985; Ryan 1988; Prestwich et al. 1989). Indeed, most studies on insects and frogs demonstrate that calling to attract mates is one of the most energetically expensive activities in which males engage

(Taigen and Wells 1985; Ryan 1988). For example, the energetic cost of advertisement calling for the gray tree frog, as measured by the rate of oxygen consumption, can be as high or higher than the amount of energy expended during forced locomotor activity, a measure commonly referred to as "$\dot{V}_{O_2}$ max" (Fig. 5.1; Taigen and Wells 1985).

One reason that these mating signals are so costly is that they are extremely inefficient to produce. The production efficiency of an acoustic signal is simply the amount of energy in the emitted sound relative to the energetic cost of its production. In most of the animals studied to date, the energetic efficiency of sound production is much less than 10% (Table 5.1 and references therein). Two factors are implicated in this low efficiency—the loss of energy as heat by the muscles used in sound production and the inefficiency of coupling acoustic energy from sound-production structures to the environment (Bradbury and Vehrenkamp 1998).

In animals that use internal sound-production structures (e.g., most birds, frogs, and mammals), the volume of air inside the vocal pathway is much smaller than the volume of air outside in the external environment. As a result, most of the sound energy that reaches the end of the vocal pathway is reflected at the boundary of the tube and will not be transmitted to the receiver.

One way to counteract this low efficiency is by the addition of radiating structures that decrease the acoustic impedance mismatch between the organism and its environment. The vocal sacs of male frogs and some pri-

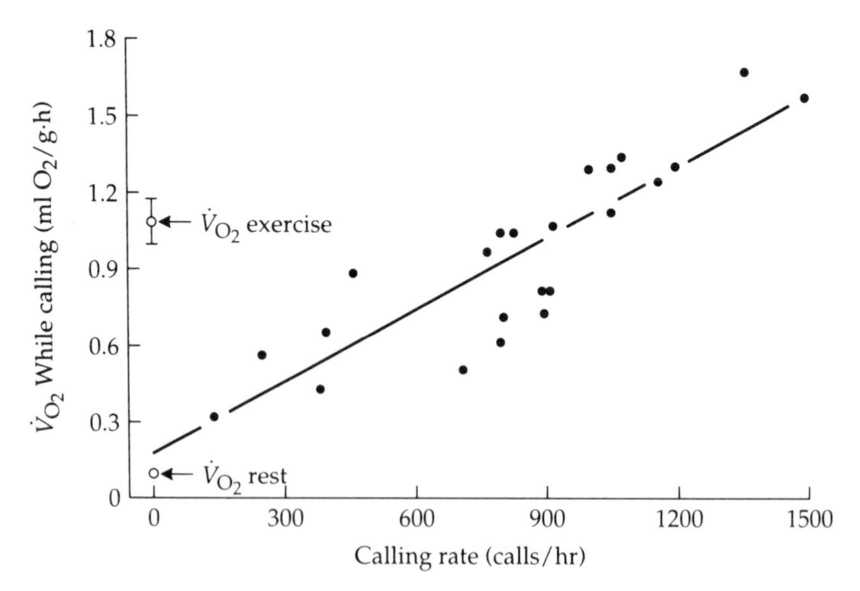

FIGURE 5.1. Rate of oxygen consumption during calling, rest, and forced locomotor activity in the gray tree frog. (From Taigen and Wells 1985.)

TABLE 5.1. Estimated efficiencies of acoustic signal production.

Species	Efficiency (%)	Reference
Insects		
Achroia grisella	0.008	Reinhold et al. 1998
Anurogryllus arboreus	0.09	Prestwich 1998
Cystosoma saundersii	0.82	MacNally and Young 1981
Gryllotalpa australis	1.05	Kavanagh 1987
Telogryllus commodus	0.5	Kavanagh 1987
Frogs		
Hyla cinerea	1.89	Prestwich et al. 1989
H. crucifer	4.9	Prestwich et al. 1989. Data from Taigen et al. 1985; Taigen unpublished
H. gratiosa	0.76	Prestwich et al. 1989
H. squirella	2.21	Prestwich et al. 1989
H. versicolor	3.6	Prestwich et al. 1989. Data from Taigen and Wells 1985; Wells and Taigen 1986
Physalaemus pustulosus	0.5–1.2	Ryan 1985
Domestic chicken, *Gallus domesticus*	1.6	Brackenbury 1977
Human	~1%	Wood 1962

mates are examples of such radiators (Martin 1972; Schön-Ybarra 1988; Rand and Dudley 1993; Fitch and Hauser, Chapter 3). The sound frequencies that can be efficiently coupled to the environment via a radiator depend on its size—larger radiators allow lower dominant frequencies to be coupled to the environment with greater efficiency. However, even with such structures, the efficiency of sound production among animals is still very low, probably because the frequencies used for communication are below the effective cutoff frequency for the radiator (Ryan 1985a; Table 5.1). In the frog *Physalaemus pustulosus*, for example, if the entire male frog were conservatively assumed to radiate the call, its effective cutoff frequency would be 3,500 Hz. Male frogs produce calls that sweep from about 900 to about 400 Hz, far below the limits of maximum radiation efficiency (Ryan 1985b).

2.1.2. Constraints on Signal Evolution

The high cost and low efficiency of calling can constrain the evolution of acoustic signals. Increasing either the length of individual calls or the rate at which calls are produced results in increased energy expenditures (Fig. 5.1; Taigen and Wells 1985; Prestwich et al. 1989; Wells and Taigen 1989). Call rates or lengths may thus be limited by an upper asymptote of energy availability. Furthermore, in some species, these two aspects of calling may be involved in an energetic trade-off. In gray tree frogs, males respond to the calls of other males by increasing call length, but they maintain calling effort at a constant level by decreasing the rate at which these longer calls are produced (Wells and Taigen 1986; Klump and Gerhardt 1987). This trade-off between call rate and length has been taken as evidence that overall calling effort is constrained by the energetic cost of signal production.

Energy limitations may also restrict calling behavior over a longer time scale, such as the amount of time that a male can spend calling over one or several nights. Some studies on the calling behavior of frogs have suggested that energetic demands limit male calling to a certain proportion of nights (Murphy 1994a; Marler and Ryan 1995) or to a restricted period during a given night (Wells and Taigen 1986; but see Murphy 1999). Male frogs usually call on less than 30% of available nights, and most males of species with prolonged breeding seasons appear in chorus only a few times during the season (Bevier 1997). Choruses usually do not last long after midnight, and large proportions of glycogen reserves of the trunk muscles can be depleted after only a few hours of calling (Wells et al. 1995; Bevier 1997).

The relative mating success of male frogs and toads is strongly tied to the number of nights in attendance at a chorus (Murphy 1994b; Wagner and Sullivan 1995). In addition, females often prefer to mate with males who produce longer mating calls or calls produced at faster rates (reviewed in Ryan and Keddy-Hector 1992). Such preferences can extend beyond the normal range of male variation (Gerhardt 1991). The origin for such preferences is unclear—it may be because of inherent sensory biases for greater neural stimulation or because the preference results in matings with males in better condition (Ryan and Keddy-Hector 1992). Whatever the cause of the preference, the response of male calling behavior to sexual selection may be constrained by the energetic cost of calling.

2.2. *Body Size and Wavelength*

The mass of the sound-producing structure plays a large role in determining the frequency of communication signals; structures with greater mass can produce lower-frequency signals more efficiently. Consequently, in many frogs, birds, and mammals, the frequency of communication signals is correlated with body size (Fig. 5.2). This relationship between body size and signal frequency often holds for comparisons among groups of species (e.g., frogs: Ryan 2001; birds: Morton 1977; Wallschager 1980; Bowman 1983; Ryan and Brenowitz 1985; Wiley 1991; mammals: Fitch and Hauser, Chapter 3). The same relationship holds among individuals within a single species or population for many frogs (e.g., Ryan 1985b; Gerhardt 1994; Howard and Young 1998). But Fitch and Hauser (Chapter 3) suggest that within species of birds and mammals, the expected correlation of body size and sound frequency is not as strong as expected.

One of the most comprehensive and enlightening studies on this subject was conducted by Martin (1972). He dissected various components of the

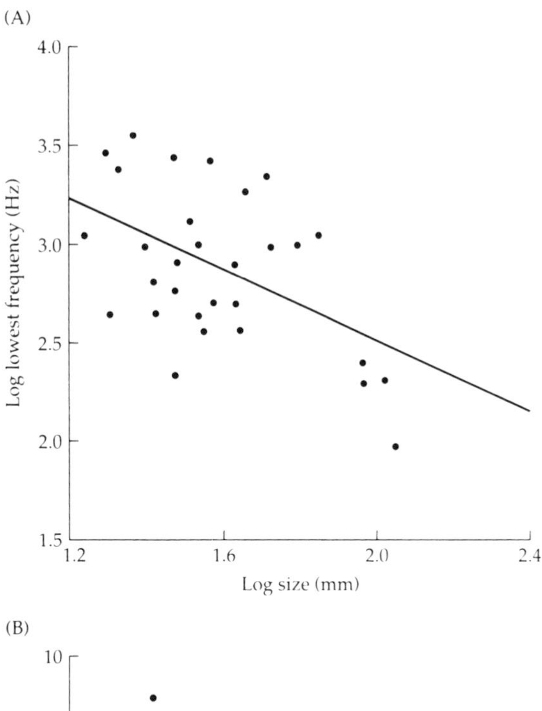

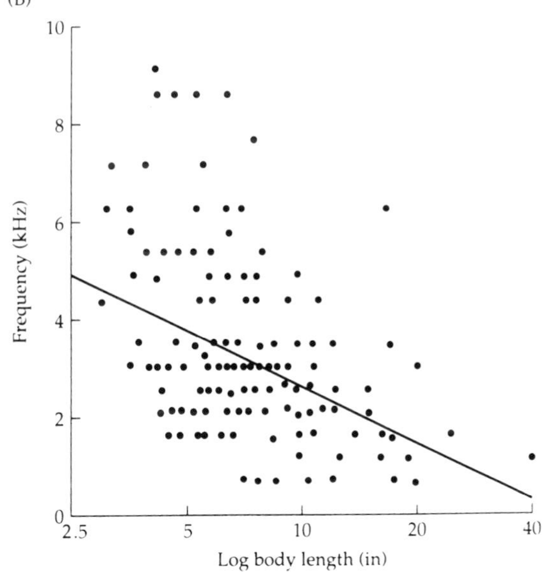

FIGURE 5.2. Relationship between body size and (**A**) the logarithm of the lowest frequency present in calls produced by leptodactylid frogs and (**B**) the emphasized song frequency of Panamanian birds. (From Bradbury and Vehrencamp 1998, modified from Ryan 1985; Ryan and Brenowitz 1985.)

toad vocal system, systematically altering them to study their effects on signal production. He showed that larger toads generally have larger vocal cords and that these larger vocal cords produce sounds with lower dominant frequencies. Of course, the relationship between body size and frequency is not perfect—Martin (1972) also showed that adding mass to vocal cords independent of body size can lower the dominant frequency of a call. In addition, call frequency can be changed by changing the tension on the vocal cords—the frequency-modulated calls of many frogs are an example of this process.

The interaction of calling efficiency and body size constrains the dominant frequency of acoustic signals. As the frequency of a signal decreases relative to the body size of the sender, the efficiency of signal production drops (e.g., Section 2.1 of this chapter; Bradbury and Vehrenkamp 1998). Taken together, the constraints of body size and energetic efficiency indicate that small animals should use the more efficiently produced higher frequencies for communication. However, as we will see in the following sections, natural and sexual selection can act on signal structure in ways that oppose this influence of sender morphology.

3. Environmental Constraints

During transmission, acoustic signals must travel through the external environment. Over distance, these signals will be altered in ways that can influence the response of receivers; all signals will eventually become so degraded that potential receivers fail to recognize them altogether. The temporal or spectral structure of signals, however, can influence the amount of change they experience and thus the distance over which they can be used. Selection can therefore act on the form of long-distance communication signals to decrease signal degradation and increase transmission distance. In this section, we will first review some of the basic properties of signal transmission in the atmosphere (see Bass and Clark, Chapter 2, for a discussion of underwater acoustics). We will then examine some of the evidence that long-distance communication signals have evolved in response to selection for increased propagation efficacy.

3.1. Signal Design for Maximum Range

The effects of transmission on acoustic signals have generally been partitioned into two main categories—loss of amplitude and loss of fidelity. Both contribute to the degradation of a signal's influence with distance. Because of spherical spreading alone, signal amplitude will decrease, or attenuate, by 6dB for each doubling of distance, even in an ideal environment. Absorption and scattering of sound waves by the air, ground, and vegetation cause additional, or "excess," attenuation in most natural environments (Wiley and Richards 1978). Attenuation of a long-distance communication signal can reduce the chances that it will be detected by a receiver either because the amplitude of the received signal falls below the auditory threshold of the receiver or because the signal-to-noise ratio has decreased to a level at which the receiver no longer recognizes the signal.

The temporal and spectral structures of a signal will also be altered as it travels from sender to receiver (Wiley and Richards 1978; Richards and Wiley 1980; Michelsen and Larsen 1983). For example, reverberations from objects in the environment or irregular amplitude fluctuations caused by atmospheric turbulence can contribute to the degradation of a signal's temporal structure. Some frequencies attenuate more rapidly than others do; this "frequency-dependent attenuation" can alter the spectral form and thus the perception of a signal. Over long distances or in harsh environments, such loss of fidelity can render a signal unrecognizable to the receiver.

All sounds experience these changes during transmission, but the amount of change is partially determined by the spectral and temporal structures of the signal itself. Selection can therefore act on the form of a signal to increase the distance over which it can be heard and recognized (Morton 1975; Sorjonen 1986; Wiley 1991; Endler 1992). Endler (1992) described 12 "rules" for the design of communication signals, most of which were considerations of the effects of the environment on sound transmission. We present a modified list of design rules in Table 5.2. Although these guidelines are applicable to many communication signals, they should be especially important for calls that must travel over long distances.

The frequency of a signal has a large effect on its propagation distance. Sound-transmission experiments in various habitats have repeatedly demonstrated that lower-frequency sounds generally experience less attenuation than higher-frequency sounds (e.g., Morton 1975; Marten and Marler 1977; Marten et al. 1977; Waser and Brown 1986). This is because higher-frequency sounds are more susceptible to absorption by objects in

Table 5.2. Guidelines for acoustic signal production. (Modified from Endler 1992.)

1. Use frequencies lower than 2 kHz in order to minimize reverberation, attenuation, and scattering during transmission.
2. For animals that must call near the ground, use frequencies above 0.5–1 kHz to avoid ground attenuation.
3. Use species-specific frequency bands and tuned receptors to minimize interference from other species and abiotic sounds.
4. Use greater-amplitude signals to maximize the signal-to-noise ratio, increase transmission distance, and increase the probability of detection.
5. Use frequency modulation rather than amplitude modulation to encode information. Reverberations and air turbulence alter amplitude more than frequency.
6. Send signals from above the ground to minimize ground attenuation and the effects of vegetation and wind and temperature gradients.
7. Use redundant signals to offset the effects of discontinuous background noise and the effects of reverberations and amplitude fluctuations.
8. Call in locations or during times that minimize turbulence and/or background noise.
9. Use alerting signals to attract the receiver's attention before sending the main signal.

the environment and to disruption by atmospheric turbulence. Near ground level, however, sounds with very low frequencies will also experience increased levels of attenuation because of destructive interference between direct waves and waves reflected from the ground (Wiley and Richards 1978). Thus, near ground level in some environments, a "sound window" of minimal excess attenuation exists for frequencies in the range of about 1–3 kHz (Morton 1975; Marten et al. 1977; Waser and Brown 1986).

The temporal structure of a signal can also influence the amount of degradation it experiences (Richards and Wiley 1980; Ryan and Sullivan 1989; Mathevon et al. 1996). Reverberations and air turbulence can blur amplitude modulation contained within signals, favoring tonal signals over amplitude-modulated signals (Richards and Wiley 1980). The frequency of the signal also influences these effects—sounds between 2 and 8 kHz are less subject to reverberations from the vegetation and ground and are thus favored with respect to the maintenance of temporal fidelity (Wiley and Richards 1978).

3.2. Testing the Acoustic Adaptation Hypothesis

Like signals, all habitats are not equal with respect to signal transmission. Selection on call structure may be stronger, and thus elicit a stronger evolutionary response, in one type of environment versus another (Morton 1975; Ryan et al. 1990). Alternatively, the optimal call structure for long-distance communication can vary among habitats or microhabitats. For example, it has been shown that amplitude-modulated calls are favored for communication in open environments, whereas tonal calls are favored in forested environments (Morton 1975; Sorjonen 1986). Similarly, the presence and shape of the frequency sound window for low excess attenuation appears to differ among environments as well as between different heights within the same environment (Morton 1975; Marten and Marler 1977; Waser and Brown 1986). The frequency window may be present in some locations and absent in others, or the range of frequencies experiencing low excess attenuation may differ among habitats.

Given that different habitats can impose differing selection on the structure of calls used for long-distance communication, one might expect predictable divergence among the signals used by the species living in these habitats. These predictions can be used to test the hypothesis that long-distance communication signals have evolved in response to selection generated by habitat acoustics. Adaptation to the acoustic environment is one possible explanation when the observed trends in call characteristics match the predictions. Following this logic, a number of studies have used comparative methods to test the hypothesis that the signals used for long-distance communication have evolved in response to selection for decreased degradation and increased transmission distance within their home environment.

3.2.1. Community Studies

One of the first and most often cited tests of the acoustic adaptation hypothesis was performed by Eugene Morton, who compared the songs of birds that reside in open and forested habitats in Panama (Morton 1975). Morton first determined the amount of excess attenuation experienced by tones of varying frequency transmitted at different heights within different environments. In all locations, higher frequencies generally experienced greater attenuation than lower frequencies. However, near the ground in forest there was a "sound window" of low excess attenuation between 1,500 and 2,500 Hz. This sound window did not exist at higher heights within forest or in edge and grassland environments. Morton then compared the results of these transmission studies with data on the song frequencies of different species of birds living in these habitats. He found that birds that call in low forest, but not birds that call in grassland or above the ground in forest, have mean call frequencies in the range of the sound window. Morton concluded that the songs of these forest species have evolved in response to selection for decreased attenuation and thus increased transmission distance.

Since Morton's original study, a number of similar comparative studies have been conducted using the calls of birds and anurans. These studies evaluated a number of different aspects of temporal and spectral fidelity in addition to attenuation and frequency-dependent attenuation. Although there is some disparity in the details of the analyses, community-level studies in temperate and tropical birds generally support Morton's findings, at least for some signal characteristics (e.g., Richards and Wiley 1980; Sorjonen 1986), but community-level studies in frogs do not (Zimmerman 1983; Penna and Solis 1996; Kime et al. 2000).

The interpretation of multispecies comparisons such as Morton's can, however, be easily confounded by other determinants of call structure such as phylogenetic relationship or body size. As Ryan and Brenowitz (1985) pointed out, the frequency differences that Morton found in his original study could also be explained by differences in the body size of birds living in the different locations or by differences in the background-noise composition of the different environments. Correcting for body size in temperate birds, Wiley (1991) failed to find differences in dominant frequency among the songs of temperate-zone oscine birds in open and forested environments. For frogs, Zimmerman (1983) showed that the phylogenetic relationships among species in a community of tropical frogs were a better predictor of signal structure than was habitat acoustics.

3.2.2. Studies of Single or Closely Related Species

Some of the more convincing tests of the acoustic adaptation hypothesis compare signal structure among populations of a single species or among closely related species with habitat shifts. At this level, it is often easier to control for the effects of body size and other morphological or phylogenetic constraints. Confounding variables are thus less problematic for studies of closely related species than for the community-level analyses described above.

The rufous-collared sparrow of Argentina can be found over a large altitudinal range. The song of males has a final trill, which varies in rate over its geographical range (King 1972; Nottebohm 1975). Although the trill rate remains constant over large areas of continuous habitat, it changes significantly with the habitat changes associated with increases in altitude (King 1972; Nottebohm 1975; Handford 1981, 1988; reviewed in Catchpole and Slater 1995; Fig. 5.3). Trill rate is generally higher in open areas and lower in forested areas because there is less scattering to mask patterns of amplitude modulation in the open environment.

In anurans, some evidence for acoustic adaptation becomes apparent when the phylogenetic scale of analysis is reduced from the level of the community to a single species. In the frog *Acris crepitans*, two subspecies

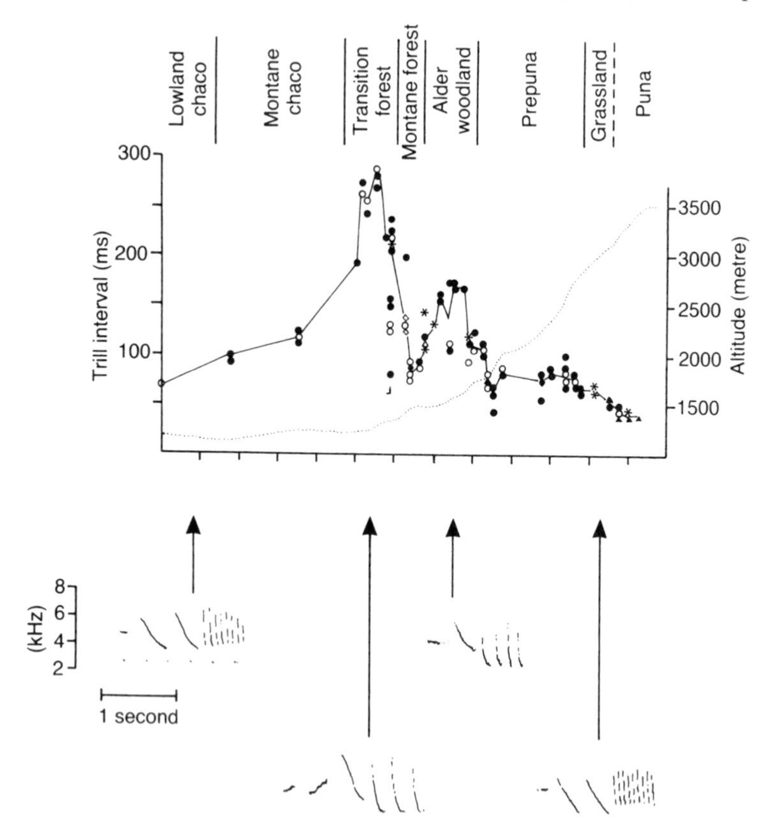

FIGURE 5.3. (Top) Changes in trill interval with vegetation type (noted on the upper x-axis) and altitude (dashed line, right y-axis) in rufous-collared sparrows. (Bottom) sonograms of song examples from four vegetation types. In each case, the trill is the last portion of the song. (From Catchpole and Slater 1995 after Handford 1988. Reprinted with the permission of Cambridge University Press.)

have slightly different mating calls. One subspecies, *A. c. blanchardi*, lives in open habitats in central Texas. The second subspecies, *A. c. crepitans*, resides in forested habitats to the east. For both subspecies, calls transmitted through an open environment experience less degradation than calls transmitted through forest. However, regardless of the environment, the calls of *A. c. crepitans* are always less subject to degradation than the calls of *A. c. blanchardi*. In addition, the difference in call degradation between the open and forested habitats is less for *A. c. crepitans* than for *A. c. blanchardi* (Ryan et al. 1990). Ryan et al. (1990) suggested that the calls of the forest subspecies, *A. c. crepitans*, have evolved in response to relatively strong selection for transmission efficiency in forest and thus transmit with less degradation in both environments. They further suggested that the calls of *A. c. blanchardi*, which lives in more open habitats, have not been under as strong environmental selection and have instead been more influenced by selection in other contexts.

Another convincing argument in favor of Morton's hypothesis comes from comparisons of short- and long-distance communication signals within species, especially in primates. As we have been discussing, long-distance communication signals should be under strong selection for increased transmission efficacy. But signals used over shorter ranges should not be under as strong selection and in some cases may be selected to degrade rapidly over distance in order to reduce eavesdropping by predators and other unintended receivers (Endler 1992). As predicted, the whoopgobble of mangabeys, a call used for intergroup communication, transmits farther and with less attenuation than calls used for intragroup communication. Three other primate species exhibit similar differences in attenuation between calls used for long- and short-distance communication (Waser and Waser 1977).

3.3. Environmental Selection on Signaling Behavior

For a given signal, both the location from which it is broadcast and the time of day during which it is transmitted can influence the amount of degradation it experiences. In addition to the temporal and spectral properties of acoustic signals, selection can also act on the behavior of the signaler.

3.3.1. Caller Height

Sounds transmitted from above the ground almost always experience less degradation than sounds transmitted near the ground (Marten and Marler 1977; Waser and Waser 1977; Henwood and Fabrick 1979; Brenowitz et al. 1984; Mathevon et al. 1996). Senders of long-distance communication signals can thus increase propagation distance simply by calling from above the ground. Likewise, the receivers of the signals can benefit from being above the ground regardless of the position of the sender (Dabelsteen et al. 1993). Birds are the most obvious example of how sound transmission might influence behavior—males of most species broadcast songs from perches in trees or while in flight in spite of the energetic costs and the chances of increased predation (Mathevon et al. 1996). Many species of frogs and insects also call from well above the ground, although species of frogs that call from ground level across water often do not experience high levels of call attenuation, perhaps because the water acts as a waveguide (Venator 1999; Bass and Clark, Chapter 2).

3.3.2. Timing of Signaling

The dawn chorus of birds and evening choruses of insects and frogs are nearly ubiquitous. Why do these species restrict calling to certain times of the day, and why to early morning and evening? Meteorological conditions vary widely over the course of the day, causing variations in signal transmission. In addition, the composition of biotic and abiotic noise in the environment differs at different times of the day. Calling at times when the meteorological conditions favor sound transmission or when other species are not calling can increase the active space of a transmitted signal. In environments ranging from desert to tropical forest, early morning and evening hours usually have the best meteorological conditions for sound transmission (Waser and Waser 1977; Henwood and Fabrick 1979).

3.4. Acoustic Niche Partitioning

In addition to the effects of the physical environment, calling structure and behavior may be influenced by the other species within a community. Selection on individual species to match the transmission characteristics of their calling habitat would lead to convergence among sympatric species. However, we can easily hear that the calls of each species have distinctive features. Such differences among sympatric species can reduce the level of interference among the calls of different species, resulting in a partitioning of the "acoustic niche." In addition, when these signals are used for the recognition of conspecific potential mates over heterospecifics, differences among calls or calling behavior can be strongly advantageous, preventing costly mismatings among incompatible individuals.

Some anuran studies provide suggestive evidence for divergence in call structure or calling behavior among species that call in the same area. Frogs sometimes maintain different calling sites within a habitat, and these differences in location can reduce the chances of matings among heterospecifics (Duellman 1967; Hödl 1977). Temporal differences in calling behavior may also reduce interference and promote reproductive isolation (Duellman 1967). Even so, many species of frogs can often be observed calling at the same time and in the same place. In these cases, different species seldom have similar calls; some combinations of temporal or spectral properties of calls will differ among species, allowing for the discrimination of conspecifics from heterospecifics (Hödl 1977; Drewry and Rand 1983).

As with habitat acoustics, however, these community-level analyses are subject to a number of different confounds, such as differences in body size and evolutionary history. These confounds render the evidence for acoustic niche partitioning of frog and insect calls inconclusive.

4. Predator- and Parasite-Generated Selection

The most fundamental communication interaction is dyadic, involving a signal and a receiver. Most communication channels, however, are not private, and there can be eavesdropping from unintended receivers. When these unintended receivers are conspecifics, they can use the information gained to compromise the fitness of the communicators. When the eavesdroppers are parasites or predators, the fitness consequence can be even greater. The parasites are usually searching for a host on which their young can develop while predators are looking for food. Long-distance signals are the most risky because they are more likely to be detected by eavesdroppers than are close-range signals.

Zuk and Kolluru (1998) recently reviewed the literature on exploitation of sexual signals by predators and parasites. They list 19 known cases. In nine of the cases, the eavesdroppers are insects; seven of those cases involve a tachinid fly. In all but one case, the insect eavesdroppers are attracted to signals produced by other insects, usually crickets. The one exception is a chaoborid fly being attracted to a tree frog. Zuk and Kolluru (1998) cite ten examples of vertebrates that eavesdrop on signals. Often these are bats homing in on signals of frogs and insects, although turtles, lizards, birds, and other mammals do so as well. Acoustic cues are more likely to be used by parasites and predators to find a victim than are visual cues, although they are used less commonly than olfactory cues.

4.1.1. Acoustically Orienting Parasites and Calling in Crickets

The best-known example of parasite/predator-generated selection on acoustic signaling comes from the classic studies of Cade (1975, 1981) on a tachinid fly, *Euphasiopteryx depleta* (= *Ormia depleta*), that parasitizes the field crickets *Gryllus rubens*, *G. integer*, and *G. lineaticeps*. As do many crickets, males of these species use long-distance acoustic cues to attract females for the purpose of mating. Female crickets are not the only ones attending to these cues, however; the tachinid flies orient toward these calls and deposit larvae on the singing male. The larvae develop inside the male, use the male as a source of nutrition, and emerge in 7–10 days (Cade 1975).

Selection on male calling from these parasites has resulted in at least two adaptations by the crickets. The most dramatic is the evolution of variation in the amount of time males spend calling. In many species, there are often alternative mating strategies in which one type of male signals for mates while the other does not and instead attempts to sneak copulations with females (Cade 1981). In crickets, these differences among males are attributable to genetic differences. Cade's studies of crickets are an illustration of how the costs of acoustic signaling can result in the evolution of genetically influenced variation in male mating behavior.

Risk of parasitism and predation does not only vary with the amount of signaling but with other environmental variables as well, and these are taken into account by female crickets when responding to calls. Hedrick and Dill (1993) showed that female response to attractive versus unattractive calls is influenced by the amount of cover in the environment. In situations that afford cover from parasites and predators, female crickets are more likely to respond to attractive calls perceived to be farther away. When the females are more exposed, they choose the closer, although what in other circumstances would be less attractive, call.

There are also adaptations on the side of the eavesdropper in this system. Robert et al. (1992) show that the tuning characteristics of the fly's hearing match the spectral energy of the cricket's call. Thus, there has been an evolutionary convergence between the fly's hearing and the cricket's call. The hearing sensitivity of the parasitoid fly appears to be an evolutionarily derived character that evolved because of the benefits derived by female flies in locating a host for their larvae.

4.1.2. The Frog-Eating Bat and Calling Frogs

A vertebrate analog to the fly–cricket system is that of the frog-eating bat and calling frogs. *Trachops cirrhosus* is a neotropical bat that feeds on a variety of frogs and uses the frog's advertisement call as a localization cue (Fig. 5.4; Tuttle and Ryan 1981). As with the fly–cricket system, *Trachops* has a number of effects on the communication system of frogs. Male túngara frogs (*Physalaemus pustulosus*) produce a whine-like advertisement call that is both necessary and sufficient to attract females. Males can add up to six chucks to the call; chucks are short duration (35 msec), harmonically rich (fundamental frequency of about 250 Hz, with up to 15 harmonics) sounds. The addition of chucks increases the call's attractiveness. Although *Trachops* are attracted to and able to localize calls without chucks, when given a choice they prefer calls with chucks. Thus, the interaction of sexual selection for calls with more chucks and natural selection for calls with fewer chucks seems to have resulted in the evolution of the túngara frog's advertisement call of facultatively varying complexity (Ryan et al. 1982).

Bat predation also influences how and when males call. Male túngara frogs exhibit more evasive behavior, such as submerging under the water, and are less likely to resume calling after the close approach of a bat (Tuttle et al. 1982). *Smilisca sila* males produce calls that overlap in time with one

FIGURE 5.4. A frog-eating bat, *Trachops cirrhosus*, about to capture a túngara frog, *Physalaemus pustulosus*. (Photo courtesy of M.D. Tuttle, Bat Conservation International.)

another and decrease bat predation risk relative to nonoverlapping calls (Tuttle and Ryan 1982). These frogs also modulate the number of syllables in the call in response to predation risk; they produce calls with more syllables under conditions of higher ambient light in which they can see an approaching bat (Tuttle and Ryan 1982). As do crickets and *S. sila* frogs, female túngara frogs assess risk when responding to male calls; they are more likely to approach attractive calls perceived as being farther away over less attractive calls perceived as being produced closer when ambient light levels are lower (Rand et al. 1997).

There was a surprising aspect of the finding that bats eavesdrop on frog calls. Bats are well-known for their reliance on the returning echo of ultrasonic (50–100 kHz in *Trachops*) signals for target localization (Barclay et al. 1981). Frog calls, on the other hand, are relatively low-frequency signals, with most of the spectral energy usually below 5 kHz (Ryan et al. 1983). *Trachops*, however, shows enhanced behavioral sensitivity (movement of the pinnae toward a sound source) to pure tones as they decrease from 15 kHz to 5 kHz; thus, they have heightened behavioral sensitivity to the frequencies that characterize frog calls relative to higher sonic (<15 kHz) frequencies (Ryan et al. 1983). Furthermore, these bats show what appears to be a suite of neuroanatomical specializations for low-frequency hearing in the inner ear, which either increase sensitivity to low-frequency sounds or extend the range of hearing into the low frequencies. These include the largest number of cochlear neurons for any mammal, the second-highest density of cochlear neurons for any mammal, and three peaks of cochlear neuron density rather than the two peaks typical of other bats or the one typical of all other mammals. The third peak is in the apical portion of the cochlea, where low-frequency sounds are detected (Bruns et al. 1989).

Although there are a number of striking cases of predator- and parasite-generated selection on long-distance acoustic signals, the impression from being in the field with these calling animals is that we probably know of only a relatively small proportion of these cases. Furthermore, it seems that there will be many more cases of adaptations on the part of eavesdroppers to detect and localize long-distance cues of prey. We predict that many of these interactions between signalers and unintended receivers will be found in the tropics, where the diversity of predator–prey interactions tends to be greater.

5. Long-Distance Signals, Speciation, and Sexual Selection

It is not only the difficulties of production and transmission of long-distance signals that have made them a central focus of so many research programs. Their involvement in the reproductive biology of many species has made these communication systems critical to the development of evolutionary theories of speciation and sexual selection.

5.1. Speciation

The middle of the twentieth century saw a rebirth of Darwinism referred to as the Modern Synthesis. This synthesis combined research traditions from paleontology, systematics, population genetics, and behavior to accomplish a grand, unified theory of evolution by natural selection (Mayr 1982). A critical issue in accomplishing this synthesis was understanding the origin of species. It was in this realm that behavior made its most lasting contribution to evolutionary theory. If a biological species is defined as a group of potentially reproductively interacting individuals, then there must be mechanisms that draw conspecifics toward each other and away from heterospecifics when searching for mates.

Despite the large number of acoustically signaling insects, frogs, and birds, most of the species that produce long-distance signals produce a species-typical one. We make this statement without implying species typology and lack of meaningful geographic variation within a species (Foster 1999; Foster and Endler 1999). By classifying a signal as "species-typical," we are making a statement about how signal variation is partitioned within species versus among species rather than implying unvarying species-specific stereotypy. It seems to us that in many if not all cases the variation in long-distance signals within the species is substantially less than the variation of the same kind of signals among closely related or ecologically sympatric species. Thus, the properties of the signal can be used for correct identification of the species producing it, as anyone who has tried to identify a bird, frog, or a cricket sight unseen might know. The females of the respective species are even more unerring in their identification, and it is this interaction of signal and receiver that has put long-distance communication at the center of research on speciation theory (e.g., Blair 1958; Alexander 1962; Andersson 1994; Howard et al. 1998).

Preferences of receivers for conspecific versus heterospecific signals lead to assortative mating, reduce the opportunity for reproductive interactions between incipient species, and contribute to the genetic divergence among populations that is critical for speciation. The preference for conspecific versus heterospecific signals is a species-isolating mechanism because it restricts reproduction among different species. The preference is also a premating isolating mechanism because the effect takes place prior to the act of mating, unlike a postmating isolating mechanism, whose effect on reproductive isolation takes place after the mating act; hybrid sterility would be an example of a postmating isolating mechanism.

5.1.1. Species Discrimination as an Incidental Consequence

The use of the term "mechanism" might imply that the function achieved, in this case preference for a conspecific versus a heterospecific signal, evolved under selection. Although there appears little doubt that selection can result in the evolution of such discrimination patterns (see Section 5.1.2), the preference for a conspecific signal versus many heterospecific signals may also be an incidental consequence of signal preferences already in place. For example, if an animal expands its range into a new area and encounters other species for the first time, it might have no problem recognizing the signals of its own species from that of the newly encountered heterospecifics. This particular set of preferences for conspecific versus heterospecific signals did not need to evolve; it was already present as a consequence of how this recognition mechanism—this interaction between properties of the signal and the receiver—happened to evolve in the past. It is *adaptive* (the current effect on fitness) to the animal to prefer the conspecific signal versus the novel heterospecific ones, but this particular preference is not an *adaptation* (the evolved function). The fact that a communication system results in effective species recognition need not mean it evolved for that purpose.

5.1.2. Reinforcement and Reproductive Character Displacement

A second possibility invokes direct selection in the evolution of mate-recognition signals. When two forms of a single species exist in sympatry, mating signals can diverge as the result of lowered hybrid viability or fertility. Such a process is called reinforcement (Butlin 1989) and can eventually lead to complete reproductive isolation, or speciation.

As in the case above, in areas where species overlap, mate-recognition signals must be sufficiently different to prevent incorrect matings between heterospecific individuals. As a result, selection may favor divergence among the signals of different species where they occur in sympatry. The term "reproductive character displacement" describes the outcome of such a process (Butlin 1989), where traits used in mate recognition differ more among sympatric than among allopatric populations because of the divergence of these characters in response to selection to reduce the probability of heterospecific matings. Reproductive character displacement has proved to be difficult to demonstrate empirically.

The strongest evidence to date for reproductive character displacement comes from examples of call divergence within sympatric populations of related species of anurans. The tree frogs *Litoria* (*Hyla*) *ewingi* and *L. verrauxi* can be found in largely disparate areas of southern Australia but do overlap in some regions. In allopatry, the advertisement calls of the two species are very similar. In sympatric populations, however, the calls of *L. ewingi* and *L. verrauxi* are quite different. In these populations, the pulse repetition rate and number of pulses per note in the calls of *L. verrauxi* are shifted away from the values for *L. ewingi* (Littlejohn 1965). There is evidence that these call characters are important in mate discrimination—females of these sympatric populations show strong preferences for pulse rates typical of conspecifics over pulse rates similar to those of heterospecifics (Loftus-Hills and Littlejohn 1971). Thus, the divergence of these call characters could result in a reduced number of hybrid matings.

5.1.3. Neuroethological Mechanisms of Species Recognition

The importance of communication in speciation not only involved behavior in evolutionary theory but also implicated neuroethology. The behavioral preferences so crucial to species recognition emerge from an interaction of stimulus variation and neural and cognitive processing. Understanding how auditory systems decode and process species-specific signals focuses the process of speciation on the nervous system. Studies on crickets (e.g., Huber 1990), frogs (Capranica 1972), and birds (Margoliash 1983) have identified features of the auditory system that bias behavioral responses toward the species' own signal. Although the neuroethologists' emphasis has been on the functional significance of these properties, they are the underlying substrates to the behavior that must evolve if speciation is to occur. Some neuroethological investigations, such as studies of neural-pattern generators that could potentially link properties of the signal with signal recognition (Hoy et al. 1977), were motivated by evolutionary as well as neurobiological issues (Doherty and Hoy 1985).

5.2. Sexual Selection

Although the focus on the role of long-distance communication in speciation continues (e.g., Howard et al. 1998), even more interest has centered on its function in sexual selection (Andersson 1994). Although speciation is the process that gives us diversity through an increase in species numbers, sexual selection is a process that enhances diversity within species. It is primarily this process that makes males and females of the same species look so different from one another—humans included. It is also the process that makes some animals look so attractive and stunning, sound so charming, and behave in manners that are sometimes amusing to us.

Sexual selection favors traits that enhance reproductive success by enhancing one's ability to acquire mates. This can be achieved by a male becoming more attractive to a female, and often, but certainly not in all cases, this judgment of attractiveness is influenced by the male's long-distance signal. A large number of studies of acoustically advertising animals shows that receivers' preferences for long-distance signals will not only guide them to males of their own species but often to a subset of males within the species (Ryan and Keddy-Hector 1992; Andersson 1994). Thus, it is preferable to refer to these signals as mate-recognition signals rather than species-recognition signals or sexually selected signals (Ryan and Rand 1993). Sexual selection generated by these receiver preferences can cause the elaboration of mate-recognition signals. This has been most clearly seen in the elaborate plumage of many male birds but is no less extreme when one examines the variety of sounds that males use to coax a female into mating. The power of sexual selection is documented not only in the extreme signals that males have evolved but in the cost incurred to exhibit them. Metabolic rate can increase tenfold during calling in some frogs (Section 2 of this chapter; Wells 2001), and crickets (Cade 1975) and frogs (Tuttle and Ryan 1981) can attract deadly acoustically orienting parasitoids and frog-eating bats when signaling for females (Section 4 of this chapter).

Much of sexual selection involves communication. One is interested in a dyadic interaction between signaler and receiver, how signal variation is perceived and acted on by the receiver, and the reproductive consequences of such actions for both pairs of the communicating dyad. As with speciation theory, a major issue is how signal and receiver are functionally linked.

Behavioral studies of sexual selection have tended to concentrate on what information the signal conveys to the receiver about the sender. Some of this information might be useful in allowing a female to determine whether a male is of the appropriate species or is in control of resources critical to the female's immediate reproductive success. In such cases, selection should favor receiver preference for the signal variant that guides the female to the mate that maximizes her reproductive success. In other cases, it is suggested that females are not as interested in maximizing the number of offspring they birth but instead choose males whose signals suggest that they will endow her offspring with genes to ensure future survival. Finally, it is possible that the relationship between signal and receiver properties might be due to males responding to selection to produce sounds that females already find appealing.

Acoustic signals have been repeatedly shown to be sexually selected, most often in birds, insects, and frogs, but also in fish and mammals (Andersson 1994). Andersson (1994) reviews a number of cases of sexually selected acoustic signals in all of these taxa. Across taxa, we often find that females prefer to mate with males that produce louder, longer, more rapidly produced calls (reviewed in Ryan and Keddy-Hector 1992). As noted above, such preferences could lead to matings with physically or genetically superior males and thus be directly or indirectly selected, or the preferences could result from the increased detectability of such signals or from females' preexisting biases toward certain types of signals. In Section 6, we discuss further the evolution of female mating preferences for male traits.

5.2.1. Neuroethological Mechanisms of Sexual Selection

Neuroethology has not played as crucial a role in explaining the behavioral mechanisms that contribute to sexual selection as it has in elucidating the role of behavioral isolating mechanisms in speciation. In some ways, the successes of neuroethology in identifying species-specific decoding mechanisms might have constrained it from similar success in sexual selection. In both speciation and sexual selection, it is crucial to understand how neural mechanisms allow the receiver to sieve through substantial signal variation to identify biologically meaningful signals. But the focus of the variation, and thus the focus of researchers, can be quite different in the two types of studies. In trying to understand species recognition, receivers and researchers alike confront the variation among species. Given signal variation at this level, how can the receiver identify a conspecific signal? In sexual selection, however, it is the variation within the species that is crucial, and one must ask how and why receivers are guided to one rather than another conspecific variant. This is not the type of question that neuroethologists originally set out to address about communication.

6. Signal–Receiver Coevolution

We have been discussing a variety of factors that cause and constrain the evolution of long-distance signals. But communication is a dyadic interaction, and signal evolution will proceed only if changes in the signal are meaningful to the receiver. This problem has been a major focus of interest in both speciation and sexual-selection studies. In the former, it is necessary to understand not only how mate-recognition signals evolve among incipient species but how the receivers evolve at the same time and in the same direction to give these new signals meaning, resulting in two functional and different mate-recognition systems (Andersson 1994).

The problem in sexual selection has been a bit more complicated and controversial (e.g., Kirkpatrick and Ryan 1991; Andersson 1994; Ryan 1997). In some mating systems, the female's choice of a mate does not influence her immediate reproductive success. In these mating systems, the male signals evolve under selection generated by female mating preferences. But how can preferences, the neural and cognitive processes biasing receivers toward certain signal variants, evolve if all receivers are producing the same number of offspring?

Here, we review some of the hypotheses for maintaining congruence between the signal and the receiver during evolution. We use the term coevolution in the more general sense—when the evolution of one suite of phenotypic characters influences the evolution of another suite of characters. Evolution of the two sets of characters need not proceed simultaneously; there could be substantial lag. Furthermore, it must be remembered that not every change in a signal requires a change in the receiver. Receivers seem not to have highly tuned accept–reject filters but are more than capable of ample generalization (Enquist and Arak 1998). The three major factors we will consider that might contribute to the functional integration of signal–receiver systems are pleiotropy, genetic correlation, and sensory exploitation.

6.1. Pleiotropy

Pleiotropy is multiple phenotypic effects resulting from the same gene. This offers the simplest explanation for how signals and receivers remain functionally integrated during evolution. If signals and receivers are controlled by the same gene or tightly linked sets of genes, then genetic changes will simultaneously and similarly affect signal and receiver.

6.1.1. Scaling

Because mate-recognition functions are so important to both sender and receiver, we assume that selection plays an important role in bringing about the match between peripheral tuning and long-distance signal. But there might be other, more parsimonious processes at work.

There are two variables to which signal variation often scales in a predictable manner: body size and, in ectotherms, temperature. Dominant or carrier frequencies tend to decrease with larger size because the characteristic vibration pattern of the morphological substrate is negatively correlated with mass. This relationship has been especially well-documented in frogs (Martin 1972; this chapter, Section 2.2), birds (Bowman 1983), and mammals (Morton 1977). In many ectotherms, ambient temperature influences rates of behavior, and sound production is no exception. The "temperature cricket" is so called because of the reliability of its pulse rate in predicting ambient temperature (Ewing 1989). How do receivers respond to such scaling effects in signals?

In anurans, the tuning of one or both of the two peripheral end organs that are sensitive to airborne sound tends to match the dominant spectral characteristics of the long-distance advertisement call (Fuzessery 1988; Zakon and Wilczynski 1988). The amphibian papilla (AP) tends to be most sensitive at threshold to frequencies below about 1,200 Hz, whereas the basilar papilla (BP) is most sensitive to frequencies above 1,200 Hz. Depending on the spectral distribution of call energy, either or both end organs will exhibit a match between the frequencies to which they are most sensitive and the spectral concentrations of call energy.

The AP and BP differ in a number of ways. The AP is thought to accommodate a traveling wave and has an array of hair cells that are tonotopically organized. The BP, on the other hand, has most of its hair cells tuned to similar frequencies, and the overall frequency sensitivity of the BP might result from its resonating properties, which will be influenced by size (Keddy-Hector et al. 1992; Wilczynski et al. 1992).

Cricket frogs, *Acris crepitans*, produce long-distance calls with most of the energy above 3,000 Hz. It seems clear that the BP rather than the AP is critical in the initial processing of the call. Calls exhibit a large amount of variation in dominant frequency across the geographical range of the species, but the auditory system tends to be tuned to frequencies characteristic of (Capranica et al. 1973; Ryan and Wilczynski 1988) or slightly lower than (Ryan et al. 1992) the local population. The geographic covariation of call frequency and auditory tuning might result from pleiotropic effects of changes in body size—both factors are negatively correlated with body size (Nevo and Capranica 1985; Ryan and Wilczynski 1991; Keddy-Hector et al. 1992; Wilczynski et al. 1992). It is assumed that the effect on tuning is derived from the effect of size on the resonating properties of the BP (Fig. 5.5). These results are consistent with patterns of frequency preference for call dominant frequency (Ryan and Wilczynski 1988; Ryan et al. 1992).

There are some cautions necessary, however. These scaling effects vary among populations (Keddy-Hector et al. 1992; Wilczynski et al. 1992), the relationships are not always very tight, and the preferences for call frequency appear to be statistically weak (although they could still generate strong biases in male mating success in the field). Nevertheless, dual scaling of signal and receiver to body-size variation does offer the potential for maintaining the functional integration of signal and receiver when size diverges among species or populations, and this will be true whether or not size has a significant heritable component.

A more short-term scaling problem involves temperature effects on signal characteristics. In most insects and frogs, temporal features of the call, such as pulse repetition rate, are drastically affected by temperature variation (Zweifel 1968; Gerhardt 1978; Bauer and Helverson 1987; Ewing 1989; Wagner 1989). In general, call rates increase with temperature. Tem-

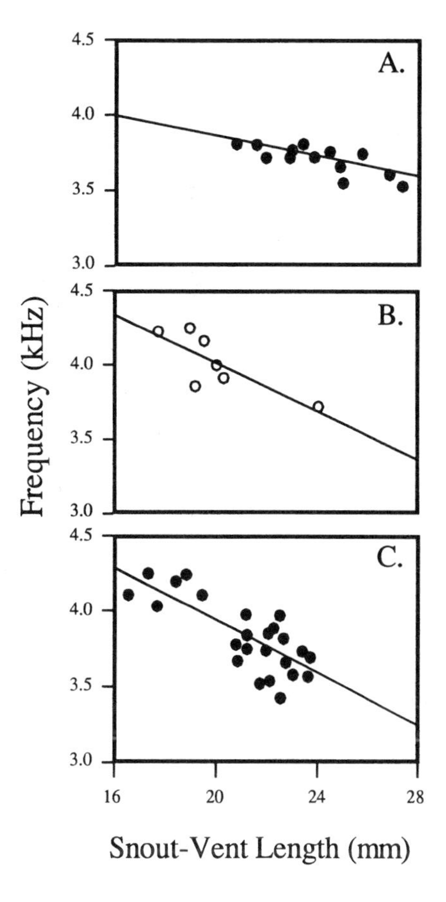

FIGURE 5.5. The relationship between body size (snout–vent length) and (**A**) the tuning of the basilar papilla in female and (**B**) male cricket frogs, *Acris crepitans*, and (**C**) the relationship between male body size and dominant frequency of the advertisement call within a single population. (Redrawn from Keddy-Hector et al. 1992.)

perature effects on spectral properties of the call are either not significant or less extreme—if there is an effect, frequency tends to be positively correlated with temperature.

One of the more interesting cases of temperature scaling involves study of the diploid-tetraploid species complex of *Hyla chrysoscelis* and *H. versicolor*. These are otherwise cryptic species that can easily be distinguished by the pulse rate of their call. *H. chrysoscelis* is diploid and has a faster pulse rate (~25–65 pulses/sec), whereas *H. versicolor* is tetraploid and has a slower pulse rate (~10–30 pulses/sec; Fig. 5.6). In both species, pulse rate is positively correlated with temperature. These two species can be sympatric, and if there is a wide enough range of temperature variation in the pond, it is conceivable that an *H. versicolor* male would have a higher pulse rate than an *H. chrysoscelis* male if, for example, the *H. versicolor* males were calling at 24°C while the *H. chrysoscelis* male was calling at 12°C (Fig. 5.6). Gerhardt (1978) showed that female preferences for pulse repetition rate scale to temperature similarly to that exhibited by conspecific calls. When challenged with calls that vary in pulse repetition rate, the female chooses that signal that would be produced by a male calling at her body tempera-

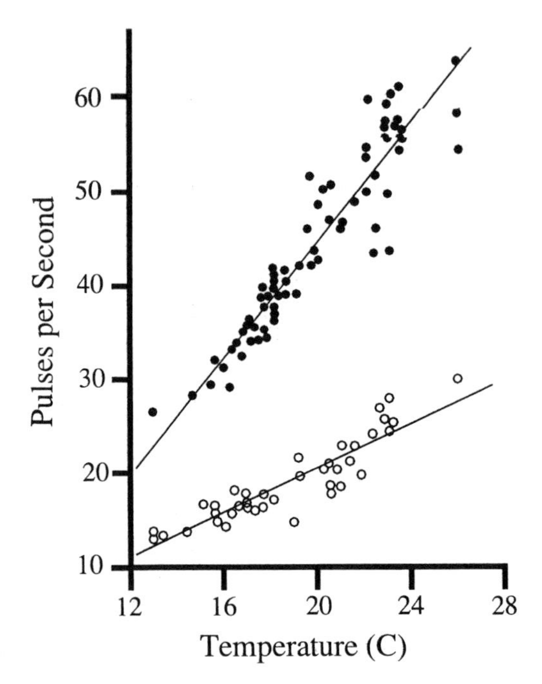

FIGURE 5.6. The relationship between pulses per second in the male advertisement calls of two species of gray tree frogs, *Hyla versicolor* (open circles) and *H. chrysoscelis* (closed circles). (Redrawn from Gerhardt 1978.)

ture. Brenowitz et al. (1985) showed that auditory neurons in the central nervous system that are sensitive to pulse rate show the same temperature scaling as the call and the preferences.

6.1.2. Pattern Generators

The dominant, or carrier, frequency of a signal is often determined by the mass of the primary vibrator that produces the sound and can be modified by radiating and resonating properties upstream from the vibrator. Temporal parameters of signals, on the other hand, are often actively regulated by behavioral–physiological properties, such as rate of stridulation or expiration. In such cases, there could be a neural oscillator that regulates the production mechanism to achieve the appropriate temporal characteristics of the signal. Receivers must decode these temporal signal characters. This could be achieved by comparing the pattern of auditory stimulation to a neural oscillator that acts as a template for temporal-pattern recognition. If the signal and receiver were both under control of the same neural oscillator or central pattern generator, any change in the temporal aspect of one component of the communication dyad would be immediately matched by the same change in the other component (Alexander 1962; Hoy et al. 1977).

A number of researchers have explored the possibility that signal–receiver systems maintain their functional integration through sharing a central pattern generator. If central pattern generators control signal–receiver variation, then hybrids should be intermediate relative to the two parental species in both temporal signal properties and response to these same properties. Studies of crickets and frogs both offered some qualified support for this hypothesis. Hoy et al. (1977) and Doherty and Gerhardt (1983) showed that in crickets and tree frogs hybrids tended to have intermediate pulse rates relative to the two parental species. Furthermore, in some but not all of the possible comparisons, hybrid females preferred the hybrid calls over the calls of at least one of the parental species.

The interpretation of these studies has been a challenge (Boake 1991). An alternative to the pleiotropy (central pattern generator) hypothesis is that signals and receivers are under separate and quantitative genetic control. If this were the case, one would still expect hybrids to be intermediate. This issue could be resolved through detailed quantitative trait locus mapping studies (Lynch and Walsh 1998). Another criticism of the central pattern generator hypothesis was offered by Bauer and Helverson (1987). If there is a single central pattern generator controlling the signal and receiver, then their response properties should be thermally linked. The authors showed, however, that if the head and thorax of a grasshopper are heated separately, the song and preference are decoupled.

One of the best-characterized genes controlling biological rhythms is the *period* locus in *Drosophila*. This locus is implicated in controlling a variety of rhythms, such as the circadian rhythms, as well as the fruit flies' love song (Hall 1994). The love song is produced by vibrating the wings, and mutants at the period locus vary in song rhythm. Previous studies had shown that mutant lines not only differed in love song pattern but in female preference for the same pattern (Kyriacou and Hall 1986). These results suggested that the female's preference for song pattern was one more rhythm under control of the *period* locus. More recent studies have suggested that this is not true, however. Greenacre et al. (1993) examined *Drosophila melanogaster* with mutations at the *period* locus, which predictably altered the rhythmic pattern of the song. Female mutants retained a preference for the wild type over the mutant song rhythm. This suggests that the song rhythm and the preference for song rhythm are under separate genetic control. Females from a *period* mutant stock that had been maintained for over ten years, however, did show preference for the mutant song rhythm. Thus, song and song preference are able to coevolve in *Drosophila*, but not through the pleiotropic effects of the period locus.

6.2. Linkage Disequilibrium

Signals and receivers could also maintain their functional integration if there were a statistical linkage between genetic variation influencing signal and receiver properties. Linkage disequilibrium is a measure of the non-random assortment of alleles at different loci. It is a process that has generated considerable interest in sexual selection theory (e.g., Andersson 1994), but its generalities can be applied to the more general problem of signal–receiver evolution. As part of this theory of runaway sexual selection, Fisher (1930) was the first to suggest that linkage disequilibrium plays an important role in the evolution of sexually selected male signal traits and female preferences for traits. It has been more recently applied to "good genes" theories of sexual selection (Pomiankowski 1988; Grafen 1990).

The crux of linkage disequilibrium is that traits that are not under selection can evolve if they are genetically correlated with traits that are under selection. In the parlance of evolutionary genetics, the trait subject to selection is said to be under direct selection. If another trait variant is more likely to be associated statistically with the trait under direct selection than expected by random chance, the two traits are genetically correlated or in linkage disequilibrium. The second trait, not subject to direct selection, is under indirect selection by virtue of its genetic correlation with the first trait and can evolve as a correlated response to the trait under direct selection.

It is the observation that a trait can evolve even if it is not subject to direct selection that makes linkage disequilibrium especially relevant to sexual-selection studies. In lekking species of animals, males gather to signal for mates and provide them with few resources besides sperm. It is assumed that in many cases, regardless of with whom the females choose to mate, the number of offspring they birth will be the same. Yet—and this is the paradox—females sometimes assiduously choose their mates (in many cases, the signal parameters that influence mate choice are well-known (Ryan and Keddy-Hector 1992; Andersson 1994)), with only a few males on the lek gathering a majority of the mating success. How could such preferences evolve if there is no difference in the reproductive success of females exerting different preferences or, for that matter, no difference in the reproductive success of females exhibiting a preference or mating at random? Two major hypotheses, runaway sexual selection and good genes selection, have been offered as solutions to this paradox, and both are dependent on indirect selection and linkage disequilibrium. Even though this problem of evolution through linkage disequilibrium is usually addressed within the context of male trait/female preference evolution, in many cases the male trait is a long-distance signal and the female preference emerges from the interaction of this stimulus and the properties of her receiver.

6.2.1. Runaway Sexual Selection

Fisher's hypothesis of runaway sexual selection is best illustrated with an example. Suppose that there is heritable variation for a male signal, such as simple song or complex song, and a female preference, such as a preference for complex song or a lack of such a preference. The genes controlling trait and preference are present in both sexes, but only the gene appropriate for each sex is expressed. For simplicity's sake, assume that the population is haploid. After the first episode of mating, males with complex songs will have garnered greater mating success. Furthermore, alleles that determine complex song will be in linkage disequilibrium with alleles that determine preference for complex song. As the frequency of complex song evolves in the population due to preference for complex song, the preference itself will "hitchhike" along with the complex song, and the preference will also evolve to be in a higher frequency in the population. The stronger the preference for complex song, the faster the rate of evolution of song and, through the genetic correlation of song and preference, the faster the rate of evolution of preference. The process can continue until the advantages of the enhanced male trait, song complexity in this example, are offset by the natural-selection costs of producing the trait, such as metabolic or predation costs, for example. At such a point, the forces of sexual selection and natural selection will be balanced, and trait and preference will reach an evolutionary equilibrium.

Although runaway sexual selection has been a popular hypothesis among population geneticists for a number of years, it has been difficult to marshal much empirical support for it (Andersson 1994; Ryan 1997). This might be because it is a transient process; once signal and receiver reach an equilibrium point, it is difficult to determine how they got there. One approach to testing the theory is to demonstrate the genetic correlation between signal and receiver either by conducting selection experiments or comparing populations of the same species that differ in trait and preference.

There are no good examples of studies in acoustic communication supporting the runaway hypothesis. But as a matter of illustration, consider an elegant study of visual signaling in stalk-eyed flies. The eyes are located at the end of long stalks in both sexes, but the stalks are much longer, and thus the eye span much greater, in males than in females. Females prefer males with greater eye spans. Wilkinson and Reillo (1994) conducted bidirectional selection experiments on male eye span and determined whether there was a correlated evolutionary response in female preference for eye span. After 13 generations, females from large eye-span lines and the unselected lines both preferred males with larger eye spans. Females in the lines for which short eye span was selected preferred males with shorter eye spans. Thus, the female preference evolved even though it was not under direct selection but instead because it was genetically correlated to the signal.

6.2.2. Good Genes Selection

Good genes selection has been viewed as an intuitively appealing and more utilitarian alternative to runaway selection. Under this scenario, females attend to signal variation to assess a male's genetic quality for survivorship. But what would keep signals honest? Why could males not cheat and evolve signals that falsely indicate high genetic quality? In some cases, signals might be constrained from doing so. For example, Hamilton and Zuk (1982) suggested that plumage brightness and song complexity in birds indicate parasite load; parasites will directly influence plumage color and, it is assumed, the energetic potential to make complex songs. If there is a genetic basis to parasite resistance, then these signals are honest indicators of some genetic quality. An alternative means for enforcing signal honesty is the handicap principle (Zahavi and Zahavi 1997). This hypothesis suggests that males evolve signals that are costly in terms of survivorship; thus, only truly healthy males can afford the handicapping signal.

Early population-genetic models of the handicap principle did not support its internal logic (reviewed in Zahavi and Zahavi 1997). Later models, however, showed that the handicap principle could work through linkage disequilibrium. In this case, however, the preference genes become correlated with the "good genes" for survivorship that are being signaled. For example, let us assume faster call rate, which will be energetically more expensive than slower call rate, indicates healthier males because they are better foragers. Once some females begin to prefer faster call rate, the alleles determining that preference will become associated with the alleles for better foraging. Natural selection will cause an increase in better foragers, and the preference for faster call rate evolves as a correlated response.

All good-genes models, and especially the handicap principle, have been controversial and difficult to support empirically. Only recently have there been data to show that female preference for male signals influences the survivorship of their offspring. Most of these studies involve visual signals (reviewed in Ryan 1997), but data involving long-distance acoustic signals have recently become available. For example, molecular paternity analysis has shown that when female great reed warblers seek extra-pair copulations, they do so from males having larger song repertoires. Hasselquist et al. (1996) speculate that the female pairs with a male with superior territories, thus ensuring the resources necessary for immediate reproductive success, but seeks extra-pair copulations from males that have "good genes"; song repertoire size is correlated with survivorship.

One of the best studies comes from anuran communication and is similar to the hypothetical example given above. Gray tree frogs, *Hyla versicolor*, produce a pulse call that can vary among males in pulse rate and pulse duration. Males can increase the energy content of the call by increasing either call rate or duration; the former is an energetically more expensive option for the males. Klump and Gerhardt (1987) showed that when given a choice between a pair of calls varying in rate and duration but similar in overall energy content, females preferred longer calls. They speculated that this energy-independent preference might be indicative of selection for good genes. This hypothesis was supported recently by Welch et al. (1998). Female gray tree frogs were mated to two males, one that produced short calls and one that produced long calls; these crosses resulted in sets of maternal half-sibs. Tadpoles were raised through metamorphosis, and a number of life-history parameters assumed to be predictive of higher survivorship were measured. The offspring of males with long calls performed significantly better than or not significantly differently from those of males with short calls. Although the data are too sparse to make any sweeping generalizations, the studies of gray tree frogs offer some of the best support for the notion that females can assess variation in long-distance acoustic signals to influence the genetic quality of their young.

6.3. Sensory Exploitation

Plants have evolved suites of adaptations to attract pollinators, much as males often use long-distance signals to attract females. In many cases, the attractions are mutualistic. The pollinator is attracted by various signals of the plant and inadvertently pollinates the plant while harvesting nectar. A number of plants, however, do not produce nectar. Orchids, for example, mimic the pheromones or the general body outline of insects; insects then attempt to mate with the flower, pollinating it during their mistaken sexual foray (Piji and Dodson 1966).

Although these interactions do not involve acoustic signals, they illustrate a relevant point. Congruence between the signal and receiver need not involve the sorts of genetic relationships within a genome envisioned in pleiotropy and linkage disequilibrium. Such phenomena cannot occur between species. In the orchid example, it appears that plants have evolved signals that exploit the insect's responses to conspecific sexual signals. The response of the insect to the orchid is not adaptive for the insect, but we assume that there is a net benefit to the receiver's biases; that is, the sum of costs and benefits of responding to plants and to sexually receptive conspecifics. Recent studies in sexual selection and communication have suggested that males evolve signals to exploit response biases of the females' receiver system (recently reviewed in Christy 1995; Endler and Basolo 1998; Ryan 1998, 1999).

6.3.1. Response Biases

Williams (1966) made a crucial point in distinguishing between an evolved function and incidental consequence in evolution. In the context of animal communication, the response properties of the receiver can be under selection to recognize the signal of a conspecific; if so, conspecific recognition is an evolved function. Depending on the recognition strategy, however, other stimuli might elicit strong receiver responses. For example, consider a receiver in an acoustic environment in which there is only one heterospecific using acoustic signals for mate recognition and its signal is much shorter in duration than the conspecific signal. A simple and effective recognition strategy would be to respond only to signals above a certain threshold in duration; this strategy, too, is an evolved function of the receiver. It seems that recognition strategies often involve such simple generalizations, as opposed to more complicated ones having sharp multivariate filters that exclude all other possible signals but those of conspecifics (Enquist and Arak 1998). Such a simple recognition strategy might have some advantages of neural economy relative to a more complicated one that relied on a multitude of signal parameters, and in the simple environment of only one heterospecific, a more generalizing strategy would be as effective as a more specific one—effective, that is, until there appeared a new heterospecific with an even longer signal and the simple recognition strategy was foiled because the receiver now made recognition errors by responding to the longer heterospecific signal. Responses to this longer heterospecific signal would be an incidental and maladaptive consequence or response bias of the original, and adaptive, mate-recognition strategy. This is bound to happen because selection cannot anticipate future situations but can only judge among alternative phenotypes in a current context. But if the costs of making a recognition error toward longer signals exceeded the benefits of such a simple recognition strategy, we assume that the recognition strategy would evolve further under selection (Dawkins and Guilford 1996; Ryan 1999).

Such response biases are the basis of another exploitative, interspecific interaction similar to the orchid–insect example reviewed above. Cuckoos do not raise their own young. They place their eggs in the nests of other species, where their young are raised by the host, often to the detriment of the host's own young. Reed warblers are one such host. Even though the host young produce begging calls to elicit feeding from their parents that are quite different from the begging calls of the cuckoo (Fig. 5.7), reed warblers feed cuckoos preferentially to their own young. The begging call of the cuckoo does, however, mimic the sounds of a group of begging reed warblers (Fig. 5.7). In a series of elegant experiments, Davies et al. (1998; see also Kilner et al. 1999) showed that it is this cuckoo signal that exploits the receiver system of the reed warbler.

In many other cases, we might not expect response biases that emerge from recognition strategies to be maladaptive. Ryan and Keddy-Hector (1992) reviewed numerous cases of female preferences based on long-distance signals. If female preference deviated from the population mean, it was usually in the direction of greater signal energy—more intense, longer signals produced at higher repetition rates. If such preferences represent general response biases of many auditory-recognition systems, they might still continue to guide females toward conspecific males and, incidentally, toward males that are in better physical condition, being able to marshal more energy to support calling. An alternative is that the general pattern of bias toward greater signal content is an evolved response to choose such males in the first place.

6.3.2. Habituation and Song Preference in Birds

Songs of many oscines are characterized by their signal complexity. One explanation for such complexity, the antimonotony hypothesis, is based on

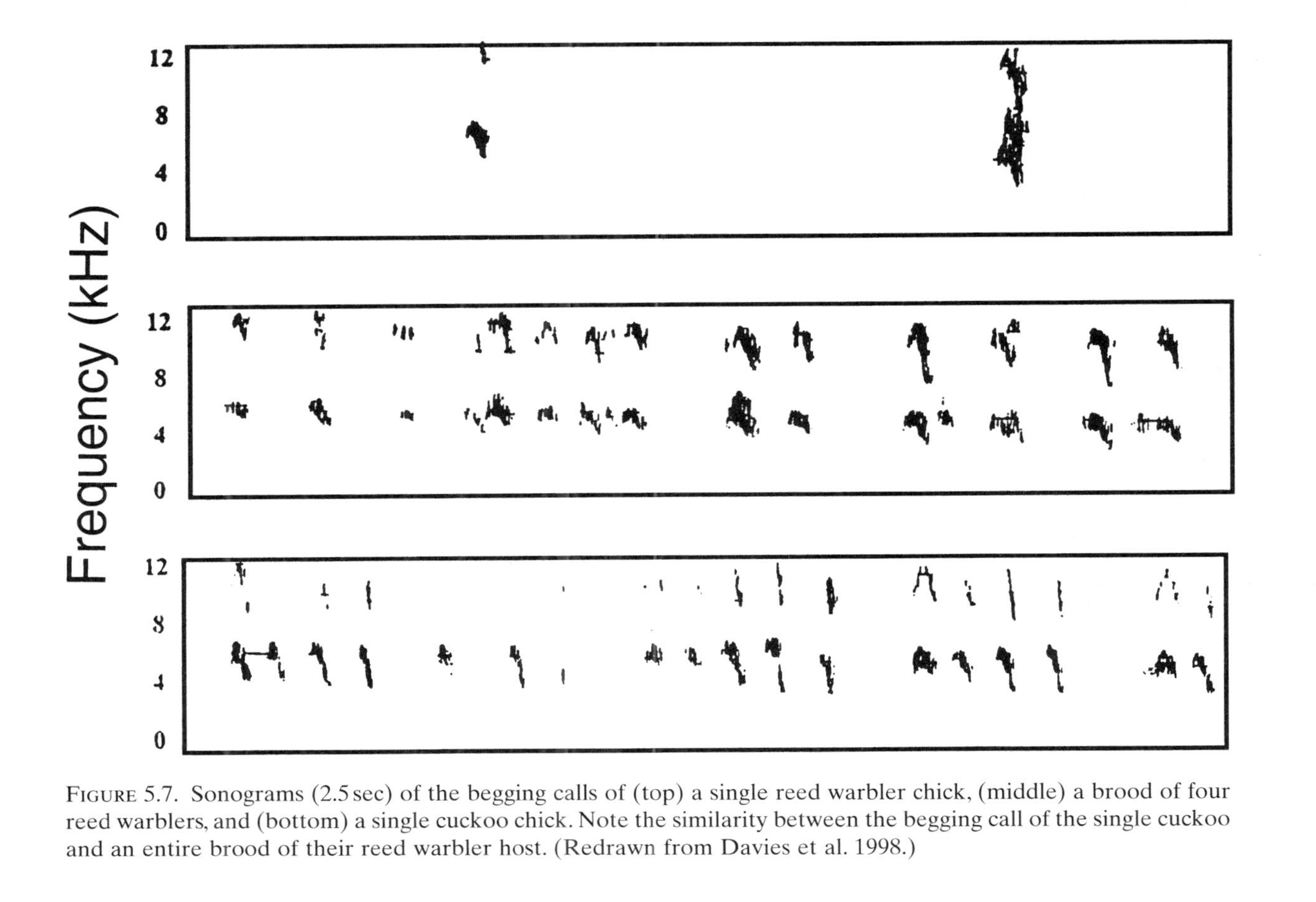

FIGURE 5.7. Sonograms (2.5 sec) of the begging calls of (top) a single reed warbler chick, (middle) a brood of four reed warblers, and (bottom) a single cuckoo chick. Note the similarity between the begging call of the single cuckoo and an entire brood of their reed warbler host. (Redrawn from Davies et al. 1998.)

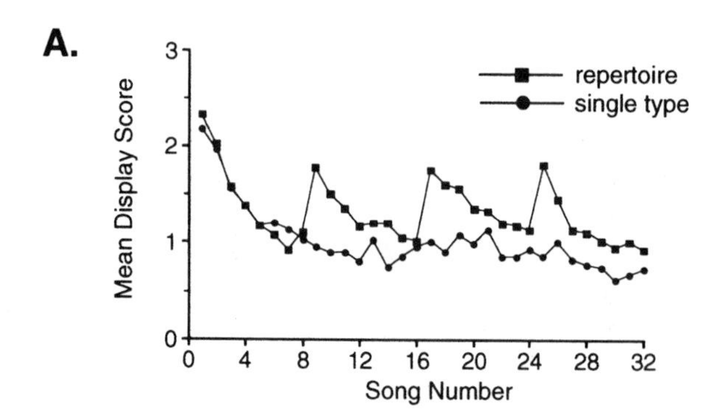
A.
Mean Display Score
Song Number
repertoire
single type
0
1
2
3
0
4
8
12
16
20
24
28
32

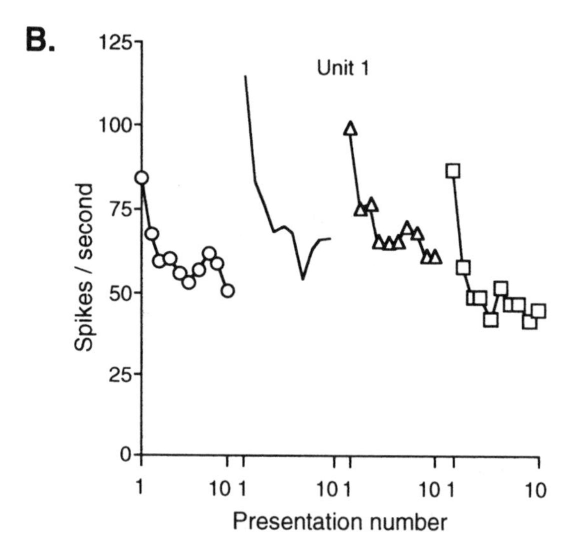
B.
Unit 1
Spikes / second
Presentation number
0
25
50
75
100
125
1
10 1
10 1
10 1
10

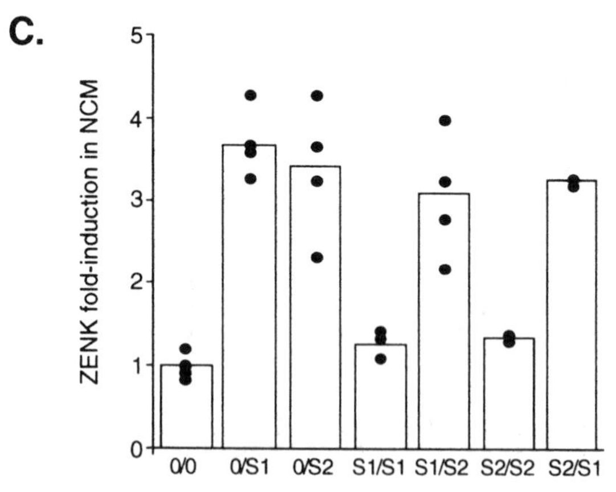
C.
ZENK fold-induction in NCM
0
1
2
3
4
5
0/0
0/S1
0/S2
S1/S1
S1/S2
S2/S2
S2/S1

a general response bias of auditory systems (Hartshorne 1956; Searcy 1992). This hypothesis suggests that song complexity per se is selected, rather than any message encoded in different song syllables, because it reduces habituation in neighboring males and courting females. Searcy presents some evidence supporting this hypothesis in studies of grackles (Fig. 5.8). These birds do not have complex repertoires, but females are attracted preferentially to artificially enhanced repertoires compared to the more monotonous song of their conspecifics. When females were presented with a song of 32 syllables, they showed more courtship solicitation displays to the song that contained eight successive repetitions of four different syllables to a song of 32 identical syllables. Interestingly, the female response decreased within the repetition of the same syllable but increased during transition between syllable types, suggesting a pattern of habituation to signals in response to the simple song but habituation and release from habituation in response to a more complicated song (Fig. 5.8, top).

Studies of zebra finches and canaries point toward some of the underlying mechanisms of this preference for more complex songs. Both electrophysiological responses (Stripling et al. 1997) of auditory neurons and expression of an early gene, *zenk*, implicated in auditory function (Mello et al. 1992), also show decreased response to repeated song stimuli and enhanced response during transition between stimulus types (Fig. 5.8; reviewed in Ryan 1998).

To the extent that habituation to signal monotony was a general phenomenon, we would expect response biases to generate selection for more complicated signals. The manner in which signal complexity could be enhanced would be constrained by the types of sounds that are efficacious in that particular system. For example, although swamp and song sparrows learn each others' song in the absence of their own, there is a strong bias to learn their own song type (reviewed in Marler 1997). Such genetic pre-

FIGURE 5.8. (**A**) Female courtship displays to complex song in grackles show higher levels in response to multiple-song repertoires than to single-song types. The single type (circles) contains 32 repeats of the same song. The repertoire (squares) contains four different song types repeated in groups of eight (e.g., 1–8, 9–6, etc.). The response to repertoire songs shows that there is habituation to repeated songs within each song type (e.g., 1–8,9–16, etc.) and release from habituation at transition between quartets (song 9, 17, etc.; Searcy 1992). (**B**) Electrophysiological responses of units in the zebra finch's causomedial neostriatum, which borders the song-control nucleus, shows decreased spike rates to repeated presentation of the same song and enhanced spike rates in response to a new song (Stripling et al. 1997). (**C**) Expression of an immediate early gene, *zenk*, is higher during transitions from no song to song (0/S1), or from one song to another song (S1/S2; S2/S1), than during absence of song (0/0) and repeated stimulation of the same song (S1/S1; S2/S2; from Mello et al. 1992). (Reprinted with permission from Ryan 1998. Copyright © 1998 American Association for the Advancement of Science.)

dispositions could bias the types of sounds incorporated into signal complexity; some species, such as mockingbirds, might be less constrained. This view of signal complexity, in general, and bird song repertoires, specifically, shifts the focus from any message that might be encoded by parts of the song to the value of complexity per se. Marler (1998), for example, has recently suggested that ". . . the song functions as affective rather than symbolic signals, and the variety is generated, not to diversify meaning, but rather to maintain the interest of anyone who is listening, and to alleviate habituation" (p. 12). Music rather than language might be the preferred analog for some types of animal communication. Such a view is consistent with the ideas of sensory biases and sensory exploitation discussed here. In fact, some recent studies in musicology have considered the proposition that characteristics of the human cochlea might dictate some aspects of music appreciation (Zentner and Kagan 1996).

6.3.3. Sensory Exploitation in Túngara Frogs

The call of the túngara frog, *Physalaemus pustulosus*, has two components: a whine and a chuck. The whine initiates the call, is always present, and may be followed by one or several chucks or can be produced alone. When chucks are added to the call, they are appended near the end of the whine. Up to six chucks can be added, although one to three chucks is the more common occurrence. In controlled laboratory experiments, a whine is both necessary and sufficient to elicit phonotaxis in female frogs. When females are given a choice between a whine only and a whine with chucks, females prefer the latter. Thus, the túngara frogs exhibit both chucks and preference for chucks (reviewed in Ryan 1985b; Ryan and Rand 1999).

Phylogenetic analysis combined with behavioral experimentation can sometimes provide insights into the historical pattern by which signals and receivers evolved. Pleiotropy and the genetic correlation hypotheses suggest that signals and receivers evolve in concert, whereas sensory exploitation suggests that signals exploit preexisting biases. These approaches were used to investigate the manner in which the chuck and the preference for chuck evolved.

The species group to which the túngara frog belongs is the *Physalaemus pustulosus* species group. This group consists of two smaller monophyletic groups, one west of the Andes mountains and the other in Middle America and east of the Andes (Cannatella et al. 1998; Fig. 5.9). Only species in the eastern group add suffixes to the call; they are lacking not only in the species in the western group but also in the more than 20 species of the genus that have been studied. This suggests that the chuck evolved after the two smaller groups within the species group diverged; examination of laryngeal correlates of the chuck support this interpretation (Ryan and Drewes 1990).

If females in the western group preferred chucks added to their conspecific whine, even though their males are incapable of producing them, it would suggest that preference for chucks was a preexisting bias exploited

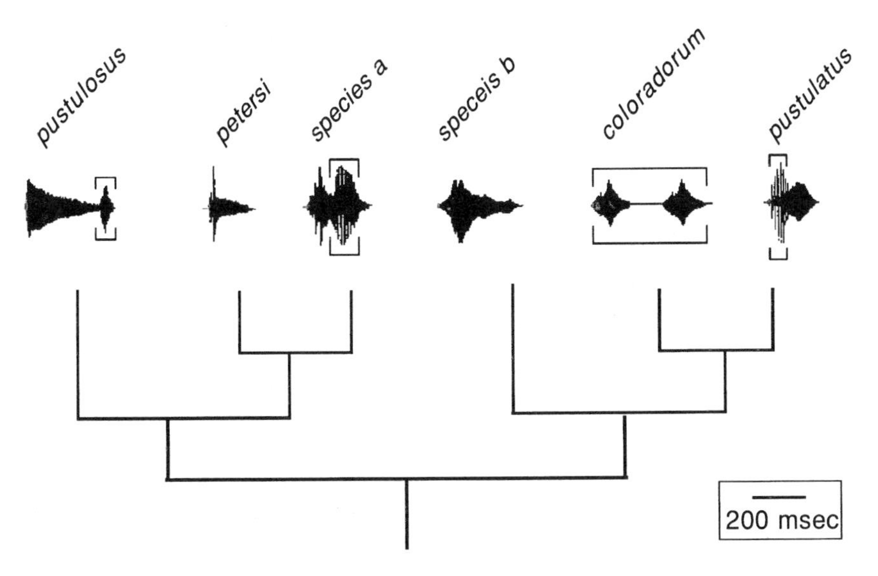

FIGURE 5.9. The phylogenetic relationships within the *Physalaemus pustulosus* species group, as determined by analysis of DNA sequences, allozymes, and morphological traits (see Cannatella et al. 1998 for details). The brackets over the calls of *P. pustulosus* and *species a* indicate call suffixes that are facultatively added. Such suffixes are absent in the sister clade and in the rest of the genus *Physalaemus*. *P. coloradorum* is the only species in the group to produce calls in doublets, while only *P. pustulatus* has an extreme amplitude-modulated component in the beginning of all its calls.

by male túngara frogs. This hypothesis would be favored over the less parsimonious one that the chuck was lost in the western group or that the preference for chucks evolved twice independently. When a chuck from a túngara frog call was digitally added to the whine call of *P. coloradorum*, a species in the western clade of the *Physalaemus pustulosus* group, females preferred calls with chucks to the normal species call, which lacks chucks (Ryan and Rand 1993). The fact that *P. pustulosus* and *P. coloradorum* both prefer calls with chucks suggests that this preference is shared through a common ancestor as opposed to the hypothesis that *P. coloradorum* females happened to evolve the same preference for traits not existing in their own males. Because the common ancestor of *P. coloradorum* and *P. pustulosus* existed before the Middle American–Amazonian and western–Andes groups diverged (Fig. 5.9), this suggests that the preference for chucks existed prior to the evolution of the chucks themselves. This supports the contention that males evolved chucks to exploit a preexisting preference for chucks.

Analogous results in a variety of other taxa suggest that sensory exploitation is not restricted to this taxon (reviewed in Ryan 1998). However, one must assume caution in interpreting these phylogenetic analyses because the most parsimonious conclusion is only a bit simpler than the alternative of multiple losses or gains of a character (Ryan 1996).

6.4. Historical Effects on Receiver Biases

6.4.1. Evolution of Communication from Prey Localization and Predator Avoidance

Although we have concentrated on the receiver's role in long-distance communication, hearing or, more generally, sensing ambient vibration patterns plays a more general role in the behavior and ecology of many species. Auditory receivers can partition their functions among different tasks. As discussed above, for example, the frog-eating bat, *Trachops cirrhosus*, exhibits a suite of neuroanatomical specializations that enhance hearing of the relatively low-frequency calls of frogs while still maintaining their ability to utilize ultrasonics for echolocation (Bruns et al. 1989). Alternatively, adaptations of a receiver for a noncommunication task might bias the types of signals that can be used for communication. Two examples, one involved in finding prey and the other in avoiding predators, illustrate this point nicely.

Water mites do not "hear" in the sense that the vibrations they sense are borne on the water's surface rather than through the air. The vibratory patterns to which they are cued are produced by copepods, one of their main prey items, and a number of species of water mites use this information for prey localization. In one species, however, males mimic these vibrations and use them to lure females. Females approach the source of this vibration and then are courted by the male producing it (Proctor 1991). To show that the female's response appears to be incidental to courtship, Proctor (1991) showed that females deprived of food were more likely to approach and encounter courting males than were females who were satiated. The courtship communication system seems to have evolved in an exploitative process. Although many species of mites use copepod vibrations to hunt, only one of those species is known to mimic food for sex (Proctor 1992).

Animals use their ability to sense ambient vibrations to avoid becoming food as well as in finding food. In some moths, evolution of adaptations to avoid bat predation seems to have dictated some aspects of the acoustic courtship communication that has evolved.

Bats are major predators on moths, and many moths have responded to this selection force by evolving the ability to detect ultrasonics, and a subset of these moths also produce ultrasonics that seem to deter predation either because they are aposematic signals (the moths are distasteful) or because they interfere with the bats' echolocation system (reviewed in Fullard 1998).

There are at least two groups of moths in which these predator-avoidance adaptations also serve the purpose of acoustic communication.

Ultrasonic courtship signals, in the same range as the hearing of bats, increase mating success in both ctenuchid (Sanderford and Conner 1995; Simmons and Conner 1996) and wax moths (Jang and Greenfield 1996), and in some ctenuchids the males and females conduct an ultrasonic dialog. It seems probable that the use of ultrasonics has nothing to do with the precise message being sent but was a convenient channel to use because it already existed, albeit for another purpose.

6.4.2. Neural Networks and Response Biases

The examples above illustrate that if an auditory receiver is adapted for one function, this can bias how it then becomes adapted for another function. A similar effect could also occur across time but within the same function. As we have been discussing, long-distance cues are often important components of a species' mate-recognition system. When an ancestral species splits into two daughter species, the two daughter species have different recognition systems: the mate-recognition signals will differ between the species, and each species will be biased toward responding to the conspecific signal. Thus, at a minimum, one daughter species evolved a new recognition system (i.e., signal plus receiver properties) and the other maintained the ancestral recognition system, or both diverged from the ancestral signal and receiver.

There are probably a large number of computational strategies by which a receiver can bias its response to the conspecific signal. The strategy it chooses might be dependent on how ancestors of this receiver achieved the same task. For example, if within a lineage of animals the mate-recognition signals of species could always be discriminated by signal duration or by a more subtle multivariate comparison of a multitude of spectral parameters, we might expect the receiver to be biased toward using temporal parameters for recognition, much as the moths discussed above utilized ultrasonics for communication. This should be true as long as such a strategy could achieve the task.

Phelps and Ryan (1998) recently addressed this issue of historical biases of receivers by combining studies of artificial neural networks with their empirical studies of mate recognition in túngara frogs. Initially, they trained recurrent artificial neural networks to recognize a túngara frog call in 20 replicate populations. In each population, they retained the network that best discriminated between the call and noise. They then determined their responses to a variety of signals, such as heterospecific calls and purported ancestral calls, with which túngara frogs had been tested. Neither the networks nor the frogs had any previous experience with these signals. The responses to these signals therefore are considered response biases because they are incidental rather than being the target of selection. There was a strong correlation between the response biases of the frogs and the networks. Thus, whatever computational strategies these two systems were employing to achieve signal recognition, they were producing similar response biases.

In a subsequent study, the authors examined the effect of history on response biases. Ryan and Rand (1995, 1999b) had used phylogenetic techniques to estimate what the calls of ancestors of túngara frogs might have sounded like. The correlation between evolutionary relationship (estimated as similarity in DNA sequences) and call similarity was not statistically significant—calls of close relatives were not more likely to sound alike than calls of more distant relatives. Nevertheless, female responses to the calls were predicted by phylogenetic relationship as well as, and independent from, overall call similarity. These results suggested that the history of the receiver can influence its response biases.

Phelps and Ryan (2000) trained neural networks along three distinct histories. In the first, the mimetic history, networks were trained to the call at the root of the phylogenetic tree (Fig. 5.10). Once the networks reached recognition criteria, they were trained to the call that was the next most immediate ancestor to the túngara frog. This procedure was continued on the line of descent to the túngara frogs until the networks were trained to the túngara frog call itself. Thus, these networks had a history of first being trained to the calls of the three direct ancestors of túngara frogs before being trained to the target call, the túngara frog call. This was replicated 20 times, and the most discriminating net in each replicate was later tested. The authors conducted the same procedure for two control evolutionary histories. For one, the random history, three calls were picked at random from the sample of heterospecific and ancestral calls. The nets were trained to these calls prior to being trained to the túngara frog calls. There were 20 random histories, and the most discriminating net in each was later tested. In another control, the calls used in the mimetic history were rotated 180° in principal component space and the new calls synthesized. The path length among these "mirror" calls was identical to the path length among the calls in the mimetic history; both of these path lengths were longer than that of the random histories. There were 20 replicates and, as with the other histories, the most discriminating net in each was saved.

The best nets from the mimetic, random, and mirror histories were tested against the same set of heterospecific and ancestral calls as with the ahistoric nets. Only the networks that were trained along the mimetic history significantly predicted the response biases of the túngara frog females. These results suggest that the past history of tasks a receiver needs to accomplish influences the computational strategies it uses to accomplish current tasks.

6.5. Summary and Conclusions

Long-distance acoustic signals are prevalent as mate-recognition signals in a number of diverse taxa and are accessible for studies by behaviorists,

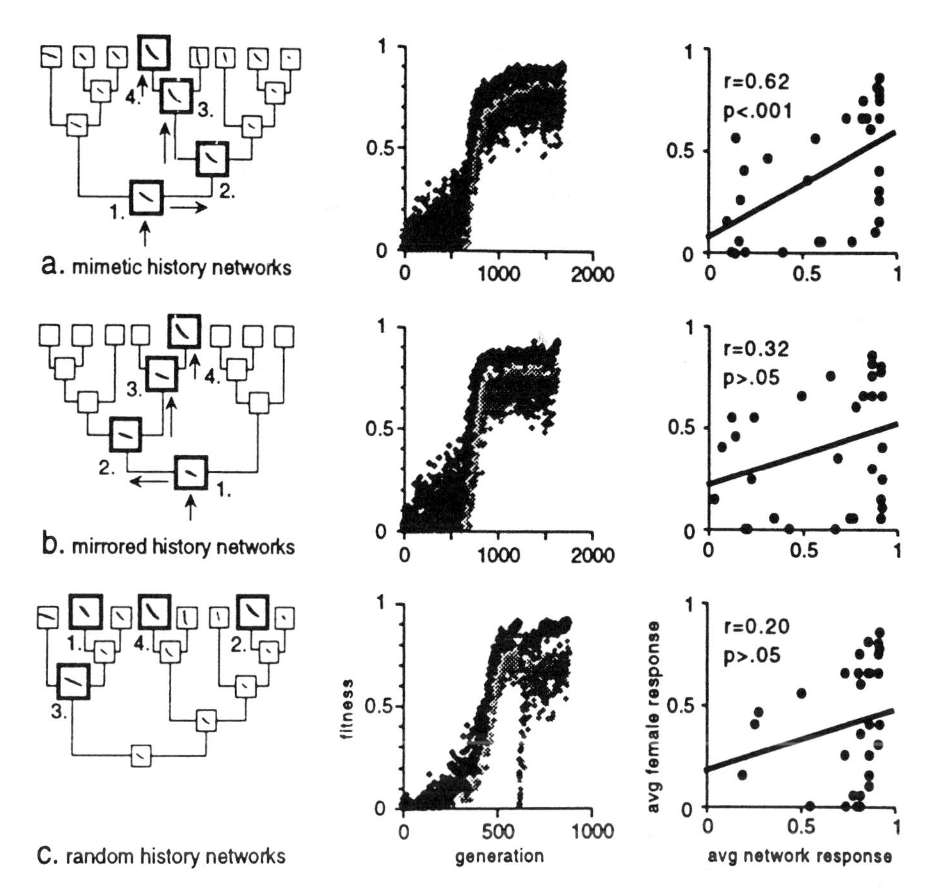

FIGURE 5.10. The left-hand columns show the various sequences of calls, or "evolutionary histories," to which artificial neural networks were trained. In all cases, the networks were trained to three calls prior to being trained on the túngara frog call. The middle column shows the relative fitness of the networks as a function of generation time. The right-hand column shows the relationship between the responses of networks and real female túngara frogs to an assortment of acoustic stimuli. Note that only networks trained along the mimetic history significantly predict the responses of real females (S.M. Phelps unpublished).

evolutionary biologists, and neuroethologists. As such, there is a wealth of information on their function, underlying morphological and physiological control, and how they might be influenced by selection. Most generalizations, however, are pieced together from data gathered in different systems. There are no model systems for which all of the issues we review here have been addressed. The advantage to such an approach is that we have a very good understanding of some of the diversity of long-distance communication systems. The disadvantage is that it is more difficult to appreciate how different constraints and selection forces that are documented in different systems might interact in a single species. The underlying theme of this chapter is that long-distance acoustic signals are subject to a variety of selection forces and constraints and that the evolutionary history of a taxon can influence the manner in which it responds to such forces. Thus, a true understanding of long-distance communication will inevitably require an increase in both breadth and depth of studies.

References

Alexander RD (1962) Evolutionary change in cricket acoustical communication. Evolution 16:443–467.

Andersson M (1994) Sexual Selection. Princeton, NJ: Princeton University Press.

Barclay RMR, Fenton B, Tuttle MD, Ryan MJ (1981) Echolocation calls produced by *Trachops cirrhosus* (Chiroptera: Phyllostomatidae) while hunting for frogs. Can J Zool 59:750–753.

Bauer M, van Helverson O (1987) Separate localization of sound recognizing and sound producing neural mechanisms in a grasshopper. J Comp Physiol 161:95–101.

Bevier CR (1997) Utilization of energy substrates during calling activity in tropical frogs. Behav Ecol Sociobiol 41:343–352.

Blair WF (1958) Mating call in the speciation of anuran amphibians. Am Nat 92:27–51.

Boake CRR (1991) Coevolution of senders and receivers of sexual signals: Genetic coupling and genetic correlations. Trends Ecol Evol 6:225–231.

Bowman RI (1983) The evolution of song in Darwin's finches. In: Bowman RI, Besson M, Leviton AE (eds) Patterns of Evolution in Galapagos Organisms. San Francisco: American Association for the Advancement of Science, pp. 237–536.

Brackenbury JH (1979) Power capabilities of the avian sound-producing system. J Exp Biol 78:163–166.

Bradbury JW, Vehrencamp SL (1998) Principles of Animal Communication. Sunderland, MA: Sinauer Associates Inc.

Brenowitz EA, Wilczynski W, Zakon HH (1984) Acoustic communication in spring peepers. Environmental and behavioral aspects. J Comp Physiol A 155:585–592.

Brenowitz EA, Rose G, Capranica RR (1985) Neural correlates of temperature coupling in the vocal communication system of the gray tree frog (*Hyla versicolor*). Brain Res 359:364–367.

Bruns V, Burda H, Ryan MJ (1989) Ear morphology of the frog-eating bat (*Trachops cirrhosus*, Family Phylostomidae): Apparent specializations for low-frequency hearing. J Morphol 199:103–118.

Bucher TL, Ryan MJ, Bartholomew G (1982) Oxygen consumption during resting, calling, and nest building in the frog *Physalaemus pustulosus*. Physiol Zool 55:10–22.

Butlin R (1989) Reinforcement of premating isolation. In: Otte D, Endler JA (eds) Speciation and its consequences. Sunderland, MA: Sinauer and Associates, pp. 158–179.

Cade WH (1975) Acoustically orienting parasites: Fly phonotaxis to cricket song. Science 190:1312–1313.

Cade WH (1981) Alternative male strategies: Genetic differences in crickets. Science 212:563–564.
Cannatella DC, Hillis DM, Chippinendale P, Weigt L, Rand AS, Ryan MJ (1998) Phylogeny of frogs of the *Physalaemus pustulosus* species group, with an examination of data incongruence. Syst Biol 47:311–335.
Capranica RR (1972) Why auditory neurophysiologists should be more interested in animal sound communication. Physiologist 15:55–60.
Capranica RR, Frishkopf LS, Nevo E (1973) Encoding of geographic dialects in the auditory system of the cricket frog. Science 182:1272–1275.
Catchpole CK, Slater PJB (1995) Bird Song, Biological Themes and Variation. Cambridge, UK: Cambridge University Press.
Christy JH (1995) Mimicry, mate choice, and the sensory trap hypothesis. Am Nat 146:171–181.
Dabelsteen T, Larsen ON, Pederson SB (1993) Habitat-induced degradation of sound signals: Quantifying the effects of communication sounds and bird location on blur ratio, excess attenuation, and signal-to-noise ratio in blackbird song. J Acoust Soc Am 93:2206–2220.
Davies NB, Kilner RM, Noble DG (1998) Nestling cuckoos, *Cuculus canorus*, exploit hosts with begging calls that mimic a brood. Proc R Soc Lond B Biol Sci 265:673–678.
Dawkins MS, Guilford T (1996) Sensory bias and adaptiveness of female choice. Am Nat 148:937–942.
Doherty JA, Gerhardt HC (1983) Hybrid tree frogs: Vocalizations of males and selective phonotaxis. Science 220:1078–1080.
Doherty JA, Hoy RR (1985) Communication in insects III. The auditory behavior of crickets: Some views of genetic coupling, song recognition, and predator detection. Q Rev Biol 60:453–472.
Drewry GE, Rand AS (1983) Characteristics of an acoustic community: Puerto Rican frogs of the genus *Eleutherodactylus*. Copeia 1983:941–953.
Duellman WE (1967) Courtship isolating mechanisms in Costa Rican hylid frogs. Herpetologica 23:169–183.
Eberhard LS (1994) Oxygen consumption during singing by male Carolina wrens (*Thryothorus ludovicianus*). Auk 111:124–130.
Endler JA (1992) Signals, signal conditions, and the direction of evolution. Am Nat 139:S125–S153.
Endler JA, Basolo AL (1998) Sensory ecology, receiver biases, and sexual selection. Trends Ecol Evol 13:415–420.
Enquist M, Arak A (1998) Neural representation and the evolution of signal form. In: Dukas R (ed) Cognitive Ecology. Chicago: University of Chicago Press, pp. 21–87.
Ewing AM (1989) Arthropod Bioacoustics, Neurobiology and Behavior. Ithaca, NY: Cornell University Press.
Fisher RA (1930) The Genetical Theory of Natural Selection. Oxford: Clarendon Press.
Foster SA (1999) The geography of behaviour: An evolutionary perspective. Trends Ecol Evol 14:190–195.
Foster SA, Endler JA (eds) (1999) Geographic Variation in Behavior. Oxford: Oxford University Press.
Fritzsch B, Ryan M, Wilczynski W, Walkowiak W, Hetherington T (eds) (1988) The Evolution of the Amphibian Auditory System. New York: John Wiley and Sons Inc.

Fullard JH (1998) The sensory coevolution of moths and bats. In: Hoy RR, Popper AN, Fay RR (eds) Comparative Hearing: Insects. New York: Springer, pp. 279–326.
Fuzessery ZM (1988) Frequency tuning in the anuran central auditory system. In: Fritzsch B, Ryan M, Wilczynski W, Walkowiak W, Hetherington T (eds) The Evolution of the Amphibian Auditory System. New York: John Wiley and Sons Inc., pp. 253–273.
Gerhardt HC (1978) Temperature coupling in the vocal communication system of the gray tree frog, *Hyla versicolor*. Science 199:992–994.
Gerhardt HC (1991) Female mate choice in tree frogs: Static and dynamic acoustic criteria. Anim Behav 42:615–635.
Gerhardt HC (1994) The evolution of vocalization in frogs and toads. Annu Rev Ecol Syst 25:293–324.
Grafen A (1990) Sexual selection unhandicapped by the Fisher process. J Theor Biol 144:473–516.
Greenacre ML, Ritchie MG, Byrne BC, Kyriacou C (1993) Female song preference and the *period* gene in *Drosophila*. Behav Genet 23:85–90.
Hall JC (1994) The mating of a fly. Science 264:1702–1714.
Hamilton WD, Zuk M (1982) Heritable true fitness and bright birds: A role for parasites? Science 218:384–386.
Handford P (1981) Vegetational correlates of variation in the song of *Zonotrichia capensis*. Behav Ecol Sociobiol 8:203–206.
Handford P (1988) Trill rate dialects in the rufous-collared sparrow, *Zonotrichia capensis*, in northwestern Argentina. Can J Zool 66:2658–2670.
Hartshorne C (1956) The monotony threshold in singing birds. Auk 95:758–760.
Hasselquist D, Bensch S, von Schantz T (1996) Correlation between male song repertoire, extra-pair paternity and offspring survival in the great reed warbler. Nature 381:229–232.
Hauser M (1996) The Evolution of Communication. Cambridge, MA: MIT Press.
Hedrick AV, Dill LM (1993) Mate choice by female crickets is influenced by predation risk. Anim Behav 46:193–196.
Henwood K, Fabrick A (1979) A quantitative analysis of the dawn chorus: Temporal selection for communicatory optimization. Am Nat 114:260–274.
Hödl W (1977) Call differences and calling site segregation in anuran species from central Amazonian floating meadows. Oecologia 28:351–363.
Howard DJ, Stewart H, Berlocher SH (eds) (1998) Endless Forms: Species and Speciation. Oxford: Oxford University Press.
Howard RD, Young JR (1998) Individual variation in male vocal traits and female mating preferences in *Bufo americanus*. Anim Behav 55:1165–1179.
Hoy RR, Hahn J, Paul RC (1977) Hybrid cricket auditory behavior: Evidence for genetic coupling in animal behavior. Science 195:82–84.
Huber F (1990) Nerve cells and insect behavior—Studies on crickets. Am Zool 30:609–627.
Jang Y, Greenfield MD (1996) Ultrasonic communication and sexual selection in wax moths: Female choice based on energy and asynchrony of male signals. Anim Behav 51:1095–1106.
Kavanagh MW (1987) The efficiency of sound production in two cricket species, *Gryllotalpa australis* and *Teleogryllus commodus* (Orthoptera: Grylloidea). J Exp Biol 130:107–119.

Keddy-Hector A, Wilczynski W, Ryan MJ (1992) Call patterns and basilar papilla tuning in cricket frogs. II. Intrapopulational variation and allometry. Brain Behav Evol 39:238–246.
Kilner RM, Noble DG, Davies NB (1999) Signals of need in parent-offspring. Communication and their exploitation by the common cuckoo. Nature 397:667–672.
Kime NM, Turner W, Ryan MJ (2000) The transmission of advertisement calls in Central American frogs. Behav Ecol. 11:71–83.
King JR (1972) Variation in the song of the rufous collared sparrow, *Zonotrichia capensis*, in northwestern Argentina. Z Tierpsychol 30:344–373.
Kirkpatrick M, Ryan MJ (1991) The paradox of the lek and the evolution of mating preferences. Nature 350:33–38.
Klump GM, Gerhardt HC (1987) Use of non-arbitrary acoustic criteria in mate choice by female gray tree frogs. Nature 326:286–288.
Kroodsma DE, Miller EH (eds) (1982) Acoustic Communication in Birds, Vols 1 and 2. New York: Academic Press.
Kroodsma DE, Miller EH (eds) (1996) Ecology and Evolution of Acoustic Communication in Birds. Ithaca, NY: Comstock Press.
Kyriacou CP, Hall JC (1986) Interspecific genetic control of courtship song production and reception in *Drosophila*. Science 232:494–497.
Littlejohn MJ (1965) Premating isolation in the *Hyla ewingi* complex (Anura: Hylidae). Evolution 19:234–243.
Littlejohn, MJ (2001) Patterns of differentiation in temporal properties of acoustic signals of anurans. In: Ryan MJ (ed) Anuran Communication. Washington DC: Smithsonian Institution Press, pp. 102–120.
Loftus-Hills JJ, Littlejohn MJ (1971) Pulse repetition rate as the basis for mating call discrimination by two sympatric species of *Hyla*. Copeia 1971:154–156.
Lynch M, Walsh B (1998) Genetic Analysis of Quantitative Traits. Sunderland, MA: Sinauer Associates.
MacNally RC, Young D (1981) Song energetics of the bladder cicada. J Exp Biol 90:185–196.
Margoliash D (1983) Acoustic parameters underlying the responses of song-specific neurons in the white-crowned sparrow. J Neurosci 3:1039–1057.
Marler CA, Ryan MJ (1995) Energetic constraints and steroid hormone correlates of male calling behaviour in the túngara frog. J Zool Lond 240:397–409.
Marler P (1997) Three models of song learning: Evidence from behavior. J Neurobiol 33:501–516.
Marler P (1998) Animal communication and human language. In: Jablonski NG, Aiello LC (eds) The Origin and Diversification of Language. Memoirs of the California Academy of Sciences, Number 24. San Francisco: California Academy of Sciences.
Marten K, Marler P (1977) Sound transmission and its significance for animal vocalization I. Temperate habitats. Behav Ecol Sociobiol 2:271–290.
Marten K, Quine D, Marler P (1977) Sound transmission and its significance for animal vocalization II. Tropical forest habitats. Behav Ecol Sociobiol 2:291–301.
Martin WF (1972) Evolution of vocalization in the genus *Bufo*. In: Blair WF (ed) Evolution in the Genus *Bufo*. Austin: University of Texas Press, pp. 279–309.
Mathevon N, Aubin T, Dabelsteen T (1996) Song degradation during propagation: Importance of song post for the wren *Troglodytes troglodytes*. Ethology 102:397–412.

Mayr E (1982) The Growth of Biological Thought. Cambridge, MA: Harvard University Press.
Mello CV, Vicario DS, Clayton DF (1992) Song presentation induces gene expression in the songbird forebrain. Proc Natl Acad Sci U S A 89:6818–6822.
Michelsen A (1998) Biophysics of sound localization in insects. In: Hoy RR, Popper AN, Fay RR (eds) Comparative Hearing: Insects. New York: Springer, pp. 18–62.
Michelsen A, Larsen ON (1983) Strategies for acoustic communication in complex environments. In: Huber F, Markl H (eds) Neuroethology and Behavioral Physiology. Berlin: Springer-Verlag, pp. 321–331.
Morton ES (1975) Ecological sources of selection on avian sounds. Am Nat 109:17–34.
Morton ES (1977) On the occurrence and significance of motivational structural rules in some bird and mammal sounds. Am Nat 111:855–869.
Murphy CG (1994a) Determinants of chorus tenure in the barking tree frogs (*Hyla gratiosa*). Behav Ecol Sociobiol 34:285–295.
Murphy CG (1994b) Chorus tenure of male barking tree frogs, *Hyla gratiosa*. Anim Behav 48:763–777.
Murphy CG (1999) Nightly timing of chorusing by male barking tree frogs (*Hyla gratiosa*): The influence of female arrival and energy. Copeia 1999: 333–347.
Nevo E, Capranica RR (1985) Evolutionary origin of ethological reproductive isolation in cricket frogs, *Acris*. Evol Biol 19:147–214.
Nottebohm F (1975) Continental patterns of song variability in *Zonotrichia capensis*: Some possible ecological correlates. Am Nat 109:116–140.
Owens DH, Morton ES (1998) Animal Vocal Communication: A New Approach. Cambridge, UK: Cambridge University Press.
Penna M, Solis R (1996) Frog call intensities and sound propagation in the South American temperate forest region. Behav Ecol Sociobiol 42:371–381.
Phelps SM, Ryan MJ (1998) Neural networks predict response biases in female túngara frogs. Proc R Soc Lond B Biol Sci 265:279–285.
Phelps SM, Ryan MJ (2000) History influences signal recognition: Neural network models of túngara frogs. Proc R Soc Lond B Biol Sci 267:1633–1639.
Pijl L, Dodson CH (1966) Orchid Flowers. Their Pollination and Evolution. Coral Gables, FL: University of Miami Press.
Pinker S (1994) The Language Instinct. New York: William Morrow Co.
Pomiankowski AN (1988) The evolution of female mate preferences for male genetic quality. Oxford Surv Evol Biol 5:136–184.
Prestwich KN (1988) Intra-specific variation in the energetic efficiency of sound production in crickets. Am Zool 28:103A.
Prestwich KN, Brugger KE, Topping M (1989) Energy and communication in three species of hylid frogs: Power input, power output and efficiency. J Exp Biol 144:53–80.
Prestwich KN, Walker TJ (1981) Energetics of singing in crickets: Effect of temperature in three trilling species (Orthoptera: Gryllidae). J Comp Physiol 143:199–212.
Proctor HC (1991) Courtship in the water mite *Neumania papillator*: Males capitalize on female adaptations for predation. Anim Behav 42:589–598.

Proctor HC (1992) Sensory exploitation and the evolution of male mating behaviour: A cladistic test using water mites (Acari: Parasitengona). Anim Behav 44:745–752.
Rand AS, Bridarolli ME, Dries L, Ryan MJ (1997) Light levels influence female choice in túngara frogs: Predation risk assessment? Copeia 1997:447–450.
Rand AS, Dudley R (1993) Frogs in helium: The anuran vocal sac is not a cavity resonator. Physiol Zool 66:793–806.
Reinhold K, Greenfield MD, Jang Y, Broce A (1998) Energetic cost of sexual attractiveness: Ultrasonic advertisement in wax moths. Anim Behav 55:905–913.
Richards DG, Wiley RH (1980) Reverberations and amplitude fluctuations in the propagation of sound in a forest: Implications for animal communication. Am Nat 115:381–399.
Robert D, Amoroso J, Hoy RR (1992) The evolutionary convergence of hearing in a parasitoid fly and its cricket host. Science 258:1135–1137.
Römer H (1998) The sensory ecology of acoustic communication in insects. In: Hoy RR, Popper AN, Fay RR (eds) Comparative Hearing: Insects. New York: Springer, pp. 63–96.
Ryan MJ (1985a) Energetic efficiency of vocalization by the frog *Physalaemus pustulosus*. J Exp Biol 116:47–52.
Ryan MJ (1985b) The Túngaraa Frog, A Study in Sexual Selection and Communication. Chicago: University of Chicago Press.
Ryan MJ (1988) Energy, calling, and selection. Am Zool 28:885–898.
Ryan MJ (1996) Phylogenetics and behavior: Some cautions and expectations. In: Martins E (ed) Phylogenies and the Comparative Method in Animal Behavior. Oxford: Oxford University Press, pp. 1–21.
Ryan MJ (1997) Sexual selection and mate choice. In: Krebs JR, Davies NB (eds) Behavioural Ecology, An Evolutionary Approach, 4th ed. Oxford: Blackwell, pp. 179–202.
Ryan MJ (1998) Receiver biases, sexual selection and the evolution of sex differences. Science 281:1999–2003.
Ryan MJ (1999) Sexual selection and sensory exploitation. Science 283:1083a.
Ryan MJ, Rand AS (2001) Feature weighting in signal recognition and discrimination by the túngara frog. pp. 86–101. In: Ryan MJ (ed) Anuran Communication. Smithsonian Institution Press, Washington DC.
Ryan MJ, Brenowitz EA (1985) The role of body size, phylogeny, and ambient noise in the evolution of bird song. Am Nat 126:87–100.
Ryan MJ, Drewes RC (1990) Vocal morphology of the *Physalaemus pustulosus* species group (Family Leptodactylidae): Morphological response to sexual selection for complex calls. Biol J Linn Soc 40:37–52.
Ryan MJ, Keddy-Hector A (1992) Directional patterns of female mate choice and the role of sensory biases. Am Nat 139:S4–S35.
Ryan MJ, Rand AS (1993) Species recognition and sexual selection as a unitary problem in animal communication. Evolution 47:647–657.
Ryan MJ, Rand AS (1995) Female responses to ancestral advertisement calls in the túngara frog. Science 269:390–392.
Ryan MJ, Rand AS (1999b) Phylogenetic influence on mating call preferences in female túngara frogs, *Physalaemus pustulosus*. Anim Behav 57:945–956.

Ryan MJ, Rand AS (1999) Phylogenetic inference and the evolution of communication in túngara frogs. In: Konishi M, Hauser M (eds) The Design of Animal Communication. Cambridge, MA: MIT Press, pp. 535–557.
Ryan MJ, Sullivan BK (1989) Transmission effects on temporal structure in the advertisement calls of two toads, *Bufo woodhousii* and *Bufo valliceps*. Ethology 80:182–189.
Ryan MJ, Wilczynski W (1988) Coevolution of sender and receiver: Effect on local mate preference in cricket frogs. Science 240:1786–1788.
Ryan MJ, Wilczynski W (1991) Evolution of intraspecific variation in the advertisement call of a cricket frog (*Acris crepitans*, Hylidae). Biol J Linn Soc 44:249–271.
Ryan MJ, Tuttle MD, Rand AS (1982) Sexual advertisement and bat predation in a Neotropical frog. Am Nat 119:136-139.
Ryan MJ, Tuttle MD, Barclay RMR (1983) Behavioral responses of the frog-eating bat, *Trachops cirrhosus*, to sonic frequencies. J Comp Physiol 150:413–418.
Ryan MJ, Cocroft RB, Wilczynski W (1990) The role of environmental selection in intraspecific divergence of mate recognition signals in the cricket frog, *Acris crepitans*. Evolution 44:1869–1872.
Ryan MJ, Perrill SA, Wilczynski W (1992) Auditory tuning and call frequency predict population-based mating preferences in the cricket frog, *Acris crepitans*. Am Nat 139:1370–1383.
Sanderford MV, Conner WE (1995) Acoustic courtship communication in *Syntomeida epilais* Wlk. (Lepidoptera: Arctiidae, Ctenuchinae). J Insect Behav 8:19–31.
Schön-Ybarra M (1988) Morphological adaptations for loud phonation in the vocal organ of howling monkeys. Primate Rec 22:19–24.
Searcy WA (1992) Song repertoire and mate choice in birds. Am Zool 32:71–80.
Simmons RB, Conner WE (1996) Ultrasonic signals in the defense and courtship of *Euchaetes egle* Drury and *E. bolteri* Stretch (Lepidoptera: Arctiidae). J Insect Behav 9:909–919.
Sorjonen J (1986) Factors affecting the structure of song and the singing behaviour of some northen European passerine birds. Behaviour 98:286–302.
Stevens ED, Josephson RK (1977) Metabolic rate and body temperature in singing katydids. Physiol Zool 50:31–42.
Stripling R, Volman SF, Clayton DF (1997) Response modulation in the zebra finch neostriatum: Relationship to nuclear gene regulation. J Neurosci 17: 3883–3893.
Taigen TL, Wells KD (1985) Energetics of vocalization by an anuran amphibian (*Hyla versicolor*). J Comp Physiol B 155:163–170.
Taigen TL, Wells KD, Marsh RL (1985) The enzymatic basis of high metabolic rates in calling frogs. Physiol Zool 58:719–726.
Tuttle MD, Ryan MJ (1981) Bat predation and the evolution of frog vocalizations in the Neotropics. Science 214:677–678.
Tuttle MD, Ryan MJ (1982) The roles of synchronized calling, ambient noise, and ambient light in the anti-bat-predator behavior of a tree frog. Behav Ecol Sociobiol 11:125–131.
Tuttle MD, Taft LK, Ryan MJ (1982) Evasive behaviour of a frog in response to bat predation. Anim Behav 30:393–397.
Venator K (1999) The influence of signal attenuation and degradation on behavior

and midbrain auditory thresholds in the cricket frog *Acris crepitans blanchardi*. Ph.D. Thesis, University of Texas, Austin.
Wagner WE Jr (1989) Social correlates of variation in male calling behavior in Blanchard's cricket frog, *Acris crepitans blanchardi*. Ethology 82:27–45.
Wagner WE Jr, Sullivan BK (1995) Sexual selection in the gulf coast toad *Bufo valliceps*: Female choice based on variable characters. Anim Behav 49:305–319.
Wallschager D (1980) Correlation of song frequency and body weight in passerine birds. Experientia (Basel) 36:412.
Waser PM, Brown CH (1986) Habitat acoustics and primate communication. Am J Primatol 10:135–154.
Waser PM, Waser MS (1977) Experimental studies of primate vocalization: Specializations for long-distance propagation. Z Tierpsychol 43:239–263.
Welch AM, Semlitsch RD, Gerhardt HC (1998) Call duration as an indicator of genetic quality in male gray tree frogs. Science 280:1928–1930.
Wells KD (2001) The energetics of calling in frogs. In: Ryan MJ (ed) Anuran Communication. Washington, DC: Smithsonian Institution Press, pp. 45–60.
Wells KD, Taigen TL (1986) The effect of social interactions on calling energetics in the gray tree frog (*Hyla versicolor*). Behav Ecol Sociobiol 19:9–18.
Wells KD, Taigen TL (1989) Calling energetics of a neotropical tree frog, *Hyla microcephala* Behav Ecol Sociobiol 25:13–22.
Wells KD, Taigen TL, Rusch SW, Robb CC (1995) Seasonal and nightly variation in glycogen reserves of calling gray tree frogs (*Hyla versicolor*). Herpetologica 51:359–368.
Wilczynski W, Ryan MJ (1999) Geographic variation in animal communication systems. In: Foster SA, Endler JA (eds) Geographic Variation in Behavior. Perspectives on Evolutionary Mechanisms. New York: Oxford University Press, pp. 234–261.
Wilczynski W, Keddy-Hector A, Ryan MJ (1992) Call patterns and basilar papilla tuning in cricket frogs. I. Differences among populations and between sexes. Brain Behav Evol 39:229–237.
Wiley RH (1991) Associations of song properties with habitats for territorial oscine birds of eastern North America. Am Nat 138:973–993.
Wiley RH, Richards DG (1978) Physical constraints on acoustic communication in the atmosphere: Implications for the evolution of animal vocalizations. Behav Ecol Sociobiol 3:69–94.
Wilkinson GS, Reillo PR (1994) Female choice response to artificial selection on an exaggerated male trait in a stalk-eyed fly. Proc R Soc Lond B Biol Sci 255: 1–6.
Williams GC (1966) Adaptation and Natural Selection. Princeton, NJ: Princeton University Press.
Wood A (1962) The Physics of Music. London: Methuen.
Zahavi A, Zahavi A (1997) The Handicap Principle. Oxford: Oxford University Press.
Zakon HH, Wilczynski W (1988) The physiology of the anuran eighth nerve. In: Fritzsch B, Ryan M, Wilczynski W, Walkowiak W, Hetherington T (eds) The Evolution of the Amphibian Auditory System. New York: John Wiley and Sons Inc., pp. 125–155.

Zelick R, Mann DA, Popper AN (1998) Acoustic communication in fishes and frogs. In: Fay RR, Popper AN (eds) Comparative Hearing: Fish and Amphibians. New York: Springer, pp. 363–412.

Zentner MR, Kagan J (1996) Perception of music by infants. Nature 383:29.

Zimmerman BL (1983) A comparison of structural features of calls of open and forest habitat frog species in the central Amazon. Herpetologica 39:235–245.

Zuk M, Kolluru GR (1998) Exploitation of sexual signals by predators and parasitoids. Q Rev Biol 73:415–438.

Zweifel RG (1968) Effects of temperature, body size, and hybridization of mating calls of toads, *Bufo a. americanus* and *Bufo woodhousi fowleri*. Copeia 1968:269–285.

6 Hormonal Mechanisms in Acoustic Communication

Ayako Yamaguchi and Darcy B. Kelley

1. Introduction

Hormones have profound effects on behavior, bridging the external and internal environments. Certain external events, such as changes in photoperiod or the social environment, are perceived by an animal and are transduced into adaptive behavioral changes via effects on hormone secretion. Hormones modulate behavior by altering cellular physiology in a global and coordinated fashion. In this chapter, we will review how hormones modulate behaviors involved in acoustic communication.

Social communication is integral to animal survival and reproduction. In many species, communication takes the form of vocal production and acoustic perception. In acoustic communication, endocrine state is an important determinant of what signal is transmitted and how it is perceived. Gonadal steroids, for example, have profound effects both on male courtship vocalizations and on the responses of females to male calls. Our aim in this chapter is to examine the physiological mechanisms whereby hormones affect acoustic signaling and perception. Most of the information available concentrates on a few species: birds, frogs, rodents, and fish. Although it is phylogenetically narrow in scope, we are beginning to form an in-depth picture of some aspects of endocrine modulation, even at the molecular level.

Hormonal influences on acoustic communication can be studied at a variety of levels. The first and critical step is to identify which, if any, hormones are involved in vocal production and acoustic perception. Establishing correlations between signal production/perception and plasma levels of hormones, and demonstration of the causal relation between a particular hormone and a communicative behavior, is required before the physiological substrates can be further studied. The next step is to identify the neural and somatic substrates for the production and perception of acoustic signals. We can then determine whether the hormone exerts its effects directly by binding to the cells that mediate the behavior and, if so, how. Current technology permits us to examine a large number of dependent variables (i.e., molecular, cellular, or systemic properties) that can change in response to changes in hormone levels. Determining which of these physiological changes is critically responsible for behavioral change is a steep challenge. Problems include identification of all of the neural and somatic cells involved in vocal production and perception and the difficulty of relating hormonally induced changes at one level of biological organization to another, higher level of organization (e.g., cellular to systems, systems to behavioral levels). Some vocal communication systems, those of frogs for example, appear less complex in this regard, especially in terms of vocal production. In this chapter, we will first review the cellular changes induced by different types of hormones and then extend our study to the endocrine basis of behavior. Abbreviations used throughout the text are listed in Table 6.1.

2. The Biological Effect of Hormones

2.1. Source of Hormones

Hormones are chemical messengers released from cells in a variety of secretory glands and the central nervous system. The types of hormones discussed in this chapter are restricted to gonadal steroids and a few neurohypophysial hormones released from the pituitary gland (Fig. 6.1). Gonadal steroids such as androgens and estrogens are primarily released from the male testis and female ovary, respectively (Fig. 6.1A). The synthesis and release of these gonadal steroids are regulated by the plasma levels of luetenizing hormone (LH) released from the adenohypophysis (anterior pituitary; Fig. 6.1B). LH synthesis and secretion, in turn, are regulated by gonadotrophin-releasing hormones (GnRH) released from the hypothalamus. Thus, the neuronal activity of GnRH-synthesizing neurons in the hypothalamus determines the levels of LH released from the pituitary and eventually controls plasma levels of gonadal steroids.

Neuropeptides such as arginine vasopressin (AVP) or its nonmammalian homolog, arginine vasotocin (AVT in birds, reptiles, amphibians, and fishes), and oxytocin (OXT) are released from the neurohypophysis of the pituitary gland (posterior pituitary; Fig. 6.1C). These hormones are synthesized by neurons whose cell bodies lie in two hypothalamic nuclei—the supraoptic and the paraventricular—and released from the axon terminals of these cells in the neurohypophysis (Fig. 6.1C). The primary function of systemic AVT/AVP is to regulate osmolarity and blood pressure, and that of OXT is to regulate milk release and uterine contraction. Recent studies reveal that OXT- and AVT/AVP-synthesizing neurons project to a variety of brain regions aside from the pituitary. Neurohormones secreted directly in the brain may affect neuronal function to modify behavior in addition to their more classical, systemic effects.

TABLE 6.1. Abbreviations used throughout the text.

11-ketoT	11-ketotestosterone
AChE	acetylcholinesterase
APOA	anterior preoptic area
AVP	arginine vasopressine
AVT	arginine vasotocin
cAMP	cyclic adenosine monophosphate
CAT	choline acetyltransferase
CNS	central nervous system
DBD	DNA-binding domain
DHT	dihydrotestosterone
DIC	diacylglycerol
DLM	medial dorsolateral nucleus of the thalamus
DTAM	pretrigeminal nucleus of the dorsal tegmental area of the medulla
E_2	estradiol
GnRH	gonadotropin-releasing hormone
G-protein	guanine nucleotide binding proteins
HRE	hormone response element
HVc	higher vocal center
IP	inositol phospholipid
IP3	inositol triphosphate
LBD	ligand binding domain
LH	leutenizing hormone
LM	myosin heavy-chain isoform
MAN	magnocellular nucleus
MPOA	medial preoptic area
n.IX-X	laryngeal motor nucleus
nXIIts	tracheosyringeal portion of hypoglossal nerve nucleus
OXT	oxytocin
P	progesterone
PG	prostaglandin
PGE_2	prostaglandin E_2
PIP2	phosphatidyl inositol phosphate
PKC	protein kinase C
PLC	phospholipase C
POA	preoptic area
RA	robustus archistriatalis
SDA	sexually dimorphic area
SDApc	sexually dimorphic area pars compacta
T	testosterone
TS	torus semicircularis
VH	ventral hypothalamus

2.2. *Cellular Response to Hormones*

Secreted hormones can be transported by the bloodstream throughout the body and modify the biological function of target cells distant from the site of secretion. Target cells express receptors that have a selective affinity for a particular hormone. The receptor–hormone complex in target cells triggers a cascade of intracellular biochemical reactions.

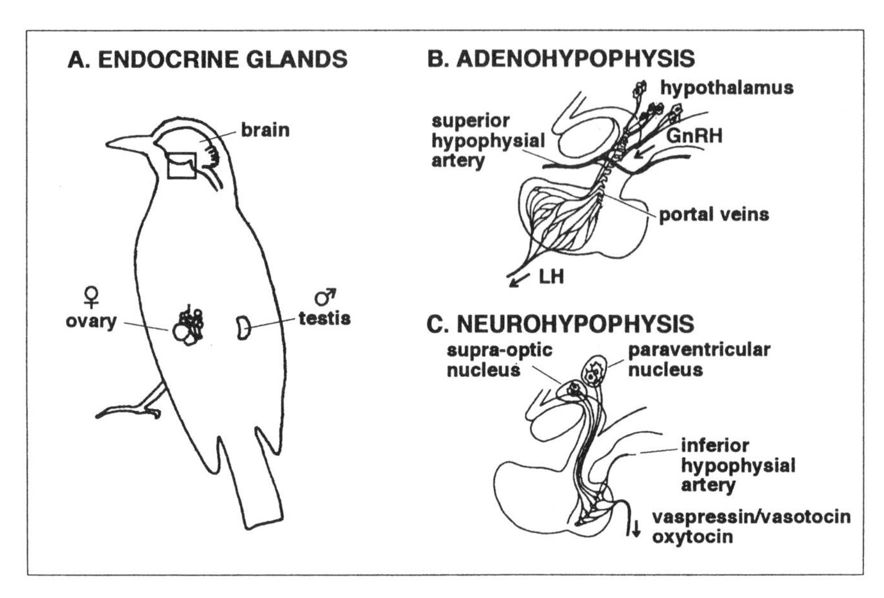

FIGURE 6.1. Sources of hormones. (**A**) Major endocrine glands that are relevant to acoustic communication, the brain, and the sex-specific gonad (testis for males and ovary for females) are depicted in a bird. The pituitary gland and the hypothalamus are enclosed in a square and are enlarged in B and C. (**B**) Adenohypophysis, the anterior pituitary gland. Gonadotrophin-releasing hormone (GnRH) released from hypothalamic neurons is transported to the adenohypophysis via the portal vein and capillaries of the superior hypophysial artery and stimulates the production of leutenizing hormone (LH), which in turn circulates through the rest of the body. (**C**) Neurohypophysis, the posterior pituitary gland. Neurons in the supraoptic and paraventricular nucleus send projections to the neurohypophysis, where vasopressin/vasotocin and oxytocin are leased. These hormones are carried to the rest of the body via the inferior hypophysial artery. (Modified from Guillemin and Burgus 1972 by kind permission of Carol Donner. © 1972 Carol Donner.)

Hormones involved in acoustic communication include steroids and peptides. Steroid hormones, such as androgens and estrogens released from the gonads, are small lipophilic molecules that readily diffuse through cell membranes. In contrast, peptide hormones, such as AVP released from hypothalamic neurons, are hydrophilic molecules that cannot readily pass through cell membranes. Consequently, receptors for these two types of hormones are localized differently: steroid receptors are found in the cytoplasm and/or nucleus of a cell, whereas peptide receptors are in the cell membrane with an extracellular hormone-binding domain.

2.2.1. Steroid Hormones

The primary action of steroid hormones is to modify gene expression in target cells. Steroid receptors are proteins localized in the cytoplasm/

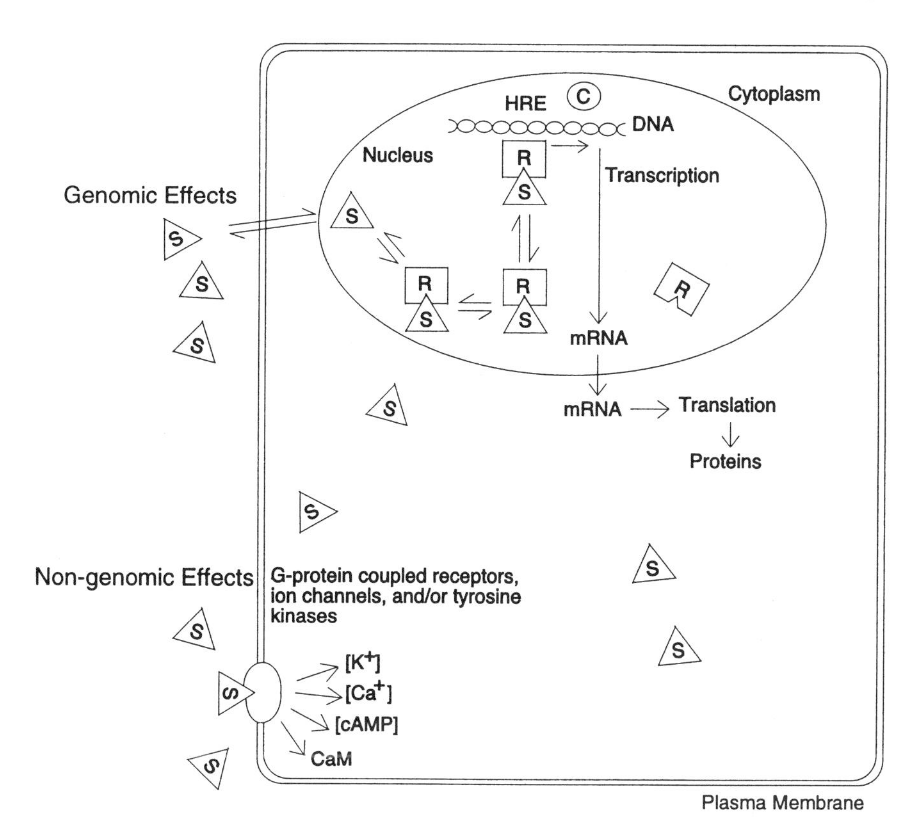

FIGURE 6.2. Genomic and nongenomic effects of steroid hormones. Genomic effects: Steroid hormones (S) diffuse into the nucleus of a target cell, where they bind to their receptors (R). The receptor–steroid complex, in the presence of a coactivator/repressor (C), acts on the hormone response element (HRE) of DNA to induce gene transcription. Nongenomic effects: Circulating steroid hormones bind to their receptors located on the cytoplasmic membrane to induce changes in the concentration of ions and second messengers within the cell. (Modified from Endocrinology 4/e by Hadley, © 1996. Reprinted by permission of Pearson Education, Inc., Upper Saddle River, NJ.)

nucleus of target cells (Fig. 6.2). Receptors contain a domain that binds to the hormone (ligand-binding domain, or LBD) and a domain that binds to DNA (DNA-binding domain, or DBD). Ligand-activated steroid receptors interact with DNA sequences called hormone response elements (HRE) located in the regulatory regions of target genes and thus modify gene transcription (Fig. 6.2). For gene transcription to be modified by a steroid–receptor complex, additional proteins, called steroid co-activators or co-repressors, must also be present (reviewed by Shibata et al. 1997). Without the presence of co-activators/co-repressors, steroids cannot modify gene expression in target cells even if they bind to receptors. Protein-synthesis inhibitors such as anisomycin and actinomycin-D (Rainbow et al. 1982) can block this transcriptional effect, the so-called genomic effect of steroids. One characteristic of the genomic effect of steroids is that its time frame of action is typically slow, occurring over hours or days.

Not all effects of steroids are mediated by modification of gene expression. Recently, more rapid effects of steroids have been identified in neurons and other tissues. *In vivo* and *in vitro* application of steroids induces rapid electrophysiological changes in steroid-sensitive neurons, including hyperpolarization of the resting membrane potential and decreased firing rate, with a latency of milliseconds to minutes (Pfaff et al. 1971; Hua and Chen 1989; Orchinik et al. 1991; reviewed by Wong et al. 1996). Because these effects are very rapid, can be observed using cytosol-free membrane patches, and cannot be blocked by protein-synthesis inhibitors, it is suspected that steroids can modulate the electrophysiological properties of the target cell nongenomically by acting directly on the membrane (reviewed by Moore and Evans 1999). Accumulating evidence suggests that circulating steroids can bind to proteinaceous receptors located on the extracellular domain of the plasma membrane in target cells and elicit immediate changes in the opening properties of ion channels (Zheng and Ramirez 1994; Fig. 6.2). In some extreme cases, the ion channel itself acts as a receptor for steroids. For example, it was recently demonstrated that estradiol binds directly to the calcium-activated K^+ channel and increases its conductance to rapidly hyperpolarize the cell (Valverde et al. 1999). In conclusion, steroids can influence the biological function of target cells both by acting on an intracellular receptor to change gene expression and by acting on a membrane receptor to alter the electrophysiological properties of a cell. The genomic actions of steroids have relatively slow latencies and long-lasting effects, whereas nongenomic actions have short latencies and effects that wear off more rapidly.

2.2.2. Peptide Hormones

Circulating peptide hormones typically modulate the function of target cells by changing membrane permeability to ions, activity of enzymes, rate of protein synthesis, and/or the cytoplasmic concentration of calcium ions. Most peptide-induced changes that are not mediated by protein synthesis are of short latency (e.g., within seconds to 15 minutes for changes in membrane permeability) and are short-lasting compared with genomic changes caused by steroid hormones (e.g., days to weeks). The effects of peptide hormones are mostly mediated by guanine nucleotide binding proteins (G-proteins) and second-messenger systems that subsequently activate various proteins via phosphorylation (Fig. 6.3). In most target cells for peptide hormones, the receptors are coupled to a G-protein that transduces the extracellular signal (i.e., binding of peptide hormones to the extracellular domain of the receptor) into an intracellular event (i.e., increases in levels of second

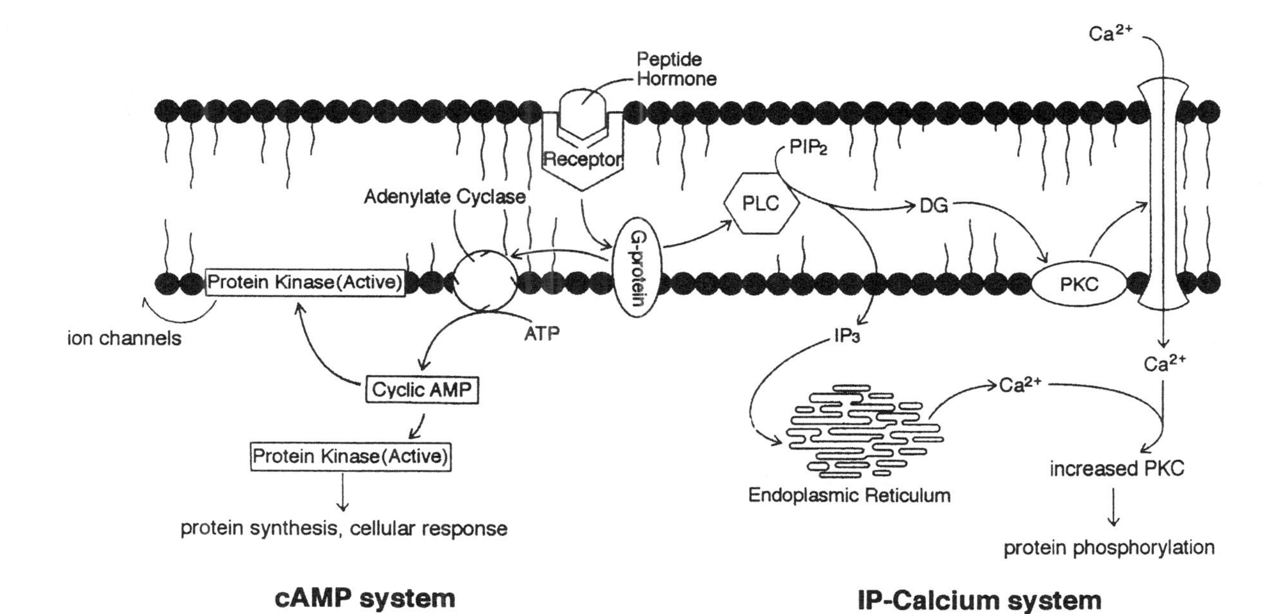

FIGURE 6.3. Intracellular response induced by peptide hormones. Two second-messenger systems, the cyclic adenosine monophosphate (cAMP) and inositol phospholipid (IP)-calcium systems, are involved. cAMP system: when a peptide hormone binds to the membrane receptor, guanine nucleotide binding proteins (G-proteins) are activated to increase the levels of cAMP via adenylate cyclase. Increased levels of cAMP induce protein kinases, which trigger a variety of intracellular reactions that modify the function of the cell. IP-calcium system: peptide hormone–receptor complex activates G-protein to increase the intracellular concentration of calcium ions (Ca^{2+}) via the action of phospholipase C (PLC). (Modified from Endocrinology 4/e by Hadley, © 1996. Reprinted by permission of Pearson Education, Inc., Upper Saddle River, NJ.)

messengers). The exact cascade of biochemical reactions that follows G-protein activation depends on the type of second-messenger system activated. One of the most ubiquitous second messengers, cyclic adenosine monophosphate (cAMP), is produced by the enzyme adenylate cyclase (Fig. 6.3). Activated G-protein regulates adenylate cyclase to increase the levels of cAMP. cAMP activates cellular enzymes called protein kinases that are localized both in the membrane and in the cytoplasm. Activated protein kinases in the membrane phosphorylate structural proteins embedded in the membrane and modify the ultimate physiological function of the cell by changing the permeability of membrane ion channels. Protein kinases in the cytoplasm, when in an activated state, bind to the response elements of some genes and alter protein synthesis. In addition to cAMP, other second-messenger systems, such as the inositol phospholipid (IP)-calcium system, are also activated by G-proteins and change calcium ion concentrations in the cytoplasm. In the IP-calcium system, an active G-protein activates enzyme phospholipase C (PLC) in the cell membrane. PLC has important effects on intracellular concentrations of Ca^{2+}, which plays an important role in determining the efficacy of neurotransmission in neurons. PLC first converts phosphatidyl inositol phosphate (PIP_2) into two components, diacylglycerol (DAG) in the membrane and inositol triphosphate (IP_3) released into the cytoplasm. Cytoplasmic IP_3 modifies the intracellular concentration of Ca^{2+} by releasing intracellular stores of calcium, and DAG brings in extracellular Ca^{2+} by changing the permeability of Ca^{2+} channels via protein kinase C (PKC). Free intracellular Ca^{2+}, in turn, increases soluble PKC to phosphorylate additional proteins. Peptide hormones therefore can trigger diverse cascades of intracellular events by utilizing different kinds of second-messenger systems.

2.3. *Systemic and Organismal Effects of Hormones*

The primary (*intra*cellular) biochemical responses induced by steroid and peptide hormones in target cells described above can lead to secondary and tertiary biochemical responses in those cells and ultimately result in more global changes at systemic and organismal levels. Hormone-induced protein synthesis may lead to modifications in enzymatic activity, rates at which other proteins are synthesized, cell morphology, and overall number of hormone receptors and ion channels. Hormone-induced changes in membrane permeability in neurons or neurosecretory cells can modify the release of neurotransmitters, neuromodulators, and/or hormones. These secondary changes influence the biological function of nontarget cells via *inter*cellular communication. As we will see later, the steepest challenge the field of behavioral endocrinology faces is interpreting hormone-induced secondary and tertiary physiological and structural changes in terms of behavior.

2.4. *Time Frame of Hormonal Action: Organizational and Activational Effects*

The impact of hormones on a biological system depends on the timing of exposure. Within a defined period during development, the critical or sensitive period, hormones can exert dramatic and irreversible (organizational) effects by regulating the growth of the CNS and somatic tissues and can permanently modify the function of the system. Hormones released during adulthood usually induce transient changes on a much smaller scale (activational) that are reversible (Phoenix et al. 1959). Although the distinction between organizational and activational effects is not always clear-cut (Arnold and Breedlove 1985), the two categories are particularly useful when describing the hormonal regulation of sexual differentiation of the nervous system and the behavior of animals. For example, steroid hormones play a critical organizational role in differentiating the physiology and morphology of male and female vertebrates during development and later play an activational role in eliciting male- and female-typical patterns of reproductive behavior in adulthood. In this chapter, major emphasis will be placed on activational effects of hormones reflecting the volume of research in the field. Organizational effects of hormones will be discussed in the section on sexually differentiated vocal production.

2.5. *Physiological Substrates for Acoustic Communication Are Targets for Hormones*

Acoustic communication systems present useful models for investigating the action of hormones on behavior. In these systems, the key elements include the sound-production apparatus used by the communicator or sender and the auditory apparatus of the receiver. Sound production is driven by central nervous system elements that include motor neurons and their afferents. Sound reception includes cells of the inner ear and CNS nuclei involved in decoding acoustic information. Many of these elements have been shown to express receptors for hormones; examples include forebrain vocal centers in songbirds, the auditory midbrain, vocal organ, and motor neurons of frogs. Thus, vocal neuroeffectors in both the CNS and the periphery can be direct targets of circulating hormones.

3. Activational Effects of Hormones on Acoustic Signal Production

Animals typically vocalize when they are in a particular physiological state. For example, sexually active males produce courtship vocalizations to attract females for successful reproduction. Thus, the internal endocrine environment activates vocalizations appropriate for a particular social context (see Crews and Moore 1985 for a variety of hormone-behavior relations). In this section, we will review endocrine control of four major types of vocalizations (courtship, unreceptive, aggressive, and alarm) produced by four focus taxa: birds, frogs, rodents, and fish. A review of hormone-induced laryngeal modifications in primates can be found in Chapter 3 by Fitch and Hauser.

3.1. Courtship Vocalizations

3.1.1. Birds

3.1.1.1. Hormone–Behavior Relation

In most bird species in the temperate zone, courtship vocalizations are given predominantly by males. Male songbirds (order Passeriformes) sing to attract females (Eriksson and Wallin 1986), male budgerigars, *Melopsittacus undulatus*, warble during courtship (Brockway 1964), male ring doves, *Streptopelia risoria*, give nest-coo vocalizations in the presence of sexually receptive females (Lehrman 1965), and male Japanese quail, *Coturnix coturnix japonica*, crow when they are visually isolated from their mates after pair formation (Potash 1975). Females of these species, in contrast, either vocalize much less frequently than males do, or give no courtship vocalization at all. Instead, these females largely play the role of signal receiver, at least in the acoustic domain.

What hormone(s) activate these courtship vocalizations? A classical approach is to establish reliable correlations between plasma levels of endogenous hormones and vocal production. The causal relation between the hormone in question and the behavior is next examined by experimental manipulation of the hormones (endocrinectomy and hormone replacement) while the behavior is observed. In most species in the temperate zone, seasonal testicular growth elevates circulating levels of endogenous testosterone (T) during the breeding season. Levels of plasma T correlate very well with vocal production both in captivity and in the wild (Marler et al. 1987, 1988; Nottebohm et al. 1987; Morton et al. 1990; Johnsen 1998). Administering exogenous T to males with low endogenous levels of T, or to females, can induce courtship vocalization (Kern and King 1972; Nottebohm 1980; Baptista 1987; Nowicki and Ball 1989; Nespor et al. 1996; Romero et al. 1998). Furthermore, in many species, castrating males often abolishes courtship vocalization, and T replacement reinstates the behavior (Pröve 1974; Arnold 1975; Adkins-Regan 1981; Harding et al. 1983, 1988). Thus, courtship vocalizations in male birds are typically activated by elevated levels of T.

In some cases, circulating T is metabolized into different active forms (Fig. 6.4), which then activate male vocalizations. T can be metabolized into 5-α-dihydrotestosterone (DHT) by the enzyme 5-α-reductase, or can be aromatized into estradiol (E_2) by aromatases (Fig. 6.4). Androgen reduc-

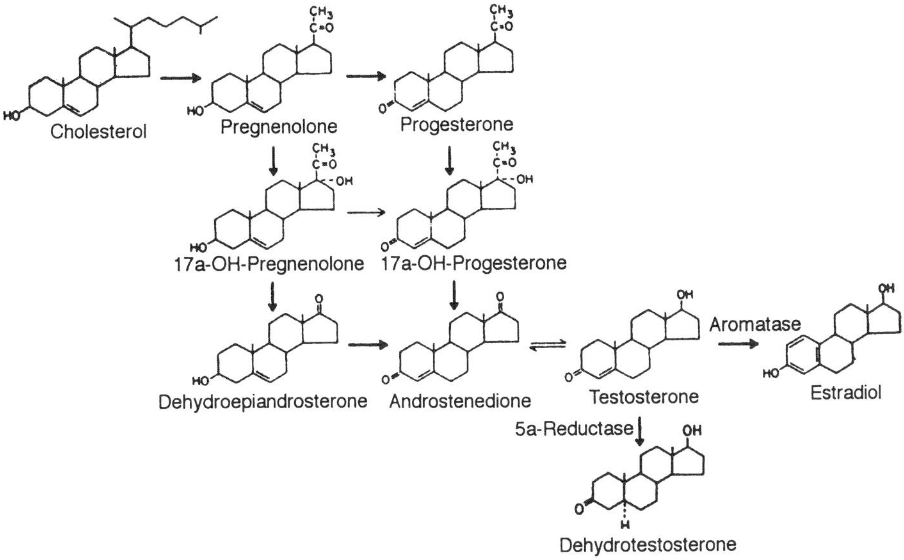

FIGURE 6.4. Biosynthesis and conversion of gonadal steroids. Gonadal steroids are synthesized from cholesterol in biological systems. (Modified from Endocrinology 4/e by Hadley, © 1996. Reprinted by permission of Pearson Education, Inc., Upper Saddle River, NJ.)

tion and aromatization are nonreversible. In Japanese quail, 5-α-DHT is more potent than T in activating crowing in castrated males, whereas other testosterone metabolites, 5-β-DHT or estradiol (E_2), have no effect on vocalizations (Adkins 1977; Adkins and Pniewski 1978; Balthazart et al. 1984). Because 5-α-reductase is present in the brain of male quail (Balthazart 1991), it is possible that the reduction of circulating T to 5-α-DHT in the brain leads to activation of crowing.

What hormones regulate the production of courtship vocalizations in female birds? Although courtship vocalizations are typically reserved for males in many species of birds, females of some species are known to emit courtship vocalizations. Female ring doves, for example, emit nest-coo vocalizations during courtship. In these females, nest-cooing is regulated by estrogen, a class of ovarian steroids that includes E_2, estrone, and estriol; ovariectomy abolishes the nest-coos, and estrogen implants reinstate their production (Cheng 1973a, 1973b). Thus, in both males and females, gonadal steroids that are necessary for egg and sperm production also play activational roles in inducing courtship vocalizations.

Although the role of gonadal steroids in activating courtship vocalization is well-established, the time course and the magnitude of the hormonal induction of vocalizations vary greatly from species to species. For example, in the tropical bush shrike, *Laniarius funebris*, a species in which the male and female within a pair engage in antiphonal duetting throughout the reproductive period, the correlation between circulating levels of gonadal steroids and singing is poor in both sexes (Schwabl and Sonnenschein 1992). In this species, behavioral interactions between mates, rather than the endocrine state of the individual, may play a larger role in activating singing behavior (Schwabl and Sonnenschein 1992). In extreme cases, courtship singing may be completely independent of T. In male European robins, blocking the surge of T during the spring by implanting flutamide, an androgen antagonist, did not influence the song production (Schwabl and Kriner 1991). The results suggest that T and its androgenic metabolites do not activate courtship vocalizations in this species, although we cannot rule out the possibility that T may exert its effect via aromatization into E_2. In conclusion, the gonadal steroids of males and females activate courtship vocalizations in both sexes although the exact steroid metabolites involved, and their potency with respect to vocal activation, differ across species.

3.1.1.2. Hormonal Modification of Vocal Organs

Circulating gonadal steroids modify the peripheral organ involved in vocal production. In gray partridges, *Perdix perdix*, administering T to both sexes results in thickening of the external membrane of their vocal organ, the syrinx, which modifies the acoustic properties of the call (Beani et al. 1995). The syrinx of adult songbirds is sexually distinct in weight and in the activity of two cholinergic enzymes required for synaptic transmission, choline acetyltransferase (CAT) and acetylcholinesterase (AChE). These anatomical and molecular differences are regulated by circulating T. Castrating adult male zebra finches and canaries decreases syringeal weights and activity of CAT and AChE, whereas administering T or other androgenic metabolites for 1–4 weeks restores weight and activity of AChE but not of CAT (Luine et al. 1980, 1983). Ovariectomizing adult female zebra finches has no effect on the syrinx, but administering T to ovariectomized females for a month masculinizes the syrinx by increasing syringeal weight and activity of AChE (Luine et al. 1980, 1983). In addition, the number of acetylcholine receptors in syringeal muscle positively correlates with circulating levels of T, indicating that neuromuscular transmission at syringeal synapses is enhanced by T (Bleisch et al. 1984). Thus, the anatomical structure and physiological function of the vocal organ can be readily modified by changes in the circulating levels of androgen in adulthood. Whether these changes in the periphery are necessary for song production in songbirds is not known.

3.1.1.3. Hormonal Modification of the CNS Associated with Activation of Courtship Vocalizations in Birds

Hormones that activate courtship vocalizations are shown to bind to the neurons in the vocal pathway in the CNS and modify their function. The specific brain regions that mediate vocal production have been identified particularly well in songbirds (Fig. 6.5). The song-control system consists of

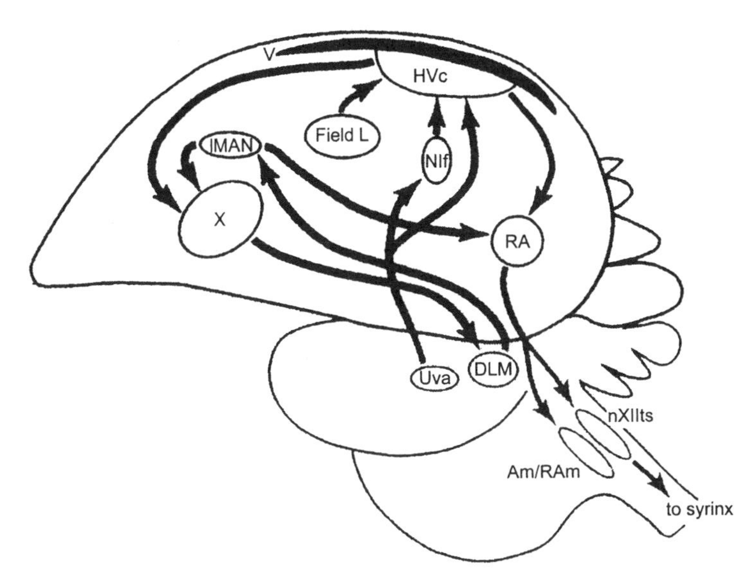

FIGURE 6.5. Song-control nuclei of songbirds. Auditory input to the vocal motor pathways comes into HVc from field L, an avian equivalent of the primary auditory cortex. HVc = higher vocal center, DLM = the medial portion of the dorsolateral nucleus of the anterior thalamus, lMAN = lateral magnocellular nucleus of the anterior neostriatum, NIf = nucleus interfacialis, nXIIts = tracheosyringeal portion of the hypoglossal nerve nucleus, RA = robust nucleus of the archistriatum, Uva = nucleus uvaeformis, X = area X. (From Brenowitz et al. 1997. Copyright © 1997 John Wiley & Sons, Inc. Reprinted by permission of Wiley-Liss, Inc., a subsidiary of John Wiley & Sons, Inc.)

a series of interconnected forebrain nuclei (cytoarchitectonically distinct brain regions) that access the vocal motor neurons projecting to the syrinx (vocal organ) via projection neurons in nucleus RA (robustus archistriatalis). Forebrain nuclei include the high vocal center (HVc), the anterior forebrain nucleus X (area X), and the magnocellular nucleus of the anterior neostriatum (MAN). Several of these brain nuclei as well as others in the forebrain, midbrain, and hindbrain, the so-called song-control nuclei, appear to be direct targets of androgens. Androgen secreted from the testis binds directly to receptors in the song-control nuclei, and these brain regions increase their sensitivity to androgens by increased receptor expression. Earlier studies using the autoradiographic method showed that androgens accumulate in many of the neurons in the song-control nuclei, including HVc, the medial dorsolateral nucleus of the thalamus (DLM), MAN, and RA (Zigmond et al. 1973, 1980; Arnold et al. 1976; Arnold and Saltiel 1979; Arnold 1980b). In addition, neurons in motor nuclei that innervate syringeal muscles accumulate androgens (Arnold et al. 1976; Arnold and Saltiel 1979; Arnold 1980b). The degree of androgen accumulation in vocal nuclei differs between males and females. In zebra finches, in which song production is monopolized by males, a larger proportion of neurons accumulate androgen in HVc and MAN of males than in females (Arnold and Saltiel 1979; Arnold 1980a, 1980b). Recent studies directly localizing intracellular androgen receptors using immunocytochemistry largely agree with previous autoradiographic studies (Balthazart et al. 1992). Moreover, these studies reveal seasonal changes in the number and density of cells expressing androgen receptors in HVc (Soma et al. 1999a). The results suggest that, in response to elevated levels of circulating androgens during the breeding season, song-control nuclei show enhanced sensitivity to androgens by expressing a larger number of receptors.

Although a wealth of studies clearly demonstrate the binding of androgens to intracellular receptors in the neurons of song-control nuclei, it is not clear whether binding actually induces novel patterns of gene expression in these neurons and, if so, what genes are affected. Whether androgens bind to plasma-membrane receptors to trigger nongenomic responses in these neurons has not yet been explored.

Although the intracellular responses that immediately follow hormone exposure have not yet been identified in neurons of song-control nuclei, longer-term androgen-induced modifications at the cellular and intercellular levels have been well-characterized. The overall volume of song-control nuclei changes in an androgen-dependent manner. In male canaries, RA is larger in the spring, when circulating levels of T are elevated and song production is at its peak; in the fall, when T levels are low, RA becomes smaller while the overall size of the brain remains constant (Nottebohm 1981). Seasonal changes in HVc, RA, area X, and the tracheosyringeal portion of the hypoglossal nerve nucleus (nXIIts) are observed in many species of songbirds, including white-crowned sparrows (*Zonotrichia leucophrys*) (Smith et al. 1995; Brenowitz et al. 1998), rufous-sided towhees (*Pipilo erythrophthalmus*) (Brenowitz et al. 1991; Smith 1996), dark-eyed juncos (*Junco hyemalis*) (Gulledge and Deviche 1997), European starlings (*Sturnus vulgaris*) (Bernard and Ball 1995), orange bishops (*Euplectes franciscanus*) (Arai et al. 1989), and red-winged blackbirds (*Agelaius phoeniceus*) (Kirn et al. 1989). Although the original studies demonstrating seasonal fluctuations in the volume of song-control nuclei were criticized on methodological grounds (Gahr 1990), subsequent studies using multiple staining methods have largely confirmed the original results (Johnson and Bottjer 1993, 1995; Bernard and Ball 1995; Smith et al. 1997a, 1997b; Soma et al. 1999a).

In addition to seasonal fluctuations in size, comparisons between male and female brains also support the androgen dependence of song-nuclei volume. In canaries and zebra finches, for example, the volume of all song-control nuclei is larger in males than in females (Nottebohm and Arnold 1976; Nottebohm 1981; Gurney 1982; Brenowitz and Arnold 1986). A similar sex difference is also apparent in the vocal motor nucleus, nucleus XII, that contains syringeal motoneurons (DeVoogd et al. 1991). Although the most dramatic sex differences in neuronal structures are due to the developmental differentiation process, sexually distinct levels of plasma androgen in adulthood also contribute to the differences in the song-nuclei volume. Implanting T into adult female canaries increases the size of HVc and RA (Nottebohm 1980; DeVoogd et al. 1985), and the same treatment in adult male zebra finches increases the volume of nXIIts and DLM (Gurney and Konishi 1980; Arnold 1980b; Gurney 1982). Thus, circulating androgen can bind to receptors in the neurons of some song-control nuclei and subsequently increase their volume in adults, although exactly how these changes occur, in terms of gene expression, is not yet clear.

How are androgen-dependent increases in the volume of song-control nuclei achieved at the cellular level? Fluctuations in the volume of vocal nuclei derive from changes in the number, density, and size of neurons. In a few species examined, the number and density of neurons in song-control nuclei increase during the spring and decrease in the fall, and the somal size and dendritic arborization of each neuron also increase during the spring in adult males, suggesting that these parameters are controlled by circulating androgen (DeVoogd et al. 1985; Clower et al. 1989; Brenowitz et al. 1991; Hill and DeVoogd 1991; Smith et al. 1995). In canaries, the cell bodies of male RA neurons are larger and their dendrites are more numerous and cover a larger area than those of female neurons, whereas the number of neurons does not differ between the two sexes (DeVoogd and Nottebohm 1981a). Treating gonadectomized female canaries with T results in growth of the dendritic field of RA neurons (DeVoogd and Nottebohm 1981b). Taken together, androgen-dependent increases in the volume of song-control nuclei result from the increases in the number, density, and/or the size of neurons within each nucleus.

What subcellular changes are associated with androgen binding in neurons of song-control nuclei? One clear example occurs at the synaptic level. DeVoogd (DeVoogd et al. 1985) found that injecting T into adult female canaries increases the number of synapses within RA. The synapses of T-treated females had shapes characteristic of stronger synapses, and the number of synaptic vesicles found in the synaptic terminal was dramatically increased compared with control females. This result strongly suggests that new and stronger synapses are formed in response to T in this vocal nucleus.

So far, we have seen that androgen activates courtship singing in male songbirds, binds to neurons in song-control nuclei, increases the volume of the nuclei by increasing the number, size, and density of neurons, and modifies the strength of synapses within song nuclei. These pieces of information make up an extended catalog of bivariate observations with hormone as an independent variable and song production or neuronal structure as dependent variables. However, it is difficult to identify which of the structural changes in the CNS are actually involved in vocal production, and how these changes mediate expression of behavior (Fig. 6.6). What we have to keep in mind is that not all physiological and structural changes induced by activating hormones in the CNS or the vocal organ can be directly related

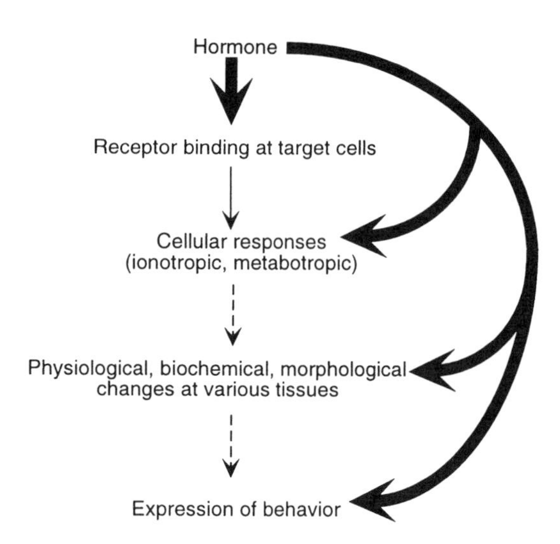

FIGURE 6.6. The challenge of understanding how hormones modify behavior. Target tissues of hormones can be identified, and causal relations between plasma levels of hormones and cellular, systemic, and behavioral changes can be established (indicated by thick arrows). However, how these cellular and systemic changes generate behaviors is not well-understood.

to vocal production. This point is well-illustrated in the relation between activation of song and volume of song-control nuclei.

As we have seen, circulating levels of T reliably correlate with male song, and androgen controls the volume of song-control nuclei (Smith et al. 1995, 1997b; Bernard et al. 1996; Brenowitz et al. 1998). The size of song nuclei also correlates well with song production. For example, in two different morphs of white-throated sparrows that differ in the rate of song production, the sizes of two song-control nuclei also covary (DeVoogd et al. 1995). Because of the remarkable correlation between size of song nuclei and singing rates across species, it is plausible to deduce a causal relation between them. However, some studies suggest that increased volume is not required for song production in all species. Castrating male zebra finches decreases the singing rate without changing the size of song-control nuclei (Arnold 1975, 1980b). Male white-crowned sparrows exposed to different photoperiods have different singing rates but maintain the same size of song-control nuclei (Baker et al. 1984; Smith et al. 1995). Adolescent male dark-eyed juncos begin to increase singing rates as they mature without increasing the volume of vocal nuclei (Gulledge and Deviche 1998). Thus, androgen-dependent size modulation of vocal control nuclei is not a general requirement for activation of song despite the reliable correlation observed in many species.

The difficulty of relating physiological and structural changes in the CNS to behavior arises because our understanding of how the brain produces behavior is still rudimentary. A direct approach is to block the change in a single parameter while holding all other parameters constant and observe concomitant changes in a behavior. However, this approach is not available for many neural characteristics, such as anatomical parameters (e.g., one cannot selectively block the changes in the number of synapses without changing the dendritic arborization). In these cases, comparison across species as well as sexes and individuals will be a fruitful approach to identify which of the neural characters holds the key to the vocal activation (Brenowitz 1997). Alternatively, one could take a bottom-up approach and focus on neurons with an identified function, such as motor neurons. By identifying molecular and cellular changes induced by activating hormones in these lower-order neurons in the vocal pathway that directly control muscles, rather than higher-order cells such as telencephalic neurons, we can obtain a better handle on how particular changes translate to behavior.

3.1.2. Frogs

Many species of frogs vocalize in the context of courtship. As in birds, males typically emit advertisement calls to attract females as well as to defend their territories during the breeding season. Advertisement calls of males are largely dependent on T and/or its metabolites. Circulating levels of T fluctuate seasonally in frogs in the temperate zone (Licht et al. 1983; Pierantoni et al. 1984). In these species, calling rates correlate well with seasonal elevations of plasma androgen (Townsent and Moger 1987; Solis and Penna 1997). Castrating males abolishes calling behavior, whereas administering T or DHT reinstates calling (Schmidt 1966; Palka and Gorbrnan 1973; Kelley and Pfaff 1976; Wetzel and Kelley 1983). Moreover, exogenous testosterone induces male-like courtship vocalizations in castrated female African clawed frogs, *Xenopus laevis* (Hannigan and Kelley 1986). These results suggest that androgen is a major hormone activating courtship vocalizations in male frogs.

In contrast to these results, however, studies in some species have shown limited potency of androgen in activating advertisement calling in males. For example, circulating levels of T show poor correlation with calling behavior in male bullfrogs, *Rana catesbeiana* (Mendonça et al. 1985), and systemic administration of T fails to activate calling behavior in both castrated male leopard frogs, *Rana pipiens*, and Mexican leaf frogs, *Pachymedusa dacnicolor* (Schmidt 1966; Palka and Gorbman 1973; Wada and Gorbrnan 1977a, 1977b; Rastogi et al. 1986). These observations led Moore (1978) to hypothesize that testicular androgens are necessary, but not sufficient, for the activation of some sexual behaviors in amphibians; some nontesticular hormones, in addition to T, are required.

Arginine vasotocin (AVT), a nonmammalian homolog of arginine vasopressin (AVP), may be such a hormone. AVT is a neuropeptide synthesized by neurons in hypothalamic nuclei and released from the posterior pituitary. In addition to its classic role in water retention, recent studies suggest that AVT influences reproductive and aggressive behaviors. Anatomical evidence suggests that AVT-synthesizing neurons project to various regions of the brain other than the posterior pituitary to modulate neuronal functions underlying reproductive and aggressive behaviors. In rough-skinned newts, *Taricha granulosa*, the concentration of AVT fluctuates seasonally in brain regions that regulate the expression of sexual behavior (Zoeller and Moore 1986; Deviche et al. 1990). Systemic administration of AVT induces advertisement calling in males of several species (Penna et al. 1992; Boyd 1994a; Marler et al. 1995; Propper and Dixon 1997; Chu et al. 1998; Semsar et al. 1998). AVT-induced calling is inhibited by application of an AVT receptor antagonist, Manning compound (Propper and Dixon 1997), in male Great Plains toads, *Bufo cognatus*, suggesting a direct role for AVT in activating calling behavior. Interestingly, AVT enhances the rate of advertisement calling without influencing the rate of other types of vocalizations. In gray tree frogs, *Hyla versicolor*, two acoustically distinct calls, advertisement and aggressive calls, are given by males during the breeding season (Fellers 1979). Exogenous injection of AVT increases the rate of advertisement calling, whereas the rate of aggressive calling is unchanged (Semsar et al. 1998). In male cricket frogs, *Acris crepitans*, whose advertisement and aggressive calls differ in dominant frequency (Wagner 1989), AVT changes acoustic properties of calls to resemble advertisement calling rather than aggressive calling (Marler et al. 1995; Chu et al. 1998). Taken together, these results suggest that AVT selectively increases rates of courtship vocalization without enhancing overall vocal activity.

Although the robust effect of AVT in activating advertisement calling has been fully recognized, it is becoming apparent that its potency depends on circulating levels of T (Moore and Zoeller 1979; Moore 1987). For example, when castrated male green tree frogs, *Hyla cinerea*, were treated with AVT, only T-implanted animals were induced to call (Penna et al. 1992). T may maintain a fundamental sexual state in males during the breeding season, whereas AVT actually triggers vocalizations (Moore 1987; Semsar et al. 1998). The complex interplay between steroids and AVT is not yet fully understood and is a focus of current research.

In contrast to activation of vocalizations, there is also evidence that a hormone can inhibit call production. Injection of prostaglandin (PG) F2 α inhibits male American toads, *Bufo americanus*, from advertisement calling in response to playback of conspecific calls (Schmidt and Kemnitz 1989). PG is an eicosanoid hormone (synthesized from cell membrane phospholipids via arachidonic acid) released from a variety of cells, and its primary function is to control the vascular smooth muscle activity. Neural correlates of electrically induced advertisement calling in an isolated brain preparation also disappear upon bath application of PG without affecting neural correlates of pulmonary respiration (Schmidt and Kemnitz 1989). This finding suggests that PG may act centrally to selectively suppress the activation of the vocal pathway for advertisement calling, although it is not clear whether the effects of PG are physiological or pharmacological (i.e., whether endogenous, not exogenous, levels of PG inhibit calling *in vivo*).

In some frogs, females also vocalize during courtship. Examples include midwife toads of the genus *Alytes* (Marquez and Verrell 1991), *Rana blythi* (Emerson 1992), and the carpenter frog, *Rana virgatipes* (Given 1993). A particularly dramatic example of female advertisement is rapping in *Xenopus laevis* (Tobias et al. 1998). When about to oviposit, females give a rapid series of clicks that act as an acoustic aphrodisiac for males, provoking heightened vocal activity, phonotaxis, and clasping. Oviposition is stimulated by gonadotropin, which also raises circulating estrogen levels, but estrogen itself is not sufficient to induce females to rap (Wu et al. 2001). Endocrine or neural activity associated with the transit of eggs through the oviduct is a candidate for stimulation of rapping.

Where do androgens act to activate advertisement calling in males? Studies in a variety of anurans have revealed intracellular receptors for androgen in the vocal organ, the larynx, or (in terrestrial frogs) in muscles associated with expiration (Segil et al. 1987; Kelley et al. 1989; Fischer et al. 1993; Boyd et al. 1999; Emerson et al. 1999). In *Xenopus laevis*, many neurons in regions of the CNS that subserve vocal behaviors also express androgen receptors (Kelley 1980; reviewed in Kelley 1996; Fig. 6.7). Thus, both the vocal organ and CNS vocal circuitry are potential activational targets for androgens. Data from *X. laevis* suggest that the activational effects of steroids are confined to the CNS. This conclusion arises from studies of the isolated larynx, a preparation in which vocal behaviors can be evoked by nerve stimulation *in vitro* (Tobias and Kelley 1987). Stimulation of male laryngeal nerves with the pattern of the advertisement call (a pattern actually generated by the male CNS: Yamaguchi and Kelley 2000; and see below) results in production of calls both in intact males and in males castrated for

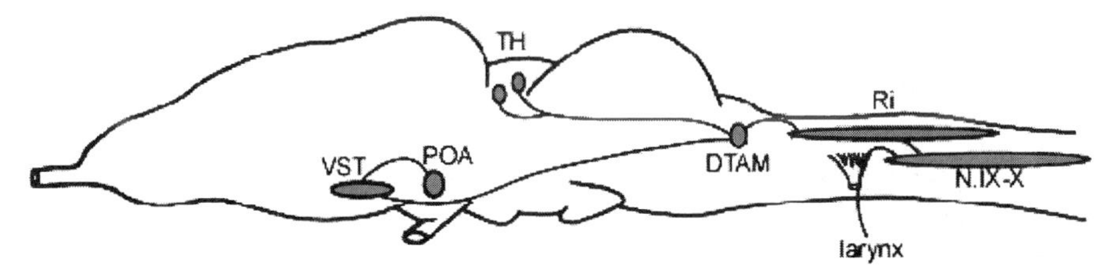

FIGURE 6.7. Schematic diagram of *Xenopus* brain nuclei involved in vocal communication and their interconnections. Gray nuclei accumulate steroids, suggesting that their function can be modified by hormones. POA = preoptic area, DTAM = pretrigeminal nucleus of the dorsal tegmental area of the medulla, n.IX–X = laryngeal motor nucleus, Ri = inferior reticular formation, T = thalamus, VST = ventral striatum. (Modified from Wetzel et al., 1985.)

over 3 years (Watson and Kelley 1992; Watson et al. 1993). Castrated males maintain a masculine laryngeal weight and complement of all fast-twitch muscle fibers. Castrated males stop calling even though their larynx is capable of producing male-typical fast trills. The activational effects of androgen must thus be on tissues other than the larynx, and the most likely candidates are the androgen-target neurons in the vocal pathway. Two of the major targets of androgens in the CNS are neurons of the laryngeal motor nucleus (n. IX–X) and its major afferent nucleus, DTAM (Kelley 1980, 1981; Pérez et al. 1996). The preoptic area is known to play an important role in male reproductive behaviors in a variety of vertebrates (Kelley and Pfaff 1978; and see below in Section 3.1.3 on rodents). However, the POA expresses estrogen receptors exclusively; because estrogen does not activate advertisement calling, the POA is unlikely to be a site for this activational effect (reviewed in Kelley 1996). Exactly what these activational effects are, at the cellular and molecular levels, is presently not known.

Where does AVT act to mediate the expression of vocalizations? The most likely target is, again, the CNS. Immunocytochemical studies reveal that AVT is localized in many brain regions that are involved in the expression of vocal behaviors, including the pretrigeminal nucleus (DTAM of *Xenopus*), the major input to vocal motor neurons (reviewed in Boyd 1997). Interestingly, the concentration and the spatial distribution of AVT within the brain are sexually distinct in bullfrogs (Boyd and Moore 1992; Boyd et al. 1992; Boyd 1994b), corresponding to sex differences in vocal behavior (i.e., advertisement calls are unique to males). Recent studies report that the firing frequency of medullary neurons can be modified by direct application of AVT into brains (Rose et al. 1995). Together, these findings suggest that AVT acts directly on the brain of male frogs by modifying the electrophysiological activity of neurons involved in the expression of vocalizations. Future studies will further identify how neuronal properties are modified by the presence of AVT, and attempts should be made to understand how the modification of these higher-order neurons orchestrates the expression of behavior.

3.1.3. Rodents

Many species of rodents produce ultrasonic vocalizations (≥20 kHz) in the context of courtship prior to copulation (e.g., Floody 1979; Holman 1980; Barfield and Thomas 1986). Following classic patterns of hormone induction of courtship vocalizations, male ultrasonic vocalizations are largely androgen-dependent; in Mongolian gerbils (*Meriones unguiculatus*) (Holman and Hutchison 1982), deer mice (*Peromyscus maniculatus bairdi*) (Pomerantz et al. 1983), golden hamsters (*Mesocricetus auratus*) (Floody 1979), Long–Evans rats (*Rattus norvegicus*) (Barfield and Thomas 1986; Matochik and Barfield 1991; Nyby et al. 1992), and house mice (*Mus musculus*) (Nunez et al. 1978; Nunez and Tan 1984; Bean et al. 1986), castrating males gradually decreases vocal production, whereas T replacement restores calling behavior. Estrous females of some rodent species also produce ultrasonic vocalizations to attract males. Female ultrasonic vocalizations are regulated by synergistic activity of two gonadal steroids, estradiol (E_2) and progesterone (P). A peak of E_2 followed by P is typical of the ovulatory menstrual cycles of female mammals. Gonadectomizing adult females decreases calling rates, and injection of E_2 followed by P reinstates calling behavior in rats (Matochik et al. 1992) and hamsters (Floody 1979; Floody and Comerci 1987). Thus, in both sexes, gonadal steroids play a potent activational role in the production of courtship vocalizations.

The activational effects of T may be mediated by its metabolites, as we have seen in other vertebrates. For example, E_2 implants in castrated males can restore ultrasonic vocalizations in mice and gerbils (Holman et al. 1991; Nyby et al. 1992). In hamsters, the aromatase inhibitor, 1,4,6-androstatriene-3,17-dione, blocks effects of T in reinstating vocal activity in castrated males (Floody and Petropoulos 1987). These results suggest that aromatization of endogenous T to E_2 is involved in the activation of vocalizations in males. In rats, there is evidence that both T and E_2 receptors are required for the activation of vocalizations. Using recently developed androgen and estrogen receptor blockers, hydroxyflutamide and RU58668, Vagell and McGinnis (1998) demonstrated that both types of intracellular receptors need to be bound to activate vocalizations in castrated males with T implants; blocking of either type of receptor prevented reinstatement of the vocal behavior. Taken together, these studies in rats suggest that ultrasonic vocalizations of males are activated when some of the circulating T binds to androgen receptors and the remaining T is aromatized to E_2 and binds to estrogen receptors. Hormonal modifications of the larynx in rodents have not received much attention.

In addition to steroid hormones, recent reports reveal a role for peptide hormones in activation of vocalizations. In hamsters, injection of oxytocin (OXT) into the hypothalamus of naturally estrous females increases the rate of ultrasonic vocalizations within 30 minutes (Floody et al. 1998). As discussed in the section on AVT-induced vocal production in frogs, OXT may interact with gonadal steroids to activate vocalizations. It will be of interest to examine whether elevated levels of estrogen and progesterone are required for OXT to activate vocal production. Involvement of OXT in vocal production is particularly interesting in light of recent revelations that AVP and OXT play important roles in controlling the social behavior of rodents (e.g., Young et al. 1998). In summary, courtship vocalizations of male and female rodents are largely activated by gonadal steroids, although the exact combination of metabolites required differs across species. In addition, neuropeptides, such as OXT, in combination with steroid hormones may play potent roles in activating vocalizations.

Although the neural substrates for vocal production encompass a number of brain regions, most research has been focused on the medial preoptic area, MPOA. An intact MPOA is necessary for the expression of sexual behavior in rodents, including ultrasonic vocalizations (reviewed by Hart and Leedy 1985).

The MPOA serves as a direct target for gonadal steroid hormones; neurons in the MPOA express receptors for gonadal steroids and their metabolizing enzymes. Intracellular androgen and estrogen receptors are concentrated in the MPOA in many species (rats: Handa et al. 1986; McGinnis and Dreifuss 1989; Roselli et al. 1989; Simerly et al. 1990; Lauber et al. 1991; gerbils: Commins and Yahr 1985; hamsters: Li et al. 1993; Woods and Newman 1993), and aromatase is also concentrated in the nucleus as measured by aromatase mRNA expression (Roselli et al. 1998). Thus, the MPOA may act as a direct link between hormones and courtship vocal behavior in rodents. Direct application of appropriate steroid hormones to the MPOA, but not to other regions of the brain, restores vocal activity in castrated males (Holman et al. 1991; Nyby et al. 1992; Matochik et al. 1994). Moreover, OXT application to the MPOA results in rapid induction of vocalizations in female hamsters (Floody et al. 1998). Thus, it is likely that activating hormones, such as steroids and OXT, directly change the function of neurons in the MPOA to induce changes in vocal production.

What are the intracellular responses that follow hormone binding? Recent studies reveal a causal relation between protein synthesis in the MPOA and vocal production. By directly applying the protein synthesis inhibitor anisomycin to the MPOA, McGinnis and Kahn (1997) demonstrated that T fails to increase vocal behavior in castrated males when protein synthesis is inhibited in the MPOA. Determining which proteins are synthesized in response to binding of T to its receptor, and how these proteins modulate neuronal function and morphology, will greatly enhance our understanding of hormonal mediation of behavioral activation.

What subsequent changes in the MPOA are observed? Similar to findings obtained in songbirds, some structural modification of the MPOA is observed in response to vocal activating hormones. For example, one nucleus included in the MPOA, the sexually dimorphic area (SDA), shows androgen-dependent changes in its volume in gerbils (Ulibarri and Yahr 1993). Interestingly, the size of SDA is lateralized in this species, and the volume and the number of cells in the left nucleus, but not the right nucleus, correlates well with rates of vocalization (Holman and Hutchison 1993; Holman and Janus 1998). However, as we discussed in Section 3.1.1.3, how these hormone-dependent anatomical modifications bring about the expression of vocal behavior is not clear.

3.1.4. Fish

Some species of teleost fishes produce courtship vocalizations underwater. Male weakfish, *Cynoscion regalis*, for example, produce a drumming sound to attract gravid females (Connaughton and Taylor 1996). The expression of drumming behavior by males is seasonal and is reliably correlated with elevated levels of both testosterone and 11-ketotestosterone (11-ketoT), a nonaromatizable testosterone found only in teleosts (Connaughton and Taylor 1994, 1995a). However, hormonal manipulation studies have not been carried out to determine whether either or both of these androgens is involved in activating the vocal behavior in this species.

A role for 11-ketoT in activating courtship vocalizations is also postulated in another sonic fish, the plainfin midshipman fish, *Porichthys notatus*. Males of this species produce long-duration advertisement calls, called hums, during the breeding season (Bass 1992; Brantley and Bass 1994). In this species, males come in two morphs, type I and type II, that show alternative reproductive strategies. About 90% of males are type I with a large body size; they build a nest and produce hums. About 10% of males are type II with a smaller body size resembling females; they neither build nests nor hum. Instead, type II males sneak into type I males' nests along with gravid females that are attracted to hums and sneak or satellite spawn to inseminate eggs as they are laid (Brantley and Bass 1994). By comparing the circulating levels of androgen in courting type I males and noncourting type II males, it was deduced that 11-ketoT plays a role in activating courtship vocalizations in this species. In type I males, 11-ketoT is the predominant plasma androgen, whereas testosterone is the predominant androgen in type II males (Brantley et al. 1993a). In addition, females that do not show courtship behavior have hormonal profiles that resemble type II males, including elevated levels of T (a feature shared with females of other teleost fishes). Thus, 11-ketoT, but not T, probably activates courtship vocalizations in midshipman fish. Similar correlations are found between elevated levels of 11-ketoT and courtship displays in six other species of fishes with courting and noncourting male morphs (reviewed by Brantley et al. 1993b). Although artificial manipulation of 11-ketoT to determine the causal relation between the hormone and vocal production has not been carried out, these reliable correlations offer support for an activational role of 11-ketoT in the expression of courtship vocalizations in fishes.

Both weakfish and plainfin midshipman fish produce vocalizations by contracting and relaxing sonic muscles attached to the lateral walls of the swim bladder (Cohen and Winn 1967; Demski et al. 1973; Bass and Marchaterre 1989). Corresponding to vocal differences between male and female weakfish, and between the two male morphs of midshipman fish, the muscle mass of vocalizing males is larger than that of nonvocalizing individuals even when scaled to body mass (Bass and Marchaterre 1989; Connaughton and Taylor 1995b). In male weakfish, sonic muscle mass changes seasonally in an androgen-dependent manner (Connaughton et al. 1997). Endogenous fluctuations in T and 11-ketoT correlate well with the hypertrophy and atrophy of the sonic muscle (Connaughton and Taylor 1994; Connaughton et al. 1997). Exogenous T implants increase muscle mass within 3 weeks in nonbreeding males, and continuous implants of T maintain hypertrophied sonic muscle well beyond the breeding season (Connaughton and Taylor 1995b). Thus, T and/or its metabolites are involved in modifying the periphery in adulthood in male weakfish. It is not clear whether androgen-induced modification of sonic muscles is required for vocal production. Hypertrophied muscle is, however, necessary to masculinize some acoustic properties of calls. Hypertrophied sonic muscles of males produce vocalizations with increased amplitude by generating a greater burst of force (Connaughton et al. 1997). Thus, the amplitude of male calls encodes the circulating levels of androgen in this species.

Although circulating androgens have a dramatic effect on sonic muscles in the male weakfish, the smaller sonic muscles of females are not due to low levels of androgens in adulthood. Exogenous androgen can induce muscle growth in adult males, but the same treatment does not masculinize the muscle in female adults. Thus, sonic muscle is probably differentiated during development in males and females such that females lose responsiveness to androgen in adulthood (Connaughton and Taylor 1995b).

At the level of the CNS, the activational role of androgen has not been explored. However, recent reports show that AVT and vasoactive intestinal polypeptide (VIP) play inhibitory roles in humming vocalization in plainfin midshipman fish. Many of the brain nuclei that are involved in vocal production are shown to contain AVT and VIP immunoreactive neurons (Foran and Bass 1998; Powers and Ingleton 1998; Goodson and Bass 2000a). Infusion of AVT into preoptic area–anterior hypothalamus and VIP into the paralemniscal midbrain inhibited electrically induced fictive vocalizations in type I (humming) midshipman fish (Goodson and Bass 2000b, 2000c), suggesting that endogenous AVT and VIP may exert opposing actions on vocalization to 11-ketoT.

In summary, androgens that activate the courtship vocalizations modify the structure of the vocal organ in weakfish. These structural changes modify the acoustic quality of calls; it is not clear whether they are required for vocal production per se. Although activational effects of androgens in the CNS are not clear, AVT and VIP act directly on vocal nuclei to inhibit vocal production.

3.2. Unreceptive Vocalizations

Unreceptive vocalizations are given by many species of frogs. When nongravid female frogs are clasped by a sexually active male, they give release calls that result in termination of the clasp attempt (Schmidt 1965; Kelley and Tobias 1999). Sexually unreceptive female *X. laevis* tick; ticking functions as a release call but also suppresses male vocalizations (Tobias et al. 1998). Gonadectomy increases ticking rates, but application of estrogen combined with progesterone had no effect on calling (Weintraub et al. 1985). Testosterone, synthesized by the ovary as a precursor to E_2 and circulating during ovulation, decreases ticking rates (Kelley 1982; Hannigan and Kelley 1986). In leopard frog females, neither gonadectomy nor estrogen with or without progesterone application had any effect on release calling (Diakow 1978). Thus, it is not clear at present what hormone regulates release calling.

In bullfrogs, males and females produce release calls; during the breeding season, females decrease their rates of release calling, whereas males increase calling rates (Boyd 1992). Although the overall circulating levels of androgen and estrogen increase in both sexes during the breeding season, males have higher levels of androgens, and females have higher levels of estrogens (Licht et al. 1983; Mendonça et al. 1985). In this species, AVT plays opposing roles in regulating release calls in males and females, depending on the concomitant levels of T and E_2. Injection of AVT in the spring increases rates of release calling in males, whereas the same treatment decreases calling rates in females (Boyd 1992). Injection of AVT in the fall has no effect on release calling in either sex (Boyd 1992). AVT-induced suppression of release calling during the breeding season was also observed in female leopard frogs (Diakow 1978; Diakow and Nemiroff 1981). These findings indicate that AVT combined with high levels of estrogen suppresses release calling, whereas AVT combined with high levels of testosterone increases release calling rates; neither gonadal steroids nor AVT alone are sufficient to modify release calling rates. Thus, steroids and AVT are likely to act synergistically in regulating release calls; steroids may act as primers for the opposing action of AVT (Emerson and Boyd 1999).

Contrary to these findings, AVT has no effect on release calling rates in *Xenopus* females. Instead, endogenous prostaglandin E_2 (PGE_2) appears to play a role in regulating release calling rates in this species. Injection of PGE_2 into intact and ovariectomized females suppressed release calling immediately (within 30 sec to 3 min). Injection of the PGE_2 synthesis inhibitors indomethacin and flurbiprofen (FBP) blocked chorionic gonadotropin-induced suppression of release calling (Weintraub et al. 1985). Thus, which hormone activates release calling may be species-specific.

3.3. Aggressive Vocalizations

The songs of male birds during the breeding season typically have two functions: courtship and aggression. These courtship/aggressive vocalizations of male songbirds are androgen-dependent, as we have seen earlier.

In some species of songbirds, however, males sing throughout the year (e.g., Blanchard and Ericksson 1949). Song during the nonbreeding season is used solely in aggressive contexts to defend winter territories (e.g., Wingfield and Hahn 1994). Females of some species, such as northern cardinals, *Cardinalis cardinalis*, produce songs with an aggressive function during the breeding season. Are these aggressive female songs and winter male songs also androgen-dependent? In mockingbirds (*Mimus polyglottos*), song sparrows (*Melospiza melodidia*), European robins (*Erithacus *rubecula*), and Gambel's white-crowned sparrow (*Zonotrichia leucophrys gambelii*), males sing in the winter when circulating levels of T are low (Logan and Wingfield 1990; Schwabl and Kriner 1991; Wingfield 1994a, 1994b). Castrating wild males does not suppress winter aggression in song sparrows (Wingfield 1994b). Songs of female cardinals are produced while plasma levels of testosterone are at undetectable levels (Yamaguchi 1996). These results indicate that aggressive songs can be independent of androgen. However, a recent study suggests that very low levels of endogenous androgen and estrogen may in fact be capable of activating aggressive song production. Inhibition of androgen and estrogen using an aromatase inhibitor and an antiandrogen during the nonbreeding season, when the T level is very low, decreased the amount of singing and overall aggression in male song sparrows (Soma et al. 1999b). Because the gonads do not synthesize steroids during nonbreeding season, nongonadal steroidogenesis and up-regulation of androgen receptors may account for activation of aggressive song production during this time (Soma et al. 1999b). In summary, despite its low circulating levels, androgen may still play a role in activating aggressive vocalizations in songbirds.

In addition to steroid hormones, AVT may also play a role in activating aggressive vocalizations in some species of birds. In the field sparrow *Spizellas pusilla*, males have two acoustically distinct song types during the breeding season: a simple, multipurpose song that is used both for courtship and territorial aggression and a complex song that is used predominantly in male–male aggression. Goodson (1998) showed that the aggressive songs, but not the multipurpose songs, of males can be induced by intraventricular infusion of AVT. This finding differs from the role of AVT in frogs, where AVT selectively activates courtship vocalizations without activating aggressive vocalizations (Semsar et al. 1998).

In previous examples reviewed in this chapter, the effects of AVT are partly dependent on gonadal steroids. However, in the case of aggressive vocalizations in songbirds, AVT might activate vocalizations independent of steroids. Female Gambel's white-crowned sparrows sing in spring and winter, and the songs are used in an aggressive context. When AVT was directly infused into the third ventricle of female sparrows, their rate of singing increased dramatically within five minutes following the treatment (Maney et al. 1997). The effect of AVT on singing behavior was uniform across seasons and was not influenced by variations in circulating levels of gonadal steroids. This observation suggests that AVT is a potent activator of songs both during breeding and nonbreeding season in females of this species.

In summary, aggressive vocalizations in birds appear to be largely dependent on androgen, although the underlying mechanism of vocal activation probably changes between breeding and nonbreeding seasons. AVT may also be involved in regulating aggressive vocalizations, although further study is necessary to generalize its role.

3.4. *Alarm Calls*

Alarm calls are ubiquitous vocalizations given in response to detecting a predator. One would expect that this emergency vocalization should not require a particular endocrine profile because it should take precedence over other needs in life. However, in domestic fowl, aerial alarm calls, emitted in response to seeing a predator overhead, increase with circulating levels of T; castrating males decreases the rate of alarm calling, whereas T implants increase the calling rate (Gyger et al. 1988). In addition, male domestic fowl give more alarm calls than females do. It is not clear from this study, however, whether T activates alarm calling per se or enhances alertness or motivation to vocalize in general.

4. Organizational Effect of Hormones on Acoustic Signal Production

The physiological substrates for vocal production are organized by hormones during development. The organizational role of hormones is most striking in sexually differentiated vocal systems. Sexual dimorphism is a characteristic feature of systems that subserve courtship and aggressive vocalizations; both the vocal organ itself and CNS pathways can differ dramatically in males and in females. Some sex differences are due to hormones that circulate in adulthood. Most, however, are due to the effects of hormones during early stages of development. In the vocal periphery, androgens stimulate cell proliferation, control cell types, and masculinize organogenesis. In the central nervous system, the best-documented effect of androgens is the prevention of ontogenetic cell death. In this section, we will review how hormones organize the central vocal pathways and the peripheral vocal organs in males and females.

4.1. *Birds*

In most songbirds in the temperate zone, males produce songs, whereas females sing much less frequently or not at all. Males and females have similar numbers of muscle fibers in the syrinx and similar numbers of neurons in the vocal motor nucleus (the hypoglossus). In vocal-control regions of the telencephalon, however, dramatic sex differences are found (Nottebohm and Arnold 1976); males not only have more cells than females but also have different sets of connections between CNS vocal nuclei. These sex differences are attributable to specific developmental programs of males and females. Song nuclei of female zebra finches can be masculinized if exogenous estrogen is given around the time of hatching; masculinized females go on to produce male-like songs when adult (Gurney 1981, 1982; Adkins-Regan and Ascenzi 1987; Simpson and Vicario 1991). Most of the early work in this field was based on the premise that the gonads of young male birds secrete T, which is then aromatized to E_2 within neurons of song nuclei to steer their developmental program in the masculine direction. The demonstration that the bird forebrain has very high levels of aromatase activity, sufficient to influence circulating estrogen levels elsewhere in the body, supported this idea (Shen et al. 1992), as did the observation that estrogen receptors are found in HVc (Walters et al. 1988). However, to date, no treatment designed to decrease levels of endogenous estrogen (including antiestrogens and castration) has been effective in demasculinizing the song systems of males (Jacobs et al. 1995; Wade and Arnold 1996; Arnold 1997). At stages during which exogenous estrogen masculinizes the song system, there are no sex differences in levels of circulating androgens (the substrate for aromatase), circulating estrogens (mostly derived from brain aromatization), or levels of aromatase itself. The song nuclei themselves do not express the aromatase gene (Metzdorf et al. 1999); of the song nuclei, only HVc expresses the estrogen receptor, and the expression levels are the same in both sexes (Johnson and Bottjer 1995). If developing zebra finches are treated *in ovo* with the aromatase inhibitor fadrozole, genetic females develop testicular tissue that secretes androgens, but song nuclei are not masculinized (Wade and Arnold 1996; Wade et al. 1996). Taken together, the results suggest that sex differences in steroid hormone secretion are not sufficent to explain masculinization of the song system. Although there is a great deal of evidence that steroid hormones play some role, the native mechanisms of sexual differentiation of the vocal system are entirely obscure (Bottjer and Arnold 1997; Kelley 1997).

Despite these caveats, the observation that exogenous steroids can masculinize song nuclei has been affirmed repeatedly and used as a basis for understanding how the number of cells, their size, and their connectivity become sexually differentiated. For example, steroids interact with neurotrophic factors in the differentiation of song nuclei. These interactions can be demonstrated when the trophic connections between song nuclei are disrupted. In young zebra finches, lesions of MAN lead to the death of neurons in RA; treatment of RA with the neurotrophin BDNF suppresses neuronal death (Johnson et al. 1997). BDNF expression is higher in developing males than in females (Akutagawa and Konishi 1998), and estradiol increases BDNF expression (Dittrich et al. 1999).

4.2. Frogs

In the African clawed frog, *Xenopus laevis*, males and females produce sexually distinct vocalizations. The calls of both sexes are made up of clicks that are repeated in distinctive, sex-specific temporal patterns (Kelley and Tobias 1999). Reflecting the vocal differences, the vocal organ shows dramatic differences between males and females. The adult male larynx has more muscle and cartilage cells than the female, and the cells also differ dramatically in function and gene expression. The sound-producing elements in males but not females are surrounded by elastic cartilage. The male larynx is made up entirely of fast-twitch muscle fibers, whereas the female larynx is mostly composed of slow-twitch fibers; all male muscle fibers express a distinct myosin heavy-chain isoform, LM, found in only a subset of fibers in females (reviewed in Kelley 1996).

How do these sexually differentiated cellular features of the vocal system come about? In *X. laevis*, removal of the testis at any point during the first six months after metamorphosis blocks addition of new muscle fibers. Castration also blocks expression of the LM gene, the transcription of which appears to require androgen secretion. Transplantation of a testis into a female at any point during her development results in a masculinized program of laryngeal development and the ability to produce male songs (Watson and Kelley 1992). Administering androgen antagonists to tadpoles blocks the rescue of laryngeal motor neurons that underlies sex differences in cell number while exogenous androgens rescue dying cells in females (Kay et al. 1999). Although testicular androgens appear necessary for masculinization of the song system, they are not sufficient. Exogenous androgens do not masculinize cell numbers in juvenile females, nor can those females sing when adult (even when given activational androgens), whereas testis implants fully masculinize the muscle cell numbers in females.

Are these cellular differences that parallel vocal differences between males and females caused by the organizational effect of hormones? The control of cell number appears to result from organizational effects of androgen because castration of juvenile males does not reverse muscle fiber addition. The ability of gonadal androgen to increase muscle fiber number depends on another hormone, thyroxine. Thyroxine, a thyroid hormone, is responsible for metamorphosis in amphibians; without thyroxine, androgen cannot induce laryngeal cell proliferation (Robertson and Kelley 1996). Premature exposure to thyroxine induces premature androgen sensitivity of the larynx (Cohen and Kelley 1996). The ability of hormones to masculinize muscle fiber extends past early juvenile stages. Even adult females can achieve male-like numbers of fibers after testis transplants. Thus, the critical period for androgen regulation of muscle fiber number is opened by exposure to thyroxine. Females maintain androgen sensitivity and can respond to testis transplants even when adult. In males, the critical period ends when the full complement of muscle fibers is achieved, perhaps due to depletion of the stem cell population (myoblasts) that generates new fibers.

In addition to the number of muscle fibers, contractile properties of muscle fibers are also controlled by a developmental program. After all muscle fibers have been generated in males, slow-twitch fibers gradually start to convert to fast-twitch fibers. If animals are castrated at any point during this process, fiber type switching is halted but can be resumed if androgen is provided (Tobias et al. 1991). Female muscle fibers can be converted to fast-twitch fibers with androgen at any point during development, although the treatment time required is considerably less in juveniles than in adults (Marin et al. 1990). As is the case for muscle fiber number, masculinization of muscle fiber type by androgen relies on another hormone, prolactin in this case. Thyroxine secretion during the metamorphic climax of tadpole life triggers synthesis and secretion of prolactin from the anterior pituitary. The androgen-sensitive fast-twitch myosin, LM, requires prolactin and then androgen for its own expression (Edwards et al. 1999). Thus, the critical period for androgen regulation of muscle fiber type is opened by the secretion of prolactin from the pituitary but never closes. The effects of androgen on both cell number and cell types are irreversible: an adult male that has been castrated for 3 years retains a full complement of fast-twitch fibers (Watson et al. 1993). That the critical period does not close, however, is supported by the observation that masculinization can be reinstated in males castrated as juveniles and in females at any stage.

To make the jump from the cellular to the behavioral level requires knowing whether a particular masculinized cellular property is required for the exhibition of masculinized vocal behaviors. In this respect, the hormonal control of cell number appears to be key. Genetic females do not produce the specific temporal pattern of male advertisement calls unless laryngeal muscle fiber number has been masculinized (Watson et al. 1993). Even if every muscle fiber in the larynx is fast-twitch, a female will not call like a male unless her muscle fiber complement is also masculinized. Why is this the case? Some clues come from recent studies of how vocal behaviors are actually generated by male and female brains.

The brains of males and females are functionally differentiated such that the motor patterns produced by the vocal pathways in the CNS are dramatically different between male and female. To examine the contribution of CNS to generating vocal patterns, we developed a preparation in which both laryngeal nerve activity and electromyograms can be recorded from awake, vocalizing frogs (Yamaguchi and Kelley 2000). Recordings reveal that the CNS of the sexes produces patterned activity that closely matches each vocalization while the larynx acts as a faithful transducer of nerve activity into sound (Fig. 6.8). Thus, the CNS is the source of sexually differentiated vocalizations in *Xenopus laevis*. Neuronal activity underlying different male call types is distinct; some calls are likely to be generated by synchronous firing of motor neuron populations either of constant size or progressively larger sizes, whereas others are generated by asynchronous activity of motor neurons, a pattern shared with vocal production in females. We suggest that these distinct neuronal activity patterns in males may be subserved by two populations of motor units in males that can be distinguished by the strength of the neuromuscular synapse.

The robust sex difference in vocal output of the CNS is most likely due to a sexually differentiated program that changes the generation of patterned output from laryngeal motor neurons. Candidate substrates for the

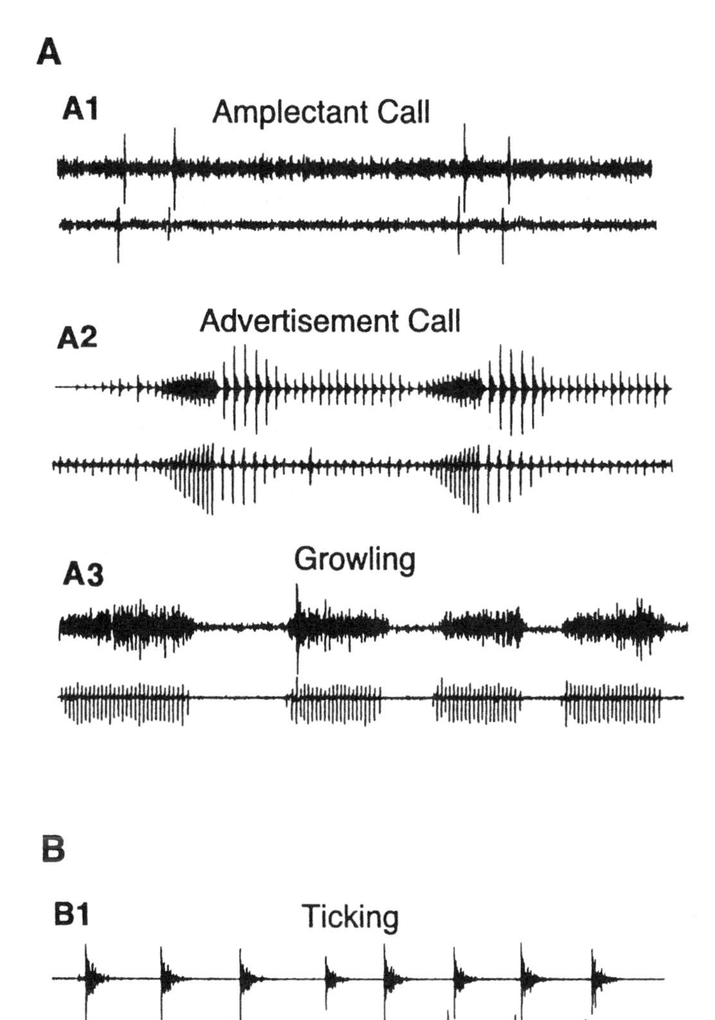

FIGURE 6.8. Extracellular recordings from laryngeal motor nerve along with sound amplitude waveforms of male and female vocalizations. The top trace of each panel is the amplitude waveform (sound amplitude over time); the bottom trace is nerve recordings (voltage over time). Each burst of activity on nerve recordings indicates the compound action potential of laryngeal motoneurons, whereas the burst of activity on sound recordings indicates a click sound. Note that each compound action potential precedes each click sound. A: Male vocalizations; A1: nerve recordings during amplectant call; A2: nerve recordings during advertisement call; A3: nerve recordings during growling; B: female vocalization; B1: nerve recordings during ticking.

behavioral effects include sex differences in number of laryngeal motor neurons, differences in number and dendritic extent of N. IX–X interneurons, and differences in synaptic connectivity; for example, the connections between DTAM and N. IX–X (reviewed in Kelley 1996; Fig. 6.7). In addition, biophysical properties of motor neurons and interneurons may differ in the sexes in ways that contribute to differences in the generation of vocal patterns.

4.3. Rodents

Some rodents, such as Mongolian gerbils, produce sexually differentiated vocalizations (Holman and Hutchison 1982). Steroid-sensitive, vocal courtship behavior is a function of a specific hypothalamic nucleus, the sexually dimorphic area pars compacta (SDApc) in the male adult gerbil. Sex-related differences in the number of neurons in this nucleus are apparent immediately after birth. The differentiation of the SDApc and subsequent vocalization result from the organizational effect of androgens. Although female Mongolian gerbils normally do not produce male-typical vocalizations, androgen treatment of a neonatal female leads to male-like vocal production. The treatment also masculinizes the structure of the brain; androgen-treated females acquire lateralized SDApc's that are larger than those of control females. Cell death and proliferation occur simultaneously in the neonatal gerbil brain. There is a lower incidence of cell death, occurring earlier in males than in females (Holman et al. 1995). The size and number of cells in the SDApc can be masculinized by androgen implantation in neonatal females. Androgens also increase laterality in the nucleus; cell number in the left SDApc is correlated with vocal behaviors (Holman and Rice 1996). Androgen thus may organize the brain of males by increasing the survival of interneurons in a laterally asymmetric hypothalamic nucleus related to courtship vocal behaviors.

4.4. Fish

The plainfin midshipman fish produces vocalizations by contracting and relaxing sonic muscles attached to the lateral walls of the swim bladder (Cohen and Winn 1967; Bass and Marchaterre 1989). Corresponding to the vocal differences between type I males and nonhumming type II males and females, the muscle mass of type I males is six times larger than that of type II males or of females when scaled to body size, and it contains 3–5 times more muscle fibers with larger diameters (Bass and Marchaterre 1989). Moreover, type I myofibrils have enlarged peripheral and central zones of sarcoplasm that are densely filled with mitochondria, a presumed adaptation for the metabolic requirement for the vocal production (Bass and Marchaterre 1989). These peripheral differences are likely to be the result of the organizational effect of androgen. When T or 11-ketoT is implanted into juvenile males (whether they will be type I or II cannot be distinguished at this stage), their sonic muscle becomes masculinized and resembles that of type I males. The treatment was ineffective in adult type II males, suggesting that the peripheral differentiation takes place during development and is irreversible in adulthood (Brantlly et al. 1993b). Thus, a morph-specific profile of circulating androgen in adulthood does not account for hypertrophy of sonic muscle in the plainfin midshipman fish. Rather, exposure to T or 11-ketoT during development is likely to organize the sonic muscle in this species.

In the juvenile plainfin midshipman fish, the differential action of aromatase may account for the differentiation of the vocal pathways in the CNS. In the vocal midbrain region, the aromatase activity of females and type II males (nonsinging morph) was much higher than that in type I males (singing morph), suggesting that aromatase may play a critical role in organizing the vocal circuitry of singing and nonsinging animals by converting circulating androgen into estrogen.

5. Hormonal Basis of Acoustical Signal Perception

In comparison to hormonal control of vocal production, the endocrine basis of acoustic perception has received less attention, perhaps because the study of perception of animals at a behavioral level is quite difficult. Acoustic perception of animals only becomes apparent when they express behavioral responses to stimuli; signals perceived without an overt behavioral response cannot be distinguished from signals that are not perceived. Thus, slight modulations in perception that might result from changes in hormonal state often are not observable.

Two techniques are commonly used to study acoustic perception: playback experiments and operant conditioning. Playbacks involve presentation of prerecorded or synthesized acoustic stimuli to animals and observation of their behavioral responses (e.g., phonotaxis, copulation postures). When animals respond differently to acoustic stimuli, one can deduce that the stimuli were perceived differently by the animal. When animals reliably show stereotyped behavioral responses to particular auditory stimuli, this technique can be a powerful tool in understanding their perception. When naturally occurring behavioral responses are difficult to elicit, operant conditioning is often used. In this experimental paradigm, the animal is first trained to respond to a sensory stimulus with a certain motor response (e.g., pecking a key in response to hearing a tone); later, the animal is presented with a test stimulus to determine whether the conditioned response will be elicited. If the animal fails to respond to a test stimulus, the training and test stimuli are regarded as having been perceived differently.

5.1. Hormonal Influences on Acoustic Perception: Behavioral Evidence

Song perception by male and female birds may be influenced by circulating hormones. Both sexes use songs to distinguish conspecifics from heterospecifics. In red-winged blackbirds, sexually receptive females typically respond to courting males with copulation solicitation displays, an avian equivalent of lordosis in female rodents, whereas males respond to songs of other males by aggressive behavior, including singing and physical attacks (Searcy and Brenowitz 1988). When male and female blackbirds were presented with conspecific male songs and imitation songs sung by the mockingbird *Mimus polyglottos* (mockingbirds mimic songs of red-winged blackbirds superbly well), female blackbirds showed a copulation solicitation response only to genuine conspecific songs, whereas males showed an aggressive response both to conspecific and imitation songs. Although the apparent sex difference in the responsiveness to conspecific and heterospecific songs may be ascribed to the differences in motivation to respond or the amount of attention paid, one explanation is that the two sexes differ in their perceptual capability to distinguish conspecific from heterospecific songs. If this is the case, steroid hormones may be involved in the perception of conspecific songs in blackbirds.

In addition to conspecific recognition, the ability to recognize individuals using songs may also show sex differences. Male songs typically show subtle intraspecific variation in acoustic properties. Female songbirds choose a male partly based on features of his courtship song (Eriksson and Wallin 1986), and males distinguish neighboring from stranger males using their songs (Falls 1982). Cynx and Nottebohm (1992a) trained male and female zebra finches to discriminate conspecific and heterospecific songs using an operant-conditioning paradigm. Males required fewer trials than females to discriminate conspecific songs, and both sexes learned to discriminate songs with fewer trials in the spring than in winter. Moreover, discrimination of various conspecific songs was enhanced when T was given to castrated males (Cynx and Nottebohm 1992b). Thus, circulating T may regulate discrimination of conspecific songs in zebra finches.

Ample evidence from human psychophysical studies supports the idea that acoustic perception is influenced by circulating hormones. For example, glucocorticoids, hormones secreted from the adrenal cortex in response to stress, modify auditory perception in humans by elevating the perceptual threshold (e.g., Henkin and Daly 1968; Fehm-Wolfdorf and Nagel 1996). Gonadal steroids may also bias acoustic perceptions. Women have greater auditory sensitivity than men in general, the auditory perception of women is slightly masculinized during the first half of the menstrual cycle, and women with twin brothers tend to have masculinized auditory perception compared with women with twin sisters (reviewed by McFadden 1998). These studies suggest that gonadal steroids could have organizational and activational effects on the auditory system. Similar effects may be present in experimental animal systems, but our behavioral techniques are not yet sensitive enough to detect them.

5.2. Electrophysiological Evidence for Hormonal Control of Acoustic Perception

Auditory pathways in the CNS are highly conserved across species. Acoustic information, conveyed via the eighth nerve, enters the medulla and is sent to the superior olive, lateral lemniscus, and midbrain (inferior colliculus in mammals and torus semicircularis in birds and frogs; see Gentner and Margoliash, Chapter 7). From the midbrain, there are projections to the thalamus and then to the forebrain. Electrophysiological recordings indicate that activity of the eighth nerve in response to sound can be sexually distinct and modulated seasonally in some species of frogs. Narins and Capranica (1976) used a combination of playback and electrophysiological methods to demonstrate sex differences in auditory perception of the Puerto Rican tree frog *Eleutherodactylus coqui*. In this frog, males and females respond to different notes of the two-note male calls (Narins and Capranica 1976). Males respond selectively to the monotonous, single-frequency "Co" note, whereas females preferentially approach the upsweep "Qui" note (Narins and Capranica 1976). This behavioral difference is accompanied by a sex difference in the tuning of primary auditory neurons at threshold. Males have a larger number of single units that respond to "Co," whereas females have a larger number of single units that respond to the "Qui" frequency (as judged by single-unit auditory tuning curves that yield "best frequencies"; Narins and Capranica 1976). Thus, males may respond to "Co" and females to "Qui" when they hear a distant calling male because of this differential acoustic sensitivity.

Seasonal fluctuations in electrophysiological properties of auditory neurons have also been described. In the fire-bellied toad (*Rana temporaria*), the grass frog (*Bombina bombinba*), and the gray tree frog (*Hyla chrysoscleis*), single-unit activity in the torus semicircularis (TS) is depressed during the winter in comparison with the summer, suggesting that seasonal variation in circulating levels of gonadal steroids may modulate activity of auditory neurons (Walkowiak 1980; Hillery 1984). Corresponding to this finding, injection of estradiol into the third ventricle of the female leopard frog produces an increase in the amplitude of auditory evoked potentials in the TS (Yovanof and Feng 1983). Although E_2 does not enhance behavioral receptiveness in females of this species (Diakow 1978), it is possible that E_2 enhances perceptual responsiveness without influencing behavioral responsiveness. Taken together, these studies suggest that the physiological functions of auditory neurons are modified by gonadal steroids and that these changes may underlie the modification of auditory perception in the context of acoustic communication.

Where do hormones exert their effects? Many of the neurons in the auditory pathway accumulate gonadal steroids and express steroid receptors in birds, frogs, and rodents (Arnold et al. 1976; Kelley et al. 1978; Kelley 1980; Simerly et al. 1990; Balthazardt et al. 1992). For example, in *Xenopus laevis*, neurons in the ventral striatum and anterior preoptic area that are considered to be auditory concentrate estrogen, whereas the TS contains receptors for both estrogens and androgens (Kelley 1980, 1981).

5.3. Beyond Acoustic Perception: Downstream Effects of Signal Perception

Subsequent to perceiving acoustic signals, animals often show a variety of physiological changes, including endocrine responses. For example, female canaries lay more eggs when exposed to complex male songs than when exposed to simpler songs (Kroodsma 1976).

Where do acoustic stimuli gain access to the neuroendocrine system that controls sebsequent steroid secretion? Anatomical projections from the TS to the ventral hypothalamus (VH) and APOA (the two anuran GnRH centers) have been described in several frogs (Neary and Wilczynski 1986; Allison and Wilczynski 1991). Auditory activity in VH and APOA varies with the circannual mating cycle (Allison 1992). This kind of anatomical pathway—from an auditory brain nucleus to one controlling GnRH release—could provide a means for acoustic stimulation to influence endocrine state. Endocrine state may then feed back on these same auditory regions.

6. Conclusion

It is clear that hormones play a critical role in triggering vocal production as well as in the development of neural and muscular substrates for vocal production. The acoustic perception of vocalizations is also likely to be influenced by hormones, although more research in this area is needed. The study of the neuroendocrine basis of acoustic communication provides an excellent opportunity to understand the effect of neuromodulation on the expression of discrete behavior. The future challenges that we face are twofold. First, extensive catalogs of pairwise parameters (hormone as independent variable and behavior and physiological phenomena as dependent variable) need to be integrated to make better sense of how hormonal changes at physiological levels lead to the expression of behavior. Comparative approaches as well as bottom-up approaches—by focusing on lower-order neurons with identified function—are expected to yield fruitful results. Second, we must recognize that the hormone–behavior relation is not unidirectional; hormones not only regulate vocal production and perception of acoustic signals, but the very act of vocalizing and/or perceiving vocalizations can modify the hormonal state. The endocrine basis of acoustic communication of animals in their natural environment—with extended interactions between individuals—is likely to be much more complex and will provide a different kind of understanding of these very powerful ways to affect behavior.

Acknowledgments. Work from the authors' laboratory was supported by a grant from the NIH (NS 23684 to DBK) and an NIH postdoctoral fellowship (NS 10881 to AY).

References

Adkins EK (1977) Effects of diverse androgens on the sexual behavior and morphology of castrated male quail. Horm Behav 8:201–207.

Adkins EK, Pniewski E (1978) Control of reproductive behavior by sex steroids in male quail. J Comp Physiol Psychol 92:1169–1178.

Adkins-Regan E (1981) Effect of sex steroids on the reproductive behavior of castrated male ring doves (*Streptopelia* sp.). Physiol Behav 26:561–566.

Adkins-Regan E, Ascenzi M (1987) Social and sexual behavior of male and female zebra finches treated with oestradiol during the nestling period. Anim Behav 35:1100–1122.

Akutagawa E, Konishi M (1998) Transient expression and transport of brain-derived neurotrophic factor in the male zebra finch's song system during vocal development. Proc Natl Acad Sci U S A 95:11429–11434.

Allison JD (1992) Acoustic modulation of neural activity in the preoptic area and ventral hypothalamus of the green treefrog (*Hyla cinerea*). J Comp Physiol A 171:387–395.

Allison JD, Wilczynski W (1991) Thalamic and midbrain auditory projections to the preoptic area and ventral hypothalamus in the green treefrog (*Hyla cinerea*). Brain Behav Evol 38:322–331.

Arai O, Taniguchi I, Saito N (1989) Correlation between the size of song control nuclei and plumage color change in orange bishop birds. Neurosci Lett 98:144–148.

Arnold AP (1975) The effects of castration and androgen replacement on song, courtship, and aggression in zebra finches (*Poephila guttata*). J Exp Zool 191: 309–326.

Arnold AP (1980a) Quantitative analysis of sex differences in hormone accumulation in the zebra finch brain: Methodological and theoretical issues. J Comp Neurol 189:421–436.

Arnold AP (1980b) Effects of androgens on volumes of sexually dimorphic brain regions in the zebra finch. Brain Res 185:441–444.

Arnold AP (1997) Sexual differentiation of the zebra finch song system: Positive evidence, negative evidence, null hypotheses, and a paradigm shift. J Neurobiol 33:572–584.

Arnold AP, Breedlove SM (1985) Organizational and activational effects of sex steroids on brain and behavior: A reanalysis. Horm Behav 19:469–498.

Arnold AP, Saltiel A (1979) Sexual difference in pattern of hormone accumulation in the brain of a songbird. Science 205:702–705.

Arnold AP, Nottebohm F, Pfaff DW (1976) Hormone concentrating cells in vocal control and other areas of the brain of the zebra finch (*Poephila guttata*). J Comp Neurol 165:487–511.

Baker M, Bottjer S, Arnold A (1984) Sexual dimorphism and lack of seasonal changes in vocal control regions of the white-crowned sparrow brain. Brain Res 295:85–89.

Balthazart J (1991) Testosterone metabolism in the avian hypothalamus. J Steroid Biochem Mol Biol 40:557–570.

Balthazart J, Schumacher M, Malacarne G (1984) Relative potencies of testosterone and 5 alpha-dihydrotestosterone on crowing and cloacal gland growth in the Japanese quail (*Coturnix coturnix japonica*). J Endocrinol 100:19–23.

Balthazart J, Foidart A, Wilson EM, Ball GF (1992) Immunocytochemical localization of androgen receptors in the male songbird and quail brain. J Comp Neurol 317:407–420.

Baptista LF (1987) Testosterone, aggression, and dominance in Gambel's white-crowned sparrows. Wilson Bull 99:86–91.

Barfield RJ, Thomas DA (1986) The role of ultrasonic vocalizations in the regulation of reproduction in rats. Ann N Y Acad Sci 474:33–43.

Bass AH (1992) Dimorphic male brains and alternative reproductive tactics in a vocalizing fish. Trends Neurosci 15:139–145.

Bass AH, Marchaterre MA (1989) Sound-generating (sonic) motor system in a teleost fish (*Porichthys notatus*): Sexual polymorphism in the ultrastructure of myofibrils. J Comp Neurol 286:141–153.

Bean NJ, Nyby J, Kerchner M, Dahinden Z (1986) Hormonal regulation of chemosignal-stimulated precopulatory behaviors in male housemice (*Mus musculus*). Horm Behav 20:390–404.

Beani L, Panzica G, Briganti F, Persichella P, Dessi-Fulgheri F (1995) Testosterone-induced changes of call structure, midbrain and syrinx anatomy in partridges. Physiol Behav 58:1149–1157.

Bernard DJ, Ball GF (1995) Two histological markers reveal a similar photoperiodic difference in the volume of the high vocal center in male European starlings. J Comp Neurol 360:726–734.

Bernard DJ, Eens M, Ball GF (1996) Age- and behavior-related variation in volumes of song control nuclei in male European starlings. J Neurobiol 30:329–339.

Blanchard BD, Erickson MM (1949) The cycle in the Gambel sparrow. Univ Calif Publ Zool 47:255–318.

Bleisch W, Luine VN, Nottebohm F (1984) Modification of synapses in androgen-sensitive muscle. I. Hormonal regulation of acetylcholine receptor number in the songbird syrinx. J Neurosci 4:786–792.

Bottjer SW, Arnold AP (1997) Developmental plasticity in neural circuits for a learned behavior. Annu Rev Neurosci 20:459–481.

Boyd SK (1992) Sexual differences in hormonal control of release calls in bullfrogs. Horm Behav 26:522–535.

Boyd SK (1994a) Arginine vasotocin facilitation of advertisement calling and call phonotaxis in bullfrogs. Horm Behav 28:232–240.

Boyd SK (1994b) Gonadal steroid modulation of vasotocin concentrations in the bullfrog brain. Neuroendocrinology 60:150–156.

Boyd SK (1997) Brain vasotocin pathways and the control of sexual behaviors in the bullfrog. Brain Res Bull 44:345–350.

Boyd SK, Moore FL (1992) Sexually dimorphic concentrations of arginine vasotocin in sensory regions of the amphibian brain. Brain Res 588:304–306.

Boyd SK, Tyler CJ, De Vries GJ (1992) Sexual dimorphism in the vasotocin system of the bullfrog (*Rana catesbeiana*). J Comp Neurol 325:313–325.
Boyd SK, Wissing KD, Heinsz JE, Prins GS (1999) Androgen receptors and sexual dimorphisms in the larynx of the bullfrog. Gen Comp Endocrinol 113:59–68.
Brantley RK, Bass AH (1994) Alternative male spawning tactics and acoustic signals in the plainfin midshipman fish *Porichtys notatus* Girard (Teleostei, Batrachoididae). Ethology 96:213–232.
Brantley RK, Wingfield JC, Bass AH (1993a) Sex steroid levels in *Porichthys notatus*, a fish with alternative reproductive tactics, and a review of the hormonal bases for male dimorphism among teleost fishes. Horm Behav 27:332–347.
Brantley RK, Marchaterre MA, Bass AH (1993b) Androgen effects on vocal muscle structure in a teleost fish with inter- and intra-sexual dimorphism. J Morphol 216:305–316.
Brenowitz EA (1997) Comparative approaches to the avian song system. J Neurobiol 33:517–531.
Brenowitz EA, Arnold AP (1986) Interspecific comparisons of the size of neural song control regions and song complexity in duetting birds: Evolutionary implications. J Neurosci 6:2875–2879.
Brenowitz EA, Nalls B, Wingfield JC, Kroodsma DE (1991) Seasonal changes in avian song nuclei without seasonal changes in song repertoire. J Neurosci 11:1367–1374.
Brenowitz EA, Margoliash D, Nordeen KW (1997) An introduction to birdsong and the avian song system. J Neurobiol 33:495–500.
Brenowitz EA, Baptista LF, Lent K, Wingfield JC (1998) Seasonal plasticity of the song control system in wild Nuttal's white-crowned sparrows. J Neurobiol 34:69–82.
Brockway BF (1964) Ethological studies of the budgerigar (*Melopsittacus undulatus*): Reproductive behavior. Behaviour 22:295–324.
Cheng M-F (1973a) Effects of ovariectomy on the reproductive behavior of female ring doves (*Streptopelia risoria*). J Comp Physiol Psychol 83:221–233.
Cheng M-F (1973b) Effects of estrogen on behavior of ovariectomized ring doves (*Streptopelia risoria*). J Comp Physiol Psychol 83:234–239.
Chu J, Marler CA, Wilczynski W (1998) The effects of arginine vasotocin on the calling behavior of male cricket frogs in changing social contexts. Horm Behav 34:248–261.
Clower RP, Nixdorf BE, DeVoogd TJ (1989) Synaptic plasticity in the hypoglossal nucleus of female canaries: Structural correlates of season, hemisphere, and testosterone treatment. Behav Neural Biol 52:63–77.
Cohen M, Kelley DB (1996) Androgen-induced proliferation in the developing larynx of *Xenopus laevis* is regulated by thyroid hormone. Dev Biol 178:113–123.
Cohen M, Winn H (1967) Electrophysiological observations on hearing and sound production in the fish, *Porichthys notatus*. J Exp Zool 165:355–369.
Commins D, Yahr P (1985) Autoradiographic localizations of estrogen and androgen receptors in the sexually dimorphic areas and other regions of the gerbil brain. J Comp Neurol 231:473–489.
Connaughton MA, Taylor MH (1994) Seasonal cycles in the sonic muscles of the weakfish, *Cynoscion regalis*. Fishery Bull 92:697–703.

Connaughton MA, Taylor MH (1995a) Seasonal and daily cycles in sound production associated with spawning in the weakfish, *Cynoscion regalis*. Environ Biol Fishes 42:233–240.

Connaughton MA, Taylor MH (1995b) Effects of exogenous testosterone on sonic muscle mass in the weakfish, *Cynoscion regalis*. Gen Comp Endocrinol 100: 238–245.

Connaughton MA, Taylor MH (1996) Drumming, courtship, and spawning behavior in captive weakfish *Cynoscion regalis*. Copeia 1996:195–199.

Connaughton MA, Fine ML, Taylor MH (1997) The effects of seasonal hypertrophy and atrophy on fiber morphology, metabolic substrate concentration and sound characteristics of the weakfish sonic muscle. J Exp Biol 200:2449–2457.

Crews D, Moore MC (1985) Evolution of mechanisms controlling mating behavior. Science 231:121–125.

Cynx J, Nottebohm F (1992a) Role of gender, season, and familiarity in discrimination of conspecific song by zebra finches (*Taeniopygia guttata*). Proc Natl Acad Sci U S A 89:1368–1371.

Cynx J, Nottebohm F (1992b) Testosterone facilitates some conspecific song discriminations in castrated zebra finches (*Taeniopygia guttata*). Proc Nati Acad Sci U S K 89:1376–1378.

Demski LS, Gerald JW, Popper AN (1973) Central and peripheral mechanisms of teleost sound production. Am Zool 13:1141–1167.

Deviche P, Propper CR, Moore FL (1990) Neuroendocrine, behavioral, and morphological changes associated with the termination of the reproductive period in a natural population of male rough-skinned newts (*Taricha granulosa*). Horm Behav 24:284–300.

DeVoogd TJ, Nottebohm F (1981a) Sex differences in dendritic morphology of a song control nucleus in the canary: A quantitative Golgi study. J Comp Neurol 196:309–316.

DeVoogd TJ, Nottebohm F (1981b) Gonadal hormones induce dendritic growth in the adult avian brain. Science 214:202–204.

DeVoogd TJ, Nixdorf B, Nottebohm F (1985) Synaptogenesis and changes in synaptic morphology related to acquisition of a new behavior. Brain Res 329:304–308.

DeVoogd TJ, Pyskaty DJ, Nottebohm F (1991) Lateral asymmetries and testosterone-induced changes in the gross morphology of the hypoglossal nucleus in adult canaries. J Comp Neurol 307:65–76.

DeVoogd TJ, Houtman AM, Falls BJ (1995) White-throated sparrow morphs that differ in song production rate also differ in the anatomy of some song-related brain areas. J Neurobiol 28:202–213.

Diakow C (1978) Hormonal basis for breeding behavior in female frogs: Vasotocin inhibits the release call of *Rana pipiens*. Science 199:1456–1457.

Diakow C, Nemiroff A (1981) Vasotocin, prostaglandin, and female reproductive behavior in the frog, *Rana pipiens*. Horm Behav 15:86–93.

Dittrich F, Feng Y, Metzdorf R, Gahr M (1999) Estrogen-inducible, sex-specific expression of brain-derived neurotrophic factor mRNA in a forebrain song control nucleus of the juvenile zebra finch. Proc Natl Acad Sci U S A 96: 8241–8246.

Edwards CJ, Yamamoto K, Kikuyama S, and Kelley DB (1999) Prolactin opens the sensitive period for androgen regulation of a larynx-specific myosin heavy chain gene. J Neurobiol 41:443–451.

Emerson SB (1992) Courtship and nest-building behavior of a Bornean frog, *Rana blythi*. Copeia 1992:1123–1127.
Emerson SB, Boyd SK (1999) Mating vocalizations of female frogs: Control and evolutionary mechanisms. Brain Behav Evol 53:187–197.
Emerson SB, Greig A, Carroll L, Prins GS (1999) Androgen receptors in two androgen-mediated, sexually dimorphic characters of frogs. Gen Comp Endocrinol 11:173–180.
Eniksson D, Wallin L (1986) Male bird song attracts females—a field experiment. Behav Ecol Sociobiol 19:297–299.
Falls JB (1982) Individual recognition by sound in birds. In: Kroodsma DE, Miller EH (eds) Acoustic Communication in Birds, Vol 2. New York: Academic Press, pp. 237–278.
Fehm-Wolfdorf G, Nagel D (1996) Differential effects of glucocorticoids on human auditory perception. Biol Psychol 42:117–130.
Fellers GM (1979) Aggression, territoriality and mating behaviour in North American treefrogs. Anim Behav 27:107–119.
Fischer L, Catz D, and Kelley D (1993) An androgen receptor mRNA isoform associated with hormone-induced cell proliferation. Proc Natl Acad Sci U S A 90:8254–8258.
Floody OR (1979) Behavioral and physiological analyses of ultrasound production by female hamsters (*Mesocricetus auratus*). Am Zool 19:443–456.
Floody OR, Comerci JT (1987) Hormonal control of sex differences in ultrasound production by hamsters. Horm Behav 21:17–35.
Floody OR, Petropoulos AC (1987) Aromatase inhibition depresses ultrasound production and copulation in male hamsters. Horm Behav 21:100–104.
Floody OR, Cooper TT, Albers HE (1998) Injection of oxytocin into the medial preoptic-anterior hypothalamus increases ultrasound production by female hamsters. Peptides 19:833–839.
Foran CM, Bass AH (1998) Preoptic AVT immunoreactive neurons of a teleost fish with alternative reproductive tactics. Gen Comp Endocrinol 111: 271–282.
Gahr M (1990) Delineation of a brain nucleus: Comparisons of cytochemical, hodological, and cytoarchitectural view of the song control nucleus HVc of the adult canary. J Comp Neurol 294:30–36.
Given MF (1993) Male response to female vocalizations in the carpenter frog, *Rana virgatipes*. Anim Behav 46:1139–1149.
Goodson JL (1998) Territorial aggression and dawn song are modulated by septal vasotocin and vasoactive intestinal polypeptide in male field sparrows (*Spizella pusilla*). Horm Behav 34:67–77.
Goodson JL, Bass AH (2000a) Vasotocin innervation and modulation of vocal-acoustic circuitry in the teleost *Porichthys notatus*. J Comp Neurol 422: 363–379.
Goodson JL, Bass AH (2000b) Rhythmic midbrain-evoked vocalization is inhibited by vasoactive intestinal polypeptide in the teleost *Porichthys notatus*. Brain Res 865:107–111.
Goodson JL, Bass AH (2000c) Forebrain peptides modulate sexually polymorphic vocal circuitry. Nature 403:769–772.
Guillemin R, Burgus R (1972) The hormones of the hypothalamus. Sci Am 227: 24–33.

Gulledge CC, Deviche P (1997) Androgen control of vocal control region volumes in a wild migratory songbird (*Junco hyemalis*) is region and possibly age dependent. J Neurobiol 32:391–402.

Gulledge CC, Deviche P (1998) Photoperiod and testosterone independently affect vocal control region volumes in adolescent male songbirds. J Neurobiol 36: 550–558.

Gurney ME (1981) Hormonal control of cell form and number in the zebra finch song system. J Neurosci 1:658–673.

Gurney ME (1982) Behavioral correlates of sexual differentiation in the zebra finch brain. Brain Res 231:153–172.

Gurney ME, Konishi M (1980) Hormone-induced sexual differentiation of brain and behavior in zebra finches. Science 208:1380–1383.

Gyger M, Karakashian SJ, Dufty AM Jr, Marler P (1988) Alarm signals in birds: The role of testosterone. Horm Behav 22(3):305–314.

Hadley ME (1996) Endocrinology, 4th ed. Englewood Cliffs, NJ: Prentice–Hall.

Handa RJ, Reid DL, Resko JA (1986) Androgen receptors in brain and pituitary of female rats: Cyclic changes and comparisons with the male. Biol Reprod 34: 293–303.

Hannigan P, Kelley DB (1986) Androgen-induced alterations in vocalizations of female *Xenopus laevis*: Modifiability and constraints. J Comp Physiol A 58: 517–527.

Harding CF, Sheridan K, Walters MJ (1983) Hormonal specificity and activation of sexual behavior in male zebra finches. Horm Behav 17:111–133.

Harding CF, Walters MJ, Collado D, Sheridan K (1988) Hormonal specificity and activation of social behavior in male red-winged blackbirds. Horm Behav 22: 402–418.

Hart BL, Leedy MG (1985) Neurological basis of male sexual behavior: A comparative analysis, Vol 7. New York: Plenum Press.

Henkin RI, Daly RL (1968) Auditory detection and perception in normal man and in patients with adrenal cortical insufficiency: Effect of adrenal cortical steroids. J Clin Invest 47:1269–1280.

Hill KM, DeVoogd TJ (1991) Altered daylength affects dendritic structure in a song-related brain region in red-winged blackbirds. Behav Neural Biol 56:240–250.

Hillery CM (1984) Seasonality of two midbrain auditory responses in the treefrog, *Hyla chrysoscelis*. Copeia 1984:844–852.

Holman SD (1980) Sexually dimorphic, ultrasonic vocalizations of Mongolian gerbils. Behav Neural Biol 28:183–192.

Holman SD, Hutchison JB (1982) Pre-copulatory behaviour in the male Mongolian gerbil: I. Differences in dependence on androgen of component patterns. Anim Behav 30:221–230.

Holman SD, Hutchison JB (1993) Lateralization of a sexually dimorphic brain area associated with steroid-sensitive behavior in the male gerbil. Behav Neurosci 107:186–193.

Holman SD, Janus C (1998) Laterally asymmetrical cell number in a sexually dimorphic nucleus in the gerbil hypothalamus is correlated with vocal emission rates. Behav Neurosci 112:979–990.

Holman SD, Rice A (1996) Androgenic effects on hypothalamic asymmetry in a sexually differentiated nucleus related to vocal behavior in Mongolian gerbils. Horm Behav 30:662–672.

Holman SD, Hutchison RE, Hutchison JB (1991) Microimplants of estradiol in the sexually dimorphic area of the hypothalamus activate ultrasonic vocal behavior in male Mongolian gerbils. Horm Behav 25:531–548.

Holman SD, Collado P, Rice A, Hutchison JB (1995) Stereological estimates of postnatal structural differentiation in a sexually dimorphic hypothalamic nucleus involved in vocal control. Brain Res 694:167–176.

Hua SY, Chen YZ (1989) Membrane receptor-mediated electrophysiological effects of glucocorticoid on mammalian neurons. Endocrinology 124:687–691.

Jacobs EC, Grisham W, Arnold AP (1995) Lack of a synergistic effect between estradiol and dihydrotestosterone in the masculinization of the zebra finch song system. J Neurobiol 27:513–519.

Johnsen TS (1998) Behavioural correlates of testosterone and seasonal changes of steroids in red-winged blackbirds. Anim Behav 55:957–965.

Johnson F, Bottjer SW (1993) Hormone-induced changes in identified cell population of the higher vocal center in male canaries. J Neurobiol 24:400–418.

Johnson F, Bottjer SW (1995) Differential estrogen accumulation among populations of projection neurons in the higher vocal center in male canaries. J Neurobiol 26:87–108.

Johnson F, Hohmann SE, DiStefano PS, Bottjer SW (1997) Neurotrophins suppress apoptosis induced by deafferentation of an avian motor-cortical region. J Neurosci 17:2101–2111.

Kay JN, Hannigan P, Kelley DB (1999) Trophic effects of androgen: Development and hormonal regulation of neuron number in a sexually dimorphic vocal motor nucleus. J Neurobiol 40:375–385.

Kelley DB (1980) Auditory and vocal nuclei of frog brain concentrate sex hormones. Science 207:553–555.

Kelley DB (1981) Locations of androgen-concentrating cells in the brain of *Xenopus laevis*: Autoradiography with 3H-dihydrotestosterone. J Comp Neurol 199:221–231.

Kelley DB (1982) Female sex behaviors in the South African clawed frogs, *Xenopus laevis*: Gonadotropin-releasing, gonadotropic, and steroid hormones. Horm Behav 16:158–174.

Kelley DB (1996) Sexual differentiation in *Xenopus laevis*. In: Tinsley R, Kobel H (eds) The Biology of *Xenopus*. Oxford: Oxford University Press, pp. 143–176.

Kelley DB (1997) Generating sexually differentiated songs. Curr Opin Neurobiol 7:839–843.

Kelley DB, Pfaff DW (1976) Hormone effects on male sex behavior in adult South African clawed frogs, *Xenopus laevis*. Horm Behav 7:159–182.

Kelley DB, Pfaff DW (1978) Generalizations from comparative studies on neuroanatomical and endocrine mechanisms of sexual behavior. In: Hutchison JB (ed) Biological Determinants of Sexual Behavior. Chichester: John Wiley and Sons, pp. 225–254.

Kelley DB, Tobias ML (1999) The vocal repertoire of *Xenopus laevis*. In: Hauser M, Konishi M (eds) The Design of Animal Communication. Cambridge, MA: MIT Press, pp. 9–35.

Kelley DB, Lieberburg I, McEwen BS, Pfaff DW (1978) Autoradiographic and biochemical studies of steroid hormone-concentrating cells in the brain of *Rana pipiens*. Brain Res 140:287–305.

Kelley DB, Sassoon D, Segil N, Scudder M (1989) Development and hormone regulation of androgen receptor levels in the sexually dimorphic larynx of *Xenopus laevis*. Dev Biol 131:111–118.

Kern MD, King JR (1972) Testosterone-induced singing in female white-crowned sparrows. Condor 74:204–209.

Kirn JR, Clower RP, Kroodsma DE, DeVoogd TJ (1989) Song-related brain regions in the red-winged blackbird are affected by sex and season but not repertoire size. J Neurobiol 20:139–163.

Kroodsma DE (1976) Reproductive development in a female songbird: Differential stimulation by quality of male song. Science 192:574–575.

Lauber AH, Romano GJ, Pfaff DW (1991) Gene expression for estrogen and progesterone receptor mRNAs in rat brain and possible relations to sexually dimorphic functions. J Steroid Biochem Mol Biol 40:53–62.

Lehrman DS (1965) Interaction between internal and external environments in the regulation of the reproductive cycle of the ring dove. In: Beach FA (ed) Sex and Behavior. New York: Wiley, pp. 355–380.

Li HY, Blaustein JD, De Vries GJ, Wade GN (1993) Estrogen-receptor immunoreactivity in hamster brain: Preoptic area, hypothalamus and amygdala. Brain Res 631:304–312.

Licht P, McCreery BR, Barnes R, Pang R (1983) Seasonal and stress related changes in plasma gonadotropins, sex steroids, and corticosterone in the bullfrog, *Rana catesbeiana*. Gen Comp Endocrinol 50:124–145.

Logan CA, Wingfield JC (1990) Autumnal territorial aggression is independent of plasma testosterone in mockingbirds. Horm Behav 24:568–581.

Luine V, Nottebohm F, Harding C, McEwen BS (1980) Androgen affects cholinergic enzymes in syringeal motor neurons and muscle. Brain Res 192:89–107.

Luine VN, Harding CF, Bleisch WV (1983) Specificity of gonadal hormone modulation of cholinergic enzymes in the avian syrinx. Brain Res 279:339–342.

Maney DL, Goode CT, Wingfield JC (1997) Intraventricular infusion of arginine vasotocin induces singing in a female songbird. J Neuroendocrinol 9:487–491.

Marin ML, Tobias ML, Kelley DB (1990) Hormone-sensitive stages in the sexual differentiation of laryngeal muscle fiber number in *Xenopus laevis*. Development 110:703–711.

Marler CA, Chu J, Wilczynski W (1995) Arginine vasotocin injection increases probability of calling in cricket frogs, but causes call changes characteristic of less aggressive males. Horm Behav 29:554–570.

Marler P, Peters S, Wingfield J (1987) Correlations between song acquisition, song production, and plasma levels of testosterone and estradiol in sparrows. J Neurobiol 18:531–548.

Marler P, Peters S, Ball GF, Dufty AM, Wingfield JC (1988) The role of sex steroids in the acquisition and production of birdsong. Nature 336:770–772.

Marquez R, Verrell P (1991) The courtship and mating of the Iberian midwife toad *Alytex cisternasii* (Amphibia, Anura, Discoglossidae). J Zool 225:125–139.

Matochik JA, Barfield RJ (1991) Hormonal control of precopulatory sebaceous scent marking and ultrasonic mating vocalizations in male rats. Horm Behav 25:445–460.

Matochik JA, Barfield RJ, Nyby J (1992) Regulation of sciosexual communication in female Long–Evans rats by ovarian hormones. Horm Behav 26:545–555.

Matochik JA, Sipos ML, Nyby JG, Barfield RJ (1994) Intracranial androgenic activation of male-typical behaviors in house mice: Motivation versus performance. Behav Brain Res 60:141–149.
McFadden D (1998) Sex differences in the auditory system. Dev Neuropsychol 14:261–298.
McGinnis MY, Dreifuss RM (1989) Evidence for a role of testosterone–androgen receptor interactions in mediating masculine sexual behavior in male rats. Endocrinology 124:618–626.
McGinnis MY, Kahn DF (1997) Inhibition of male sexual behavior by intracranial implants of the protein synthesis inhibitor anisomycin into the medial preoptic area of the rat. Horm Behav 31:15–23.
Mendonça MT, Light P, Ryan MJ, Barnes R (1985) Changes in hormone levels in relation to breeding behavior in male bullfrogs (*Rana catesbeiana*) at the individual and population levels. Gen Comp Endocrinol 58:270–279.
Metzdorf R, Gahr M, Fusani L (1999) Distribution of aromatase, estrogen receptor, and androgen receptor mRNA in the forebrain of songbirds and nonsongbirds. J Comp Neurol 407:115–129.
Moore FL (1978) Differential effects of testosterone plus DHT on male courtship of castrated newts, *Taricha granulosa*. Horm Behav 11:202–208.
Moore FL (1987) Reproductive endocrinology of amphibians. In: Chester-Jones I, Ingleton PM, Phillips JG (eds) Fundamentals of Comparative Vertebrate Endocrinology. New York: Plenum Press, pp. 207–221.
Moore FL, Evans SJ (1999) Steroid hormones use non-genomic mechanisms to control brain functions and behaviors: A review of evidence. Brain Behav Evol 54:41–50.
Moore FL, Zoeller RT (1979) Endocrine control of amphibian sexual behavior: Evidence for a neurohormone–androgen interaction. Horm Behav 13:207–213.
Morton ML, Peterson LE, Burns DM, Allan N (1990) Seasonal and age-related changes in plasma testosterone levels in mountain white-crowned sparrows. Condor 92:166–173.
Narins PM, Capranica RR (1976) Sexual differences in the auditory system of the tree frog *Eleutherodactylus coqui*. Science 192:378–380.
Neary TJ, Wilczynski W (1986) Auditory pathways to the hypothalamus in ranid frogs. Neurosci Lett 71:142–146.
Nespor AA, Lukazewicz MJ, Dooling RJ, Ball GF (1996) Testosterone induction of male-like vocalizations in female budgerigars (*Melopsittacus undulatus*). Horm Behav 30:162–169.
Nottebohm F (1980) Testosterone triggers growth of brain vocal control nuclei in adult female canaries. Brain Res 189:429–436.
Nottebohm F (1981) A brain for all seasons: Cyclical anatomical changes in song control nuclei of the canary brain. Science 214:1368–1370.
Nottebohm F, Arnold AP (1976) Sexual dimorphism in vocal control areas of the songbird brain. Science 194:211–213.
Nottebohm F, Nottebohm ME, Crane LA, Wingfield JC (1987) Seasonal changes in gonadal hormone levels of adult male canaries and their relation to song. Behav Neural Biol 47:197–211.
Nowicki S, Ball GF (1989) Testosterone induction of song in photosensitive and photorefractory male sparrows. Horm Behav 23:514–525.

Nunez AA, Tan DT (1984) Courtship ultrasonic vocalizations in male Swiss–Webster mice: Effects of hormones and sexual experience. Physiol Behav 32: 717–721.

Nunez AA, Nyby J, Whitney G (1978) The effects of testosterone, estradiol, and dihydrotestosterone on male mouse (*Mus musculus*) ultrasonic vocalizations. Horm Behav 11:264–272.

Nyby J, Matochik JA, Barfield RJ (1992) Intracranial androgenic and estrogenic stimulation of male-typical behaviors in house mice (*Mus domesticus*). Horm Behav 26:24–45.

Orchinik M, Murray TF, Moore FL (1991) A corticosteroid receptor in neuronal membranes. Science 252:1848–1851.

Palka YS, Gorbman A (1973) Pituitary and testicular influenced sexual behavior in male frogs, *Rana pipiens*. Gen Comp Endocrinol 21:148–151.

Penna M, Capranica RR, Somers J (1992) Hormone-induced vocal behavior and midbrain auditory sensitivity in the green treefrog, *Hyla cinerea*. J Comp Physiol A 170:73–82.

Pérez J, Cohen MA, Kelley DB (1996) Androgen receptor mRNA expression in *Xenopus laevis* CNS; Sexual dimorphism and regulation in the laryngeal motor nucleus. J Neurobiol 30:556–568.

Pfaff DW, Silva MT, Weiss JM (1971) Telemetered recording of hormone effects on hippocampal neurons. Science 172:394–395.

Phoenix CH, Goy RW, Gerall AA, Young WC (1959) Organizing action of prenatally administered testosterone propionate on the tissues mediating mating behavior in the female guinea pig. Endocrinology 65:369–389.

Pierantoni R, Iela L, d'Istria M, Fasano S, Rastogi RK, Delrio G (1984) Seasonal testosterone profile and testicular responsiveness to pituitary factors and gonadotropin releasing hormone during two different phases of the sexual cycle of the frog (*Rana esculenta*). J Endocrinol 102:387–392.

Pomerantz SM, Fox E, Clemens LG (1983) Gonadal hormone activation of male courtship ultrasonic vocalizations and male copulatory behavior in castrated male deer mice (*Peromyscus maniculatus bairdi*). Behav Neurosci 97:462–469.

Potash LM (1975) An experimental analysis of the use of location calls by Japanese quail, *Coturnix coturnix japonica*. Behaviour 54:153–180.

Powers DM, Ingleton PM (1998) Distribution of vasoactive intestinal peptide in the brain and hypothalamo-hypophysial system of the sea bream (*Sparus aurata*). Ann N Y Acad Sci 839:356–357.

Propper CR, Dixon TB (1997) Differential effects of arginine vasotocin and gonadotropin-releasing hormone on sexual behaviors in an anuran amphibian. Horm Behav 32:99–104.

Pröve E (1974) Der einfluss von kastration und testosteronesubstitution auf das sexualverhalten mannlicher Zebrafinken (*Taeniophygia guttata castanotis* Gould). J Ornithol 115:338–347.

Rainbow TC, Parsons B, McEwen BS (1982) Sex differences in rat brain oestrogen and progestin receptors. Nature 300:648–649.

Rastogi RK, Iela L, Delrio G, Bagnara JT (1986) Reproduction in the Mexican leaf frog, *Pachymedusa dacnicolor:* II. The male. Gen Comp Endocrinol 62:23–35.

Robertson JC, Kelley DB (1996) Thyroid hormone controls the onset of androgen sensitivity in the developing larynx of *Xenopus laevis*. Dev Biol 176:108–123.

Romero LM, Soma KK, O'Reilly KM, Suydam R, Wingfield JC (1998) Hormones and territorial behavior during breeding in snow buntings (*Plectrophenax nivalis*): An arctic-breeding songbird. Horm Behav 33:40–47.

Rose JD, Kinnaird JR, Moore FL (1995) Neurophysiological effects of vasotocin and corticosterone on medullary neurons: Implications for hormonal control of amphibian courtship behavior. Neuroendocrinology 62:406–417.

Roselli CE, Handa RJ, Resko JA (1989) Quantitative distribution of nuclear androgen receptors in microdissected areas of the rat brain. Neuroendocrinology 49:449–453.

Roselli CE, Abdelgadir SE, Ronnekleiv OK, Klosterman SA (1998) Anatomic distribution and regulation of aromatase gene expression in the rat brain. Biol Reprod 58:79–87.

Schmidt RS (1965) Larynx control and call production in frogs. Copeia 1965: 143–147.

Schmidt RS (1966) Hormonal mechanisms of frog mating calling. Copeia 1966: 637–644.

Schmidt RS, Kemnitz CP (1989) Anuran mating calling circuits: Inhibition by prostaglandin. Horm Behav 23:361–367.

Schwabl H, Kriner E (1991) Territorial aggression and song of male European robins (*Erithacus rubecula*) in autumn and spring: Effects of antiandrogen treatment. Horm Behav 25:180–194.

Schwabl H, Sonnenschein E (1992) Antiphonal duetting and sex hormones in the tropical bush shrike *Laniarius funebris* (Hartlaub). Horm Behav 26: 295–307.

Searcy WA, Brenowitz EA (1988) Sexual differences in species recognition of avian song. Nature 332:152–154.

Segil N, Silverman L, Kelley DB (1987) Androgen binding levels in a sexually dimorphic muscle of *Xenopus laevis*. Gen Comp Endocrinol 66:95–101.

Semsar K, Klomberg KF, Marler CA (1998) Arginine vasotocin increases calling-site acquisition by nonresident male grey treefrogs. Anim Behav 56: 983–987.

Shen P, Schlinger BA, Campagnioni AT, Arnold AP (1992) Circulating estrigens in a male songbird originate in the brain. Proc Nati Acad Sci U S A 89:7650–7653.

Shibata H, Spencer TE, Onate SA, Jenster G, Tsai SY, Tsai MJ, O'Malley BW (1997) Role of co-activators and co-repressors in the mechanism of steroid/thyroid receptor action. Recent Prog Horm Res 52:141–165.

Simerly RB, Chang C, Muramatsu M, Swanson LW (1990) Distribution of androgen and estrogen receptor mRNA-containing cells in the rat brain: An in situ hybridization study. J Comp Neurol 294:76–95.

Simpson HB, Vicario DS (1991) Early estrogen treatment of female zebra finches masculinizes the brain pathway for learned vocalizations. J Neurobiol 22:777–793.

Smith GT (1996) Seasonal plasticity in the song nuclei of wild rufous-sided towhees. Brain Res 734:79–85.

Smith GT, Brenowitz EA, Wingfield JC, Baptista LF (1995) Seasonal changes in song nuclei and song behavior in Gambel's white-crowned sparrows. J Neurobiol 28:114–125.

Smith GT, Brenowitz EA, Wingfield JC (1997a) Roles of photoperiod and testosterone in seasonal plasticity of the avian song control system. J Neurobiol 32:426–442.

Smith GT, Brenowitz EA, Beecher MD, Wingfield JC (1997b) Seasonal changes in testosterone, neural attributes of song control nuclei, and song structure in wild songbirds. J Neurosci 17:6001–6010.
Solis R, Penna M (1997) Testosterone levels and evoked vocal responses in a natural population of the frog *Batrachyla taeniata*. Horm Behav 31:101–109.
Soma KK, Hartman VN, Wingfield JC, Brenowitz EA (1999a) Seasonal changes in androgen receptor immunoreactivity in the song nucleus HVc of a wild bird. J Comp Neurol 409:224–236.
Soma KK, Sullivan K, Wingfield JC (1999b) Combined aromatase inhibitor and antiandrogen treatment decreases territorial aggression in a wild songbird during the nonbreeding season. Gen Comp Endocrinol 115:442–453.
Tobias M, Kelley D (1987) Vocalizations of a sexually dimorphic isolated larynx: Peripheral constraints on behavioral expression. J Neurosci 7:3191–3197.
Tobias ML, Marin M, Kelley DB (1991) Temporal constraints on androgen directed laryngeal masculinization in *Xenopus laevis*. Dev Biol 147:260–270.
Tobias ML, Viswanathan S, Kelley DB (1998) Rapping, a female receptive call, initiates male/female duets in the South African clawed frog. Proc Natl Acad Sci U S A 95:1870–1875.
Townsent DS, Moger WH (1987) Plasma androgen levels during male parental care in a tropical frog (*Eleutherodactylus*). Horm Behav 21:92–99.
Ulibarri CM, Yahr P (1993) Ontogeny of the sexually dimorphic area of the gerbil hypothalamus. Dev Brain Res 74:14–24.
Vagell ME, McGinnis MY (1998) The role of gonadal steroid receptor activation in the restoration of sociosexual behavior in adult male rats. Horm Behav 33: 163–179.
Valverde MA, Rojas P, Amigo J, Cosmelli D, Orio P, Bahamonde MI, Mann GE, Vergara C, Latorre R (1999) Acute activation of Maxi-K channels (hSlo) by estradiol binding to the β subunit. Science 285:1929–1931.
Wada M, Gorbman A (1977a) Relation of mode of administration of testosterone to evocation of male sex behavior in frogs. Horm Behav 8:310–319.
Wada M, Gorbman A (1977b) Mate calling induced by electrical stimulation in freely moving leopard frogs *Rana pipiens*. Horm Behav 9:141–149.
Wade J, Arnold AP (1996) Functional testicular tissue does not masculinize development of the zebra finch song system. Proc Natl Acad Sci U S A 93:5264–5268.
Wade J, Springer ML, Wingfield JC, Arnold AP (1996) Neither testicular androgens nor embryonic aromatase activity alters morphology of the neural song system in zebra finches. Biol Reprod 55:1126–1132.
Wagner WE Jr (1989) Fighting, assessment, and frequency alteration in Blanchard's cricket frog. Behav Ecol Sociobiol 25:429–436.
Walkowiak W (1980) The coding of auditory signals in the torus semicircularis of the fire-bellied toad and the grass frog: Responses to simple stimuli and to conspecific calls. J Comp Physiol 138:131–148.
Walters MJ, McEwen BS, Harding C (1988) Estrogen receptor levels in hypothalamic and vocal control nuclei in the male zebra finch. Brain Res 459:37–43.
Watson J, Kelley D (1992) Testicular masculinization of vocal behavior in juvenile female *Xenopus laevis*: Prolonged sensitive period reveals component features of behavioral development. J Comp Physiol 171:343–350.
Watson J, Robertson J, Sachdev U, and Kelley D (1993) Laryngeal muscle and motor neuron plasticity in *Xenopus laevis*: Analysis of a sensitive period for testicular masculinization of a neuromuscular system. J Neurobiol 24:1615–1625.

Weintraub AS, Kelley DB, Bockman RS (1985) Prostaglandin E_2 induces receptive behaviors in female *Xenopus laevis*. Horm Behav 19:386–399.
Wetzel DM, Kelley DB (1983) Androgen and gonadotropin effects on male mate calls in South African clawed frogs, *Xenopus laevis*. Horm Behav 17:388–404.
Wetzel DM, Haerter UL, Kelley DB (1985) A proposed neural pathway for vocalization in South African clawed frogs, *Xenopus laevis*: Tracing afferents to laryngeal motoneurons with HRP-WGA. J Comp Physiol A 157:749–761.
Wingfield JC (1994a) Control of territorial aggression in a changing environment. Psychoneuroendocrinology 19:709–721.
Wingfield JC (1994b) Regulation of territorial behavior in the sedentary song sparrow, *Melospiza melodia morphna*. Horm Behav 28:1–15.
Wingfield JC, Hahn T (1994) Testosterone and territorial behavior in sedentary and migratory sparrows. Anim Behav 47:77–89.
Wong M, Thompson TL, Moss RL (1996) Nongenomic actions of estrogen in the brain: Physiological significance and cellular mechanisms. Crit Rev Neurobiol 10:189–203.
Woods RI, Newman SW (1993) Intracellular partitioning of androgen receptor immunoreactivity in the brain of the male Syrian hamster: Effects of castration and steroid replacement. J Neurobiol 24:925–938.
Wu K-H, Tobias ML, Kelley DB (2001) Estrogen and laryngeal synaptic strength in *Xenopus laevis:* Opposite effects of acute and chronic exposure. Neuroendocrinology 74:22–32.
Yamaguchi A (1996) Female bird song: Function, physiology, and development in the Northern Cardinal. Ph.D. Dissertation University of California-Davis, Davis, CA.
Yamaguchi A, Kelley DB (2000) Generating sexually differentiated vocal patterns: Laryngeal nerve and EMG recordings from vocalizing male and female African clawed frogs (*Xenopus laevis*). J Neurosci 20:1559–1567.
Young LJ, Wang Z, Insel TR (1998) Neuroendocrine bases of monogamy. Trends Neurosci 21:71–75.
Yovanof S, Feng AS (1983) Effects of estradiol on auditory evoked responses from the frog's auditory midbrain. Neurosci Lett 36:291–297.
Zheng J, Ramirez VD (1994) Purification and identification of estrogen-binding proteins from neuronal membranes of female rat brain. Soc Neurosci Abstr 20:95.
Zigmond RE, Nottebohm F, Pfaff DW (1973) Androgen-concentrating cells in the midbrain of a songbird. Science 19:1005–1007.
Zigmond RE, Nottebohm F, Pfaff DW (1980) An autoradiographic study of the localization of androgen concentrating cells in the chaffinch. Brain Res 182: 369–381.
Zoeller RT, Moore FL (1986) Correlation between immunoreactive vasotocin in optic tectum and seasonal changes in reproductive behaviors of male rough-skinned newts. Horm Behav 20:148–154.

7
The Neuroethology of Vocal Communication: Perception and Cognition

Timothy Q. Gentner and Daniel Margoliash

1. Introduction

In its most common sense, acoustic communication occurs between animals, but in special cases it may occur within individual animals in the context of autocommunication. Communication can be described as an information exchange that alters the behavior of the communicating animals. Acoustic communication signals (typically vocalizations) are shaped by the physics of the sound-producing organs, the physical media they traverse, and the physics of the receptor organs (Bass and Clark, Chapter 2; Fitch and Hauser, Chapter 3; Ryan and Kime, Chapter 5). Vocal communication signals are also shaped by the perceptual mechanisms of the receiver, by the proximate behavioral states of the senders and receivers (Boughman and Moss, Chapter 4; Yamaguchi and Kelley, Chapter 6), and by the evolutionary history of the senders and receivers, most often in the context of sexual selection. The information in vocal signals is represented by nonrandom acoustic variation that may either form discrete categories or fall along graded continua. Likewise, vocal signals may be perceived as members of discrete categories or along graded continua. Whether graded or discrete, animals must account for the statistical variation in vocal signals as they are produced, transmitted, and perceived. Ultimately, this means that acoustic behaviors are constrained by the natural variation in communication signals.

The wide diversity of behavioral constraints makes exploration of the neural mechanisms of acoustic communication both exciting and daunting. On one hand, one of the grand challenges in neurobiology has been the application of mechanistic analysis to perceptual and cognitive components of brain function, and the neuroethology of acoustic communication provides a logical and rigorous approach to such problems. At the same time, however, proper treatment of the mechanisms of acoustic communication requires a neural analysis that is sensitive to multiple levels of biological organization along with information-theoretic analyses that capture variations in behavior and signal acoustics. Despite good progress on many fronts, and a bright future, such analysis has yet to be achieved in any single system.

This chapter attempts to place the neuroethology of acoustic communication in the context of its multiple levels of analysis. Rather than striving to review the entire field, our approach is to focus on the neural and behavioral mechanisms that underlie what we consider to be a fundamental class of behaviors present in most vocal communication systems, namely vocal recognition. We have chosen specific examples to elaborate the points we make, and to some extent these choices reflect our own backgrounds. The choice of examples, however, is fundamentally immaterial because the evolutionary backdrop of neuroethology provides a logical framework for inferences drawn across phylogeny. In addition, although we are ultimately interested in questions of auditory perception and cognition, we do not incorporate the various literatures pertaining to the psychophysical basis of hearing and synthetic sound perception. Again, rather than any basic incompatibility, this reflects the practical constraints of the present format. A mature neuroethology of acoustic communication will complement the knowledge gained from research in other areas of comparative hearing and audition, which are reviewed elsewhere (e.g., Dooling et al. 2000; Feng and Ratnam 2000; see Simmons, Chapter 1).

1.1. The Neuroethological Approach

In one of his most important theoretical contributions to biology, Tinbergen (1963) pointed out that any given behavior could be understood in four primary ways. One can consider the evolutionary history of a behavior, its adaptive significance (i.e., its relation to survival or reproduction, past and present), its developmental history within an organism, and its physiological bases. The first two of these are often referred to as "ultimate" causes because they concern behavioral function on an evolutionary time scale. The latter two are often referred to as "proximate" causes of behavior because they are concerned, on a much finer time scale, with ontogenetic and physiological mechanisms within an organism. Although it is possible, and in fact common, to study behavior at only one or the other level, Tinbergen's heuristic is not meant to imply independence across levels. On the contrary, because evolutionary selection mechanisms operate on variation at the level of the individual, the natural history of a behavior, and its adaptive significance, are likely to have profound effects on underlying physiological and developmental processes. Likewise, the proximate mechanisms provide physical constraints for the evolutionary trajectory of a particular behavioral trait. The bidirectional interactions between ultimate and proximate mechanisms of behavior form the basis of neuroethology.

Neuroethology focuses on the mechanisms of ethologically analyzed behaviors by bringing a top-down approach to questions of brain function. By this, we mean that neuroethology takes a behavior whose adaptive significance has been studied in natural conditions (e.g., prey capture, food storage, vocal recognition) as the functional output of the central nervous system and then attempts to determine the underlying neural mechanisms. The neurobiological analysis may follow either a top-down approach (e.g., Moiseff and Konishi 1981) or a bottom-up approach (e.g., Heiligenberg 1991), but in either case the general methodology is predicated on prior knowledge about adaptive behavior.

Traditionally, neuroethologists interested in acoustic communication have investigated species with well-developed vocal systems and have focused on neuronal selectivity and specificity within the vocal repertoire. Given this focus, the extent to which the neural mechanisms of vocal communication in specialized animals can yield insight into more general mechanisms of audition should be considered. Historically, such consideration has led to criticisms that mimic those leveled against early ethology in that generalization is hampered by concentration on a limited component of the acoustic biotope (i.e., vocalizations) and by the choice of animals highly specialized for acoustic communication.

These criticisms fail at both the theoretical and empirical levels. Foremost, such criticisms are inconsistent with an evolutionary perspective maintaining that the perceptual world of each species is a unique consequence of its evolutionary history. As with morphological traits, similarities in behavioral phenotypes across species may result from either common origins or convergent evolution and so do not ensure a corresponding similarity in the underlying neural mechanisms. Comprehensive theories of behavior must therefore embrace, not ignore, evolution (e.g., Gallistel 1990; cf. Dickinson 1980) and must be allowed to emerge from comparative studies of many different species engaged in natural behaviors. A similar reasoning applies to the derivation of general neural mechanisms and their emergence through comparative studies. Moreover, as a matter of practical experience, the foregoing criticisms of the neuroethological approach to acoustic communication are not substantiated by the experimental literature. For example, among the most extreme cases of acoustic specialization is autocommunication in the contexts of echolocation and vocal learning. Yet many organizational features common to the vertebrate auditory (or octavolateralis) system have been usefully described—in many cases first elucidated—in relation to processing of vocalizations in echolocating bats, weakly electric fish, other fishes specialized for vocal communication, frogs, songbirds, and related systems. These principles include the forms and actions of parallel and hierarchical systems (including distorted tonotopic maps, feedforward, feedback, and lateral dynamic connections), distributed representations, single neurons with complex receptive field properties, temporal coding in single neurons and populations of neurons, sensorimotor interactions, state-dependent and dynamic receptive field properties, and sexual dimorphisms. Current studies of these systems continue to be highly productive and, if anything, reflect an expanding scope of issues successfully being addressed.

Finally, an additional question concerns the extent to which the neuroethological approach to acoustic communication can give insight into human acoustic perception. These concerns rest largely on the assumption that human perceptual processes, particularly those involved in speech and language, are unique. Nevertheless, determining which phenotypes of speech are unique and which are conserved requires comparative analysis. For example, although early work on speech processing emphasized what appeared to be unique properties of the human system, many of these initial claims evaporated in the face of later animal research (Kuhl and Miller 1975; Kluender et al. 1987). More recent theoretical perspectives emphasize that features of the speech signal, and associated neural processing mechanisms, are elaborations of general features of the mammalian and even vertebrate auditory system (see Fowler 1996; Lotto et al. 1997). Comparative studies have also demonstrated specialized sensory representations for autocommunication signals (see Popper and Fay 1995; Brenowitz et al. 1997). Such representations are likely to exist in humans because humans are well-known to be sensitive to the acoustic structure of their own vocalizations, but such representations have hardly been studied at all in humans. Furthermore, where there are specific theories of vocal learning in animals, they can lead to specific predictions regarding speech learning in humans (Margoliash 2001). These examples give confidence that insight into human speech perception can be gained by studying the perception of acoustically complex signals from a comparative perspective (Doupe and Kuhl 1999; Fitch 2000).

1.2. Structure of Animal Communication Systems

Arriving at explicit definitions for animal communication is a notoriously difficult problem. However, most will agree on a minimum description of communication as a process involving the transmission of information, via a signal, from a sender to a receiver. Where debate arises is in the extent to which various researchers attribute intent to the sender and in the degree to which the various fitness benefits for either the sender or the receiver are emphasized. Although such considerations remain a topic of continued debate among theoreticians (Dawkins and Krebs 1978; Beer 1982; Smith 1997; Bradbury and Vehrencamp 1998, 2000), from a proximate standpoint, it is the transmission of information that is of interest.

For information to be transmitted via some signal, there must be parity between the sender of that signal and the receiver. That is, the receiver must interpret at least some of the variability (i.e., information) in the signal in a predictable manner. Thus, the structure of a communication system can be considered as a behavioral feedback loop in which information flows

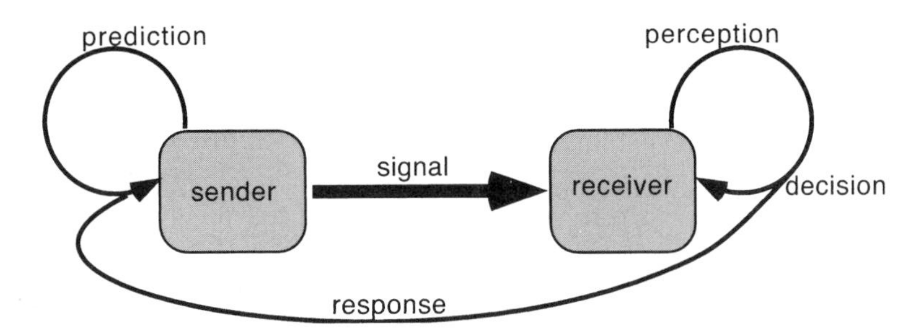

FIGURE 7.1. Diagram showing the theoretical cognitive relationship between sender and receiver in a communication system. Primary information is transmitted from the sender to the receiver in the form of a signal and from the receiver to the sender in the form of a behavioral response. The receiver must decide whether to engage in specific behaviors following the detection of some signal, and the sender must decide whether the receiver's behavior coincides with the predicted response for a given signal.

from the sender to the receiver and then back to the sender again, where the receiver's response is compared to some predicted behavior (Fig. 7.1). At the ultimate level, physiological mechanisms in the sender are under selective pressures to produce a specific variability in the signal, and those in the receiver are under pressure to perceive this variability in specific ways. Although particular circumstances may not always ensure mutual benefits to the sender and receiver, and thus that there is convergence upon specific signal characteristics, the general requirement for signal parity will hold as long as the signal remains functional. Conversely, given a functional signal, one can assume parity and thus that the production and perception mechanisms in communication systems are closely tuned to specific aspects of the acoustic variation in that signal. At the behavioral level, this is evidenced by that fact that specific functional behaviors come under the direct and strong control of acoustic variation in the signal.

One excellent example of the relationship between signal variation and behavior can be seen in female choice among songbirds. In many species of songbirds, males sing elaborate vocalizations (songs) that are often directed at females, and there is an extensive body of literature indicating that male song serves as the basis for female behavioral preference and choice in many species (reviewed in Searcy and Yasukawa 1996). Depending on the species, females choosing among individual males may attend to variation in song duration, dialect, output rate, "quality" (i.e., presence or absence of specific features), repertoire size, or measures of the acoustic complexity in a song. (Curiously, amplitude, which is so important in anuran and insect studies, has rarely been tested as a parameter for song preference and choice in female songbirds; see Dabelsteen and Pederson 1993; Searcy 1996.) Some of the behavioral variables describing male song may map simply onto neuronal mechanisms in females, whereas others may not. For example, in the case of repertoire–size preferences, it has been suggested that the acoustic entropy in the signal may drive neuronal habituation mechanisms (see Ryan 1998; Gentner et al. 2001) as opposed to more direct measures of repertoire sizes employed by researchers, such as simple song counts.

1.3. Technical Challenges

Given the attractiveness of vocal communication systems as research objects, it is somewhat surprising that the underlying neural mechanisms are relatively poorly understood. More than anything else, this lack of understanding reflects the tremendous technical challenges involved in studying these complex biological systems. Assessing perceptual responses to conspecific stimuli often requires elaborate conditions difficult to stage in the laboratory, and ethologists have only cautiously embraced conditioning procedures otherwise common in the study of perceptual phenomena (e.g., Dooling and Searcy 1980; Weary and Krebs 1992; Adret 1993). In part, such caution has been in an effort to avoid potential confusions that can arise when interpreting arbitrary conditioning paradigms in the context of natural behavior.

A related technical limitation is that the proper analysis of vocal communication will often require one or more animals to be actively engaged in communication as the experimenter records physiological activity. Ideally, this should include both field and laboratory settings. For the most part, this vision is more fanciful than real because the ability to reliably conduct single-neuron recordings in awake, behaving vertebrate animals of the small size commonly employed in neuroethological studies is only beginning to emerge (Dave et al. 1998b; Nieder and Klump 1999; Venkatachalam et al. 1999), so far mainly under relatively constrained laboratory conditions (Yu and Margoliash 1996; Dave et al. 1998a; Nieder and Klump 1999).

Yet another reason for the relative paucity of analysis of vocal communication is in some sense accidental and historical. Three of the most compelling vertebrate neuroethological systems studied from a vocal communication perspective—bats, weakly electric fish, and songbirds—involve autocommunication in the form of echolocation or feedback-mediated learning. Although some lessons and principles may emerge from the study of autocommunication that are common to sender/receiver systems, some clearly cannot. In autocommunication, the animal as the receiver has knowledge of the timing and structure of its own motor behavior (although an efference copy of the final motor output may not be directly available to the CNS—e.g., Heiligenberg 1977). An independent receiver cannot have such detailed knowledge. Thus, the signal coding and subsequent processing mechanisms, along with the role of attentional systems, are likely to vary between receivers engaging in autocommunication and those that are not.

In addition, in echolocation systems and passive localization systems, the structure of the vocal signal, or the parameters of the signal to be analyzed, are strongly and directly constrained by physical characteristics of the signal-generating mechanism, the medium for signal transmission, and the physical characteristics of the target. This means that sensory mechanisms for localization are influenced principally by natural selection. In all localization systems, there is compelling evidence for behavioral sensitivity to microsecond timing and considerable evidence for specialized neural systems to accurately represent information on such time scales. To date, there is no behavioral or physiological evidence to suggest that receiver sensitivity to microsecond timing is important outside the context of autocommunication. Vocal communication, on the other hand, generally arises in the context of reproductive behaviors such as courtship and territorial defense. Thus, vocal communication is influenced principally by sexual selection, although physical constraints can also influence the design of communication signals (Konishi 1970; Ryan and Kime, Chapter 5). The differences in selection pressures on different types of signaling may also have profound influences on CNS organization.

2. Organizational Principles for Neural Representations of Conspecific Vocalizations

Internal representations of information in the world are a function of both the form of the input and state of the neural system. Neuroethological studies using conspecific vocalizations bridge these "input" and "state" components in a biologically plausible way and have made important contributions to theories of representation and brain organization at multiple levels of analysis. In this section, we focus on the organizational principles of representation that underlie the discrimination and classification of vocal communication signals. These neuroethological analyses provide insight into behavioral mechanisms at both proximate and ultimate levels. At the proximate level, such studies relate to questions of neural coding, such as the forms of hierarchical and parallel representation, the representation of spectrally and temporally complex acoustic signals, and representational plasticity. Insights into ultimate processes that derive from neuroethological studies of conspecific vocalizations include coupling mechanisms between motor and sensory systems and the relationship between vocal recognition at inter- and intra-specific levels.

2.1. The Feature Detector Hypothesis

An understanding of the hierarchical and parallel organization of neural activity patterns is fundamental to the study of integrative brain mechanisms. Behavioral phenomena such as sensory perceptions may manifest themselves as particular states of activity in large aggregates or "assemblies" of neurons, with each neuron providing only a coarse coding of the stimulus in terms of a graded response (Hebb 1949). This is called the population hypothesis and has received particular attention in recent years with the emergence of technologies to record and analyze data from multiple electrodes in behaving animals (e.g., Eichenbaum and Davis 1998; Nicolelis 2001). Alternatively, phenomena may be encoded in the brain by the activity of small, possibly redundant populations of relatively specialized neurons. This is called the single–cell hypothesis and is closely tied to the idea of "feature detectors" (Barlow 1972). The two hypotheses are not mutually exclusive, and although the early theoretical literature rejected the single-cell hypothesis (e.g., Marr 1982), in fact both theories may be considered substantially established. The analysis of conspecific vocalizations has helped in synthesizing a unified perspective.

2.1.1. Single Cells and Distributed Representations

In the extreme case, the single-cell doctrine has been characterized, or perhaps caricatured, as the "grandmother cell" hypothesis (see Martin 1994; Barlow 1995). Imagine a cell that responds always and only whenever you perceive the face of your grandmother and is a requisite for that perception. The grandmother cell concept is defined by necessity and sufficiency arguments in relation to the percept. However, where these constraints have been proposed in other contexts of neural coding, such as motor control, they have generally not proven to be satisfactory criteria (Kupfermann and Weiss 1978; Eaton 1983). The grandmother cell hypothesis has difficulty addressing the issues of combinatorial explosion, redundancy, and coverage. For instance, consider the case for a theoretical system containing N neurons. If the response of one (and only one) neuron codes for one stimulus, the system can only represent N stimuli. If precepts are represented at high resolution, or in combination, the number of percepts can easily overwhelm the number of neurons available to code those percepts. Loss of a single neuron would represent loss of the percept. These facts combined with the uncertain support in the experimental data suggest that the limits of a single-cell coding scheme are not realized, at least in vertebrates.

The extreme case of the distributed population-coding hypothesis is equally implausible. For the theoretical system containing N neurons, a fully distributed system could represent at least 2^N stimuli (given a binary response for each cell). Many more percepts could be represented than there were neurons, and loss of any one neuron or small set of neurons would result only in gradual degradation of the information represented in the network. These are desirable properties that superficially mimic biological neural networks, and highly interconnected networks are attractive to theoreticians because they are amenable to quantitative analysis.

Nevertheless, such coding schemes are biologically unrealistic. In the distributed model, neurons would be fully interconnected (i.e., there would be no anatomical specificity of feedforward, feedback, or lateral connections). There would be no topographic representation of information. Neurons would fail to show stimulus specificity. Such anatomical and physiological patterns of organization, however, are not observed in nervous systems, which exhibit precision of connections between and within different classes of neurons, with the spatial location of a neuron typically representing aspects of the neuron's specificity for parameters that are encoded. Thus, the extreme forms of the single cell and distributed-representation hypotheses are both rejected because they fail to reflect known biological reality.

The synthesis of the single-cell and distributed-coding hypotheses arises from an appreciation of the different end states of a behavioral continuum. Perceptions of arbitrary objects are unlikely to be processed by highly object-specific neurons. This conclusion derives from constraints on proximate mechanisms as described above and constraints on ultimate mechanisms, especially the unpredictable behavioral significance of an arbitrary object. For arbitrary objects, a distributed representation is likely, where the activation of ensembles of less specialized neurons is ultimately related directly to the perceptual event. In contrast, a predictable environment may allow for either genetic fixation or learning during ontogeny (or both) to establish specialized processing for reliable objects. There is considerable evidence for the existence of such specialized hierarchical streams of processing (see below) with mnemonic cells that exist at higher levels of these hierarchies. Rather than yielding a combinatorial explosion, extension of the hierarchical organization within the constraints provided by a predictable environment may increase coding efficiency. Perception is still ultimately related to the activation of ensembles of neurons, but the size of the ensemble may be different from that in the case of an arbitrary stimulus, and the neurons that encode the predictable stimulus may be more highly specialized. Thus, there is a balance among the various computational requirements associated with different behavioral requirements (Rolls 1992). Assessing neuronal variation across such behavioral variation is at the heart of the neuroethological approach.

In the context of the proposed framework, neurons act not as single-cell indicators of complete percepts representing components of signals that in combination result in perception of objects (cf. Barlow 1972), nor as featureless cogs in a vast distributed machine, but as localized feature detectors. In this proposal, the mechanisms for combining features may be single cells, population dynamics, or both. For more arbitrary objects, the features might be described by the statistics of natural scenes (Simoncelli and Olshausen 2001). Each species also lives in its own unique perceptual environment, and this serves to delimit the more complex features and combination of features that are represented at the level of single cells.

The evidence for feature detectors of specific signals is extensive, especially in acoustical-vocal systems (see below), which have been extensively studied from the perspective of natural stimuli. The evidence in other systems, especially vision, is not as complete, in spite of pioneering work on the frog visual system (Lettvin et al. 1959; Ewert et al. 1983a). Recent work suggests that combinations of visual feature detectors (columns) are a part of the process of object recognition (Tsunoda et al. 2001). Resolution of the issue of visual feature detectors may await more consistent application of neuroethological principles to vision research. There is strong suggestive evidence and considerable dispute, for example, regarding the existence of feature detectors in the context of visual face recognition in monkeys (e.g., Fujita et al. 1992; Gross 1992; Wang et al. 2000). Yet, far more of visual processing than face and hand recognition—in the case of monkeys, for instance, for fruits and other food sources, classes of predators, and classes of habitat—may be dominated by specialized mechanisms than is generally appreciated. Visually guided behaviors for these classes of stimuli provide logic for searching for corresponding biases in the visual system. The well-documented referential call system based on classes of predators in vervet monkeys (Seyfarth et al. 1980a) suggests an interesting parallel that could guide vision research.

Finally, if neurons in some pathways act as local feature detectors for certain classes of stimuli, then it becomes important to specify how subsets of those neurons interact to produce the percept. There is considerable theoretical work on population coding and the role of synchronous neuronal activity to dynamically "bind" simpler response profiles into global percepts (Singer and Gray 1995; see deCharms and Zador 2000) and some computational work that might suggest how simple subfeatures are combined to detect objects (Lee and Seung 1999). This is an area where a neuroethological approach could be quite advantageous but so far has only begun to be applied.

2.1.2. The Organization of Feature Detectors

In early single-neuron studies of sensory systems, neuroethologists observed specializations of the auditory system related to vocalizations of interspecific predators and courtship signals (Capranica 1965; Roeder 1966). These observations were interpreted in terms of a hierarchical organizational scheme of sensory systems, with neurons higher in the hierarchy having more selective responses related directly to behavioral output. Extensive data, collected from many systems following these early studies, now support the existence of both hierarchical organization schemes and feature-detector cells. The salient observations included tuning of peripheral responses to a behaviorally relevant range of parameters and cells at higher levels of a sensory hierarchy sensitive to particular combinations of spectral and temporal components of behaviorally relevant sounds. The extent of peripheral and central specialization is species- and behavior-dependent and, at least in some cases, can also be influenced by sex (e.g., Narins and Capranica 1976). It is noteworthy that peripheral and central specializations were predicted in the vigorous debates of the early ethologists (Lehrman 1953). Obviously, each species must vocalize in a range that it can hear, but this does not predict the observed peripheral specializations. For example, in several species of anurans, different frequency ranges are represented in different peripheral auditory organs (Capranica 1978). In some species of bats, the important second-harmonic region of the echolocation signal is highly overrepresented in the cochlea (Bruns and Schmieszek 1980). Peripheral specializations have also been observed in reptiles and birds (Manley 1990).

It follows that if the evolution of vocal signals can shape the action and distribution of peripheral receptors to match the vocal signal (or vice versa), then there are likely to be central effects as well. Indeed, in numerous systems, central neurons have been observed that respond selectively to specific spectral and temporal features of conspecific or autogenous (self-produced) vocalizations or are selective for conspecific vocalizations within a repertoire of sounds (Leppelsack and Vogt 1976; Mudry et al. 1977; Suga et al. 1978, 1979; Scheich et al. 1979b; Margoliash 1983; Rose and Capranica 1983; Rose et al. 1988; Rauschecker et al. 1995; Bodnar and Bass 1997, 1999; Crawford 1997; see Ewert et al. 1983b; Feng and Schellart 1999). In many cases, the specializations involve processing for spectral combinations and temporal sequences of sounds. These are reviewed separately in the next sections.

The feature-detector concept does not require stimulus and response invariance. Feature-detector cells may exhibit dynamic modulation of response properties (Zhang et al. 1997), in some cases in response to changes in behavioral state (Dave et al. 1998a), and more permanent plastic changes in response profiles (Knudsen 1985; Doupe 1997). These dynamic properties may complicate the analysis of the feature representations, but if the dynamic process is lawful, then it does not invalidate the concept of feature detection (cf. Manley and Müller-Preuss 1978). In the best-studied cases, the dynamic changes observed have been closely correlated with changes in behavior.

2.1.2.1. Parallel Hierarchical Pathways

Feature detectors imply hierarchical organization, but there is also extensive evidence that within a sensory system several parallel pathways may exist, each hierarchically processing different sets of sensory cues associated with different behavioral tasks (e.g., different aspects of recognition or localization). Such data have provided further support of the feature-detector hypothesis. The highly specific behavioral deficits in human patients with certain brain lesions are particularly compelling in this regard because they link regional localization with profound yet restricted perceptual specializations (Damasio et al. 1990). Animal studies complement this analysis by permitting controlled experimental manipulations. In electric fish, the magnitude and phase of the electric organ discharge are processed separately by different "P" (probability) and "T" (timing) classes of receptors that encode amplitude and phase, respectively, and distinct pathways arising from these receptor classes ascend the CNS until they converge at higher levels (Heiligenberg 1991). The neurons in the highest levels can be described as "recognition" neurons, whose activity is directly predictive of behavior (Rose et al. 1988). In bats, separate ascending systems of projections and multiple cortical areas appear to mediate the differential processing of constant-frequency and frequency-modulated components of the echolocation calls (Olsen and Suga 1991a, 1991b; see Casseday and Covey 1995), and lesion studies support this idea (Riquimaroux et al. 1991). Similarly, in barn owls, sound amplitude and timing information appear to be encoded in separate ascending pathways prior to convergence at the level of the midbrain (Moiseff and Konishi 1981; Sullivan and Konishi 1984). Again, lesions of each pathway produce specific behavioral deficits associated with loss of discrimination of one dimension in the parameter space but not the other (Takahashi et al. 1984). In frogs, the different frequency pathways that arise in the periphery are eventually combined at the level of single neurons (Fuzessery and Feng 1982, 1983). A similar conclusion appears to obtain for the primate auditory system, where separate ascending pathways exhibit differential sensitivity for stimulus morphology and location (Romanski et al. 1999; Rauschecker and Tian 2000). Collectively, these and other data provide strong support for a hierarchical-, modality-, and parameter-specific organization of vertebrate sensory systems (Ulinski 1984).

2.1.3. Combination Sensitivity

Neurons can exhibit processing for complex sounds by virtue of elaborations in spectral or temporal components of their receptive fields. In audition, the sequence of discrete events is critical for perception, and neuronal specializations for detecting sequences of events can be expected. One form of neuronal sequence sensitivity is called combination sensitivity. A neuron that is combination-sensitive responds with nonlinear summation (also called facilitation) to a combination of sounds (either discrete spectral components of the sound or two or more temporally discrete elements of a sequence) as compared with the response of the neuron to subsets (typically individual components) of the sound. There is no standard for what a "sufficient" nonlinear response is for a neuron to be considered combination-sensitive as long as the additional nonlinear component reaches statistical significance. Neuronal combination sensitivity may be far more common than has been appreciated (e.g., Brosch and Schreiner 2000;

Gehr et al. 2000). Different magnitudes of the nonlinear components of the combination response, and different temporal windows of integration, may reflect different mechanisms that give rise to combination sensitivity.

2.1.3.1. Spectral Combination Sensitivity

Neurons with spectral combination sensitivity are reported in a broader range of species than are neurons with temporal combination sensitivity. This may reflect, at a neurobiological level, the fundamental nature of the process of integration of spatial information arranged across a sensory epithelium and, at a behavioral level, that sensitivity to temporal structure is probably a secondary adaptation. Neurons that require two or more spectral lines before exhibiting a facilitated, typically excitatory response have been well-characterized in bats (Suga et al. 1979), birds (Langner et al. 1981; Margoliash and Fortune 1992), cats (Sutter and Schreiner 1991; Nelken et al. 1994a, 1994b), and frogs (Mudry et al. 1977; Fuzessery and Feng 1982, 1983), and there is weaker evidence in other systems as well. The responses have almost always been linked to specific spectral components of vocalizations, whereas spectral combination sensitivity was not observed in the auditory cortex of monkeys trained on a missing-fundamental task (Schwarz and Tomlinson 1990). Spectral combination-sensitive neurons can be common throughout the auditory system in some animals (bats), or locally common within a specific neural structure (frogs, cats) or within a specific neural pathway (songbirds). Because these response properties can be difficult to identify without the adequate stimulus, the failure to find spectral (or temporal) combination-sensitive neurons cannot be taken as proof that such response properties are absent.

The mechanisms for producing spectral combination-sensitive neurons are probably best described from extracellular recordings conducted in the echolocating CF-FM mustached bat, *Pteronotus parnelli*. Members of this species produce a biosonar "pulse" for orientation composed of a constant-frequency (CF) and frequency-modulated (FM) component, each of which has four harmonics (Fig. 7.2). The returning echoes of these pulses are delayed temporally, reflecting target distance, and have a Doppler shift in frequency based on the relative velocity of the target, typically an insect. Periodic frequency (and amplitude) modulations in the echo may result from insect wing beats. Many types of combination-sensitive neurons have been described in this system, and the auditory cortex can be divided into multiple subregions based on the particular combinations of pulse and echo components that give rise to facilitated responses. For example, neurons in the CF_1/CF_3 subregion are tuned to the first harmonic of the pulse and the third harmonic of the CF component in the returning echo (see Fig. 7.2). Within this region, specific cells are tuned to specific frequency differences between the pulse and echo and/or phase-locked to frequency modulation of the harmonic echo, representing a specific relative target velocity or

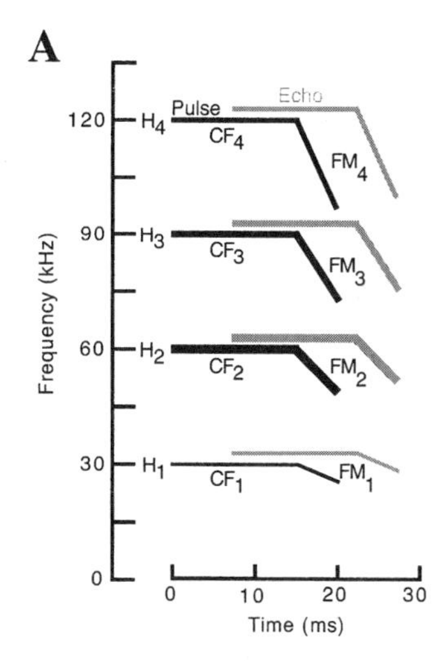

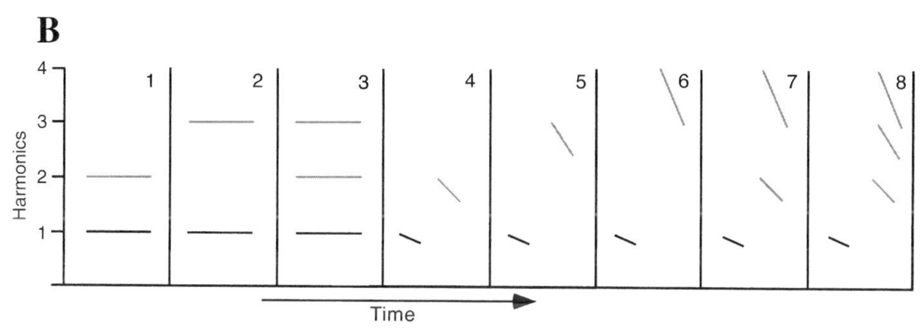

FIGURE 7.2. Schematic spectrograph (**A**) of the echolocation signal of the mustached bat, *Pteronutus parnellii*, showing the constant-frequency (CF_{1-4}) and frequency-modulated (FM_{1-4}) components for each of the four harmonics (H_{1-4}) of the pulse (black lines) and the echo (gray lines). Note that the echo is delayed in time and Doppler-shifted in frequency compared with the pulse. In (**B**), schematic representations of the signal components that facilitate common forms of spectral and temporal combination sensitivity are shown. Pulse harmonics are in black and echo harmonics in gray. (After Suga et al. 1998.)

beating wings, respectively. These multiple regions form a frequency versus frequency topographical representation that systematically maps relative target velocity. Another class of spectral combination-sensitive cells is found in the Doppler-shift constant-frequency (DSCF) region of the cortex. These neurons are tuned to very fine-grain frequency and amplitude modulations of the CF_2 component (see Fig. 7.2) and may contribute to determining target angle along with velocity (Riquimaroux et al. 1991; Kanwal et al. 1999). There are additional cortical areas in mustached bats that exhibit combination sensitivity, including some that respond to unusual combinations of stimulus components (e.g., Tsuzuki and Suga 1988;

Fitzpatrick et al. 1998). Different neural mechanisms may underlie processing of echolocation signals in different species of CF-FM bats and in different classes of bats (e.g., FM bats), but this is not reviewed here (see O'Neill 1995).

Spectral combination-sensitive neurons are also present in the ascending auditory system of mustached bats at the level of the medial geniculate body of the thalamus (MGB) (Olsen and Suga 1991a; Wenstrup 1999) but have not been found in the inferior colliculus (O'Neill 1985). This led to the hypothesis that the spectral combination sensitivity originated at the level of the MGB as a result of converging inputs from inferior collicular (IC) neurons tuned to specific frequencies in the sonar pulse and higher harmonics in the returning echoes. Recent findings, however, have challenged the idea that the IC is the primary source of converging input to the MGB (Wenstrup and Grose 1995), so additional study is clearly needed.

One potential source of combination-sensitive responses in the MGB is descending cortical projections. Corticofugal projections originating in layers V and VI of the auditory cortex project to both the IC and the MGB as well as subcollicular auditory nuclei. These projections appear to maintain tonotopic organizations such that high-frequency tuned regions of the cortex project to high-frequency tuned regions of the MGB and IC, for example. Although it is clear that these projections enhance the tuning of cortical cells through feedforward stimulation of frequency-matched subcortical units and the reduction of responses in unmatched subcortical units (Suga et al. 1997; Zhang and Suga 2000), their role in the generation of combination-sensitive responses has only recently come under investigation. The complexity of this corticofugal system is compounded by the presence of both direct excitatory connections and inhibitory connections via the thalamic reticular formation (see Wenstrup 1999).

2.1.3.2. Temporal Combination Sensitivity

Temporal combination-sensitive (TCS) neurons have been well-described in bats and birds, and there is some evidence for TCS neurons in other systems (primates: Olsen 1994; cats: Brosch and Schreiner 2000; Gehr et al. 2000). The behavioral requirements for temporal combination sensitivity differ dramatically for songbirds and bats, and the observed TCS response properties are tightly linked to the behaviors. This suggests that in systems where the animal's behavioral requirements are not sufficiently understood, it might be possible to overlook neurons with a TCS response for lack of use of the appropriate stimulus.

In mustached bats, TCS neurons have been described in the ascending auditory system, starting at the level of the inferior colliculus (Mittmann and Wenstrup 1995; Leroy and Wenstrup 2000) and continuing in the medial geniculate nucleus (Olsen and Suga 1991b; Wenstrup 1999) and in multiple cortical areas (e.g., Suga et al. 1978; Suga and Horikawa 1986; Taniguchi et al. 1986). Cortical neurons typically require FM_1–FM_N ($N = 2, 3, 4$) stimuli, that is, the fundamental FM component and one of the higher harmonics. The neurons typically are tuned to a fixed delay between the two FM components, in which case they are sensitive to a particular target distance (echo delay). In some cases, FM/FM neurons respond to systematic changes in delay between the two FM components, which could facilitate tracking an approaching target such as an insect (O'Neill and Suga 1979). As with the CF/CF neurons, FM/FM neurons are organized in a topographic manner. In the cortex, each pair of components (i.e., FM_1 paired with $FM_{\{2|3|4\}}$) is mapped in a separate subregion. The principal parameter mapped within each FM/FM subregion is delay. That is, time delays (which relate to target distance) are systematically mapped within the cortex (O'Neill 1995).

Recent data from the mustached bats suggest that TCS facilitation first arises at the level of the inferior colliculus (Portfors and Wenstrup 1999; Wenstrup et al. 1999) and is mediated via glycineric inhibitory mechanisms (Wenstrup and Leroy 2001). Neural tuning for sound duration also appears to arise via inhibitory processes in IC (Casseday et al. 1994), and neurons in IC are also tuned to periodic frequency modulations according to the rate and amount of the FM (Casseday et al. 1997). The data support a model in which duration tuning arises from the interaction among several different types of subthreshold inputs and suggest that similar mechanisms may give rise to FM tunings as well (Covey 2000). It may be the interaction of these processes at the population level that gives rise to TCS facilitation. Here again, corticofugal projections may play a role as their effect on the delay tunings for FM–FM combination-sensitive neurons has recently been shown (Yan and Suga 1999).

In songbirds, TCS neurons have been described in forebrain components of the song vocal-control system. TCS neurons in songbirds are sensitive to sequences of notes or syllables (Margoliash 1983; Margoliash and Fortune 1992; Doupe 1997; Dave and Margoliash 2000). Starting with what appears to be the first song system nucleus with auditory responses (Janata and Margoliash 1999), all forebrain areas of the song system are selective for acoustic features of the individual bird's own song. It is probable that each of these areas contains TCS neurons, although this has yet to be tested.

The extracellularly described properties of TCS neurons differ dramatically between birds and bats. Whereas in mustached bats the range of time delays represented in the neuronal population spans roughly 0.4–18 msec (corresponding to typical echo delays experienced during predation), the time course for TCS cells in birds is much longer (~80–350 msec) and can include many song syllables (Margoliash 1983; Margoliash and Fortune 1992; Dave and Margoliash 2000). Delay tuning of TCS neurons is assessed by systematically varying the interval between the first and second components that, in combination, result in the TCS response. When such tests were conducted in white-crowned sparrows (Margoliash 1983) or zebra finches

(Margoliash and Fortune 1992), the memory of TCS neurons was observed to extend up to approximately 300–500 msec. The long integration periods and high degree of stimulus specificity place the TCS cells described in songbirds among the most complex auditory neurons known.

Originally, a simple model of the TCS response in birds was proposed whereby release from inhibition resulted in the nonlinear component of the response (Margoliash 1983). Data from intracellular studies found evidence for that model but also for other interacting subthreshold mechanisms similar to those described in the bat IC, along with additional threshold nonlinearities (Lewicki and Konishi 1995). Whereas axonal delay lines may account for the shortest-time-scale TCS responses in bats (Kuwabara and Suga 1993), the much longer time scales of the TCS response in songbirds cannot be accounted for by axonal delays. In songbirds, recent data show that different subthreshold responses are associated with two distinct populations of projection neurons and a population of interneurons in the song-control nucleus HVc (Mooney 2000). TCS responses are thought to be present for at least some neurons in both classes of projection neurons and for the interneurons. Thus, as in bats, multiple mechanisms in birds may be responsible for the TCS response in different classes of neurons. One difference between TCS neurons in CF-FM bats and birds is that, in the latter, the neurons are predominately found in vocal-control areas—areas that also participate in generating motor output. This suggests the hypothesis that a relatively undifferentiated auditory input to song-system nuclei might be patterned in interaction with central pattern generators for song that are shaped by song learning (Margoliash et al. 1994). TCS responses are apparently rare in the ascending auditory system of birds, whereas they are common in the ascending auditory system of bats.

2.2. Input Constraints on Representational Systems

The search for specializations in the auditory system related to vocal behavior has often resulted in identification of complex, nonlinear neural-response properties such as the combination-sensitive neurons described above. However, practical limitations on stimulus power, stimulus and recording duration, and stationarity, coupled with the high-order nonlinearities central neurons typically exhibit, have handicapped quantitative approaches to nonlinear analysis (e.g., white-noise analysis). As a result, generalized procedures to characterize nonlinear neuronal responses have not been established, and linear techniques to analyze complex, especially natural, stimuli are being developed (e.g., spike-triggered receptive fields: Klein et al. 2000; Theunissen et al. 2000; spike-based stimulus reconstruction: Rieke et al. 1996). Information-theoretic approaches can describe nonlinear statistical properties of neurons, but these often require more data than is practical to collect. Furthermore, the ultimate utility of information-theoretic descriptions is not yet resolved. From this perspective, the use of natural stimuli in neurophysiological experiments is very valuable. Although the theory underlying their use is not yet well-developed in a formal sense, natural stimuli realistically constrain the search space to a range of input parameters that, because of their behavioral relevance, are likely to be represented by central neurons.

The use of natural stimuli to characterize neuronal responses is somewhat analogous to the application of psychophysical paradigms to characterize neuronal responses. Just as a psychophysical illusion may identify new neuronal response properties (von der Heydt et al. 1984), so may presentation of a new category of natural sounds identify new components of receptive fields, even in cases where neurons have previously been extensively studied. For example, although bat cortical neurons had been extensively studied in the context of echolocation, entirely new excitatory regions of their receptive fields that had previously been missed were uncovered when social communication calls were first presented to those neurons (Ohlemiller et al. 1996). The receptive field domain suitable for communication calls may represent the plesiomorphic condition. The existence of two separate receptive field domains highlights the degree of nonlinearity or high dimensionality (complexity) of the parameter space and emphasizes that the choice of stimulus repertoire can be a subtle decision that shapes the limitations of the experimental paradigm. Beyond this obvious caveat, procedures have been established for using natural stimuli in neurophysiological experiments.

2.2.1. Selectivity and Specificity for Conspecific Vocalizations

Two strategies have been employed to assess the responses of neurons to species-typical vocalizations. The first is to use natural vocalizations and derivative sounds to identify the potential behavioral relevance of neural responses. This is assessed in terms of a neuron's selective responses to a subset or category of natural vocalizations (i.e., its selectivity). The second strategy is to identify the specific acoustic features underlying a neuron's selective responses (i.e., its specificity) by using artificial stimuli to characterize its response properties. These two strategies are complementary, and both may be necessary to adequately describe a neuron's responses to natural stimuli. Testing for selectivity rarely unambiguously identifies the specific acoustic features that result in the selective response. Conversely, exploration of specific features of a neuron's response properties may give only limited biological insight into the significance of those features in the absence of a well-delineated behavioral context. Absent knowledge of the acoustic behaviors, neural analyses may ultimately emphasize response parameters that are not central to the behavioral decisions associated with real-world tasks such as vocal recognition.

In cases of complex vocal behavior (i.e., where multiple acoustic parameters affect the behaviors under study), an experimental approach is mandated that renders the distinction between selectivity and specificity ultimately arbitrary. Such cases require the combination of selectivity and specificity approaches. To accomplish this, the relevant natural vocalizations are systematically decomposed into simpler signals. At the same time, artificial sounds are used to synthesize increasingly closer approximations (models) of the natural vocalizations. Optimally, both decomposition and synthesis procedures are sensitive to behaviorally salient variation identified in the natural vocalizations. The goal is to bring the decomposed and synthetic stimuli to some common intersection and thus to establish a logical relationship between the variation in neuronal response and the variation in the stimuli.

Combining the approaches of modified natural and artificial stimuli need not result in convergence on the same solution—different parameter spaces may be identified using the two approaches. In addition, errors are possible, because there are practical limits to the size of a repertoire that depend on the stability of the recordings and the natural rate of stimulus repetition. In the absence of a specific model, these limits may prevent the choice of a sufficient repertoire in cases of large and complex vocal repertoires or highly selective neuronal responses. Errors in analysis may also result from experimental decisions about the appropriate stimuli that have to be made online without the benefit of retrospective analysis. These can be exceptionally challenging and exciting experiments! In the limited number of cases where this approach has been employed, data from natural and artificial stimuli have converged upon common solutions (e.g., Margoliash 1983). When the two approaches converge on a common set of parameters, this gives confidence that a uniform model of the neuron's response profile has been achieved. In this case, the derived model can account for the neuron's response in terms of a specific set of acoustic parameters in the natural vocalizations.

The success of a receptive-field model of neuronal response selectivity can be independently determined by quantitative predictions of response selectivity based on acoustic specificity that in turn can suggest specific predictions regarding cellular mechanisms that give rise to the selective responses. Achieving such quantitative predictions may itself require a significant modeling effort and is not frequently attempted. For example, the responses to songs of some avian auditory thalamic neurons can be quantitatively predicted from their responses to tone and noise bursts (Bankes and Margoliash 1993; Anderson et al. 1996). Achieving this result required extensive testing to find the appropriate model (network) architectures. Interestingly, the predictive power of the models was most sensitive to the dynamics of the neuronal responses to the artificial stimuli. Model output was less sensitive to manipulation of traditional static descriptions of neuronal response such as the frequency-amplitude response curves and rate intensity functions.

2.3. State Constraints on Representational Systems

Selection processes shape the acoustics of vocal communication signals. Constraints that result from these pressures are not restricted to the signal (i.e., input) acoustics but also the organization (i.e., state) of the representational system. In the following section, we consider a number of ways in which the organization of the representational system is constrained by both phylogenetic and ontogenetic processes. These constraints include potential links between motor and sensory representations, learning, and the relationship between species and subspecies-level recognition behaviors.

2.3.1. Motor-Sensory Linkages

2.3.1.1. Genetic Coupling and Coevolution

Signal production and perception are linked behaviorally (see Section 1; Fitch and Hauser, Chapter 3; Ryan and Kime, Chapter 5). One possible explanation for this parity is that the production of a class of sounds and perception of sounds of the same class are tightly linked within each individual. Such an arrangement could result from genetic coupling, wherein the neural mechanisms for production and perception of a signal would share common elements controlled by the same set of genes (Alexander 1962). Thus, modification of the genetic material would result in concomitant changes in production and perception. Although the simplicity of this idea is attractive, there is no guarantee that the changes will be coordinated or that the modified element will contribute the same change, in the same direction and magnitude, for both modalities. Rather than postulate a single genetic basis for separate production and perception systems, a more parsimonious solution is to postulate a single central pattern generator responsible for both production and perception (Bentley and Hoy 1972; Hoy et al. 1977). The predicted role of central pattern generators in perception represents an additional prediction of the genetic-coupling hypothesis.

Alternatively, the match between perception and production could result from coevolution (von Helversen and von Helversen 1994). By this account, the mechanisms of production and perception of communication signals are genetically independent but evolve under reciprocal selective pressure toward representation of a common set of communication features. Coevolution may arise through the common effects of a shared environment. For example, males in the gray tree frog species *Hyla versicolor* and *H. chrysoscelis* both produce pulsatile calls to which gravid females are attracted (Gerhardt 1982). Although the mean pulse rate varies between species, it is also temperature-dependent, such that a male *H. chrysoscelis* calling at 20°C could have a pulse rate similar to a male *H. versicolor* calling at 15°C. In areas where the species are sympatric, this temperature-dependent property of male calls creates the potential for mating errors.

However, female preferences for calls within the conspecific frequency range are also temperature-dependent (Gerhardt and Doherty 1988). Because these temperature dependencies scale differently under similar temperature ranges, they are likely to be controlled by different neural mechanisms (see van Dijk et al. 1997). Of course, genetic coupling and coevolution are not mutually exclusive.

2.3.2. Motor-Sensory Linkages in Learned Signals

Linkages between production and perception can also result from developmental changes that selectively modify phenotypic patterns. A compelling example is the development of human speech. It has long been known that human infants can distinguish many of the phonetic contrasts in speech, even those in languages to which they have not been exposed (e.g., Lasky et al. 1975). Adult speakers of different languages, on the other hand, not only differ from one another in the location of perceptual boundaries between phonetic contrasts but also lose altogether the ability to perceive some nonnative contrasts (e.g., Best et al. 1988; Logan et al. 1989). The loss of sensitivity to selected nonnative contrasts occurs somewhere near the end of the first year (Werker and Tees 1984).

The developmental changes in speech perception are roughly matched to developmental changes in vocal production, and such findings have been taken to support the theory (Liberman et al. 1967) that speech perception is guided by a process that matches speech sounds to the vocal gestures required to produce those sounds. This is the so-called "motor" theory of speech perception. As with the genetic-coupling hypothesis, in motor theory, the same pattern generators participate in perception and production. The modern version of the motor theory of perception (Liberman and Mattingly 1985) stresses representational modularity as conceived by Fodor (1983).

Psychologists have staged a series of strong assaults on the motor theory (e.g., Lindblom 1991; Fowler 1996; Nearey 1997). Because the motor theory of speech perception appears to make clear predictions regarding the recruitment of specific motor pathways in speech perception, it would appear possible to test predictions of motor theory with neurobiological techniques. Unfortunately, instantiations of the motor theory of speech perception do not specify the neural mechanisms implied by requisite processes such as vocal gesture, information encapsulation, and so forth. Absent these precise definitions, motor theory is not falsifiable at a neurobiological level. In addition, experimental neuronal data from humans with the spatial and temporal precision required to test theories of perception such as motor theory cannot yet be obtained.

The comparative approach can help resolve questions that might otherwise be difficult to resolve with the available human data. Humans are phylogenetically isolated with regard to the central feature of interest, vocal learning. This limits the effectiveness of a phylogenetic analysis (Fitch 2000). Nevertheless, computational problems are shared by all biological systems that exhibit vocal learning, which may result in similar mechanistic solutions (Margoliash 2001). The following section describes production and perception of song in relation to various pathways in the avian song system.

2.3.3. Production-Perception Linkages in Birdsong

The bird song system provides an experimental model system amenable to detailed neurobiological approaches with which to pursue questions of production/perception linkages in the context of a learned vocalization. Comparisons of neural responses to conspecific songs with responses to a bird's own song(s) explore the hypothesis that a bird's own song holds a position of particular perceptual significance within the representational system. This hypothesis is well-supported by a variety of data. For instance, several field playback studies demonstrate that aggressive responses from territorial males are often dependent on whether the particular playback song is similar to one that the subject sings (see Owings and Morton 1998). In most species, playback of the individual's own song elicits a level of aggressive response intermediate to the weakly aggressive responses to songs of established neighbors and strongly aggressive responses to unfamiliar (strangers') songs (Stoddard 1996). The salience of a bird's own song is also apparent under more controlled conditions, as for example when examining acquisition rates or category formation in operantly trained tasks (Cynx and Nottebohm 1992; Gentner and Hulse 1998) or behavioral preferences (Pytte and Suthers 1999; Okanoya et al. 2000). Together, such data suggest that some form of coupling exists between the song-production and perception systems.

2.3.3.1. The Vocal Motor Pathway

The extensive analysis of the avian song system provides a basis for understanding how different components of the system contribute to perceptive and productive components of behavior. Several forebrain pathways have been identified as components of the song system, of which two have been extensively analyzed (Fig. 7.3). A vocal motor pathway (VMP) is required for song production, participates in moment-to-moment control of singing, and represents the descending motor outflow of the forebrain song system. An anterior forebrain pathway (AFP) shares similarities with the mammalian corticothalamo-basal ganglia loop. The AFP is required for normal song development and may be necessary for adult song maintenance, although contributing little (depending on species) to moment-to-moment control of singing. The relative contributions, if any, of the VMP and AFP to components of perceptual processing of conspecific songs would represent different schemes for establishing perception/production linkages, a

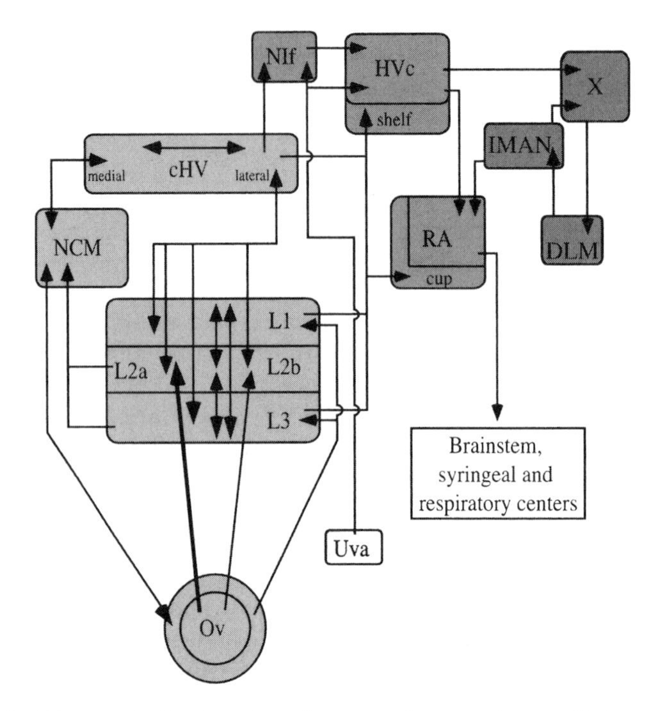

FIGURE 7.3. Diagram indicating the primary auditory pathways (light gray), vocal motor pathway (VMP, gray), and anterior forebrain pathway (AFP, dark gray) in songbirds. Ov: nucleus ovoidalis; L1–L3: field L complex; NCM: caudomedial neostriatum; cHV: caudoventral hyperstriatum; NIf: nucleus interfacialis; Uva: nucleus uvaeformis; RA: robust nucleus of the archistriatum; lMAN: lateral magnocellular nucleus of the anterior neostriatum; DLM: dorsolateral region of the medial thalamus; X: area X; HVc is used as the proper name. Ov comprises a core and a surround. HVc and RA have subjacent regions, the "shelf" and "cup," respectively.

direct connection with motor output in the case of the VMP, or a shared developmental trajectory in the case of the AFP.

In the VMP, the roles of the nuclei HVc and RA in song production have been well-established through electrolytic lesions (Nottebohm et al. 1976), electrophysiological recordings in singing birds (McCasland and Konishi 1981; Yu and Margoliash 1996), and perturbation experiments using electrical-stimulation (Vu et al. 1994). Zebra finch songs are composed of repeated motifs, which are sequences of different syllables. Each syllable typically comprises several notes. Recordings of single neurons in singing birds demonstrate that premotor activity, in what are probably HVc interneurons (see below), is organized by the syllable type. That is, repetitions of the same syllables are associated with similar activity patterns, independent of position within the song, whereas different syllables are associated with different patterns of activity (Yu and Margoliash 1996). This results even if the syllable types share some or many notes. RA neurons, which receive from HVc, exhibit patterns of physiological activity that contrast with those of HVc neurons. Premotor activity of RA neurons is organized by note type, independent of the syllable in which the note is embedded (Yu and Margoliash 1996).

The syllable is probably the largest unit of vocalization that represents a motor program. When singing birds are startled, they tend to complete the current syllable before ceasing to sing (Cynx 1990). Upon electrical stimulation of HVc during singing, birds tend to restart the song at the beginning of motifs (i.e., at large segmental boundaries representing repeated sequences of syllables). In contrast, stimulation of RA during singing disrupts the syllable morphology but not the sequence of syllables (Vu et al. 1994). Thus, information about the large-scale organization of song (sequences of motor programs) could be available to the VMP and the AFP through the two distinct classes of HVc projection neurons; HVc projection neurons target exclusively either RA or area X (Fig. 7.3).

The different classes of HVc projection neurons, and HVc interneurons, exhibit differences in their intrinsic properties and subthreshold auditory-response properties (Mooney 2000). Furthermore, recent data from electrical-stimulation studies designed to establish identities of extracellularly recorded classes of HVc neurons suggest that most of the chronic recordings were of HVc interneurons (Shea et al. 2001). Thus, it remains to be seen whether the mapping from syllable-level to note-level representations occurs within HVc (at the level of the RA-projecting HVc neurons) or in the projection of HVc to RA. The HVc to RA mapping potentially represents a transition from categories of behavior to motor output. It is likely, in any case, that the VMP and the AFP receive different mixes of premotor activity and sensory feedback during singing.

The physiological properties of other VMP nuclei contrast with those of HVc and extend the concept of a hierarchical arrangement of sensorimotor control. Neurons in NIf and Uva (Fig. 7.3), which project to HVc, are active during singing (McCasland 1987; Williams and Vicario 1993). Multiunit recordings from Uva suggest that some activity is more closely associated with the timing of entire motifs, and bilateral lesions of Uva disrupt the suprasyllabic organization of song but do not abolish singing (Williams and Vicario 1993). Recent data from zebra finches suggest that NIf is a major source of auditory input to HVc (Janata and Margoliash 1999; Boco and Margoliash 2001). Bilateral lesions of NIf have phrase-level effects on singing in Bengalese finches (Hosino and Okanoya 2000) and apparently only transient effects on singing, with full recovery, in zebra finches (Vu et al. 1995). Thus, lesions of VMP nuclei afferent to HVc affect but do not abolish singing whereas lesions of HVc or RA abolish all singing behavior.

2.3.3.2. Auditory Responses in HVc

Neurons in HVc show auditory responses; in particular, they exhibit selective responses to the bird's own song compared with conspecific or parametrically modified songs (Katz and Gurney 1981; McCasland and Konishi 1981; Margoliash 1983, 1986, 1987; Margoliash and Konishi 1985; Margoliash and Fortune 1992; Volman 1993; Lewicki 1996; Yu and Margoliash 1996; Theunissen and Doupe 1998; Mooney 2000). The selective responses of song-system neurons have motivated the hypothesis that HVc and the AFP may be involved in production-independent perceptual processing of songs that is influenced by representations of a bird's own song (Margoliash 1986, 1987). This hypothesis suggests developmental linkage of responses to a bird's own song with motor output, without invoking direct motor processing during perception.

Studies of HVc-lesioned female canaries are consistent with a role for HVc in song perception. Lesions to HVc abolish female behavioral preferences for conspecific over heterospecific song (Brenowitz 1991) and for sexually attractive song phrases over other phrases of conspecific song (Del Negro et al. 1998; and see Vallet and Kreutzer 1995). That is, although only conspecific song elicits copulation displays in prelesion females, all songs (conspecific and heterospecific) elicit solicitations following HVc lesions. Multiunit recordings from HVc in sexually receptive female canaries have identified inhibition that is restricted to presentations of sexually attractive conspecific song phrases (Del Negro et al. 2000). Thus, there may be a linkage between physiological inhibition and that observed in behavior. Because the projection of area X onto DLM (Fig. 7.3) is probably inhibitory (Luo and Perkel 1999), this suggests an overall release from inhibition in the rest of the AFP (DLM and lMAN) that is proportional to the response in HVc. In female canaries, the degree of release would be proportional to the sexual potency of the stimulus.

The role of HVc in song discrimination may be species-dependent. In both male and female starlings, HVc lesions do not affect the retention of learned conspecific song discriminations but do affect the ability to form new associations with familiar songs (Gentner et al. 2000). In contrast to the studies in canaries and starlings, in female zebra finches, *Taeniopygia guttata*, lesions to HVc have no effect on copulation solicitations (MacDougall-Shackleton et al. 1998a). Qualitative differences in sexual dimorphism between zebra finches and other species, especially the magnitude of HVc projections onto RA and area X, may help to explain these differences (Nottebohm and Arnold 1976; Gurney 1981). The data suggest that rather than providing a direct role in perceptual representation, HVc may be acting as a selective filter regulating motor control of subsequent behaviors, including copulation responses in females and song output in males. Under this hypothesis, HVc filtering, and hence the perceptual processing of song, varies across species depending on the degree of sexual dimorphism in the brain of the animals.

2.3.3.3. The Anterior Forebrain Pathway

A second pathway in the songbird forebrain involves an avian corticobasal ganglia-thalamocortical pathway (Bottjer and Johnson 1997; Farries and Perkel 2000; Perkel and Farries 2000). This three-nucleus pathway makes an obligatory contribution to song development (Bottjer et al. 1984; Sohrabji et al. 1990; Scharff and Nottebohm 1991). In zebra finches, the AFP has also been shown to contribute to song maintenance (Williams and Mehta 1999; Brainard and Doupe 2000). Although neurons in the AFP in zebra finches show premotor activity during singing (Jarvis and Nottebohm 1997; Hessler and Doupe 1999), lesions of AFP nuclei—in contrast to lesions of HVc or RA—have little disruptive effect on motor output (Nordeen and Nordeen 1993). In Bengalese finches, partial lesions of an AFP nucleus result in the transient disruption of song (Kobayashi et al. 2001). The premotor activity in the AFP has been interpreted as an efference copy signal that combines with auditory feedback to create an error signal that helps stabilize vocal output (Williams and Mehta 1999; Brainard and Doupe 2000, 2001; Solis et al. 2000). The differences between zebra and Bengalese finches may result from the apparent greater reliance on auditory feedback, and hence a greater effect of the AFP, in song maintenance in adult Bengalese finches (Okanoya and Yamaguchi 1997).

Different populations of HVc neurons project to RA in the VMP and area X in the AFP. Thus, the apparent role of HVc in some aspects of song perception does not distinguish between potential differential roles of the VMP and AFP in song perception. If the VMP is involved in song perception, this would provide strong evidence for a "motor" theory of birdsong perception (Williams and Nottebohm 1985) and its related theory for speech perception by reference to production (Liberman et al. 1967; Liberman and Mattingly 1985). Alternatively, the perceptual role of HVc could be mediated through its AFP projections. In this case, the observed linkage between production and perception would arise through a shared developmental history, not through direct moment-to-moment coupling.

The data appear to support the ontogenetic theory. In what is perhaps the most convincing evidence that the song system is involved in non-BOS acoustic stimulus recognition, lesions to lMAN in female canaries disrupt the retention of auditory discriminations (Burt et al. 2000). These lesions effect discrimination among pairs of conspecific songs, heterospecific songs, and synthetic sounds, but not among pairs of visual stimuli, and therefore suggest a general auditory processing role for the AFP rather than one specifically related to conspecific songs or a bird's own song. In contrast, the effects of lesions to the AFP in adult male zebra finches appear to be restricted to conspecific discriminations involving a bird's own song (Scharff et al. 1998). However, the later study only examined acquisition rates, not retention, and used a small stimulus set. Thus, it remains to be seen whether general auditory effects are observable in male songbirds, where the AFP shows clear selectivity to BOS (Doupe 1997).

Volumetric studies are consistent with the lesion work in suggesting a role for the AFP in song perception, particularly among females. In adult female brown-headed cowbirds, the volume and number of neurons in lMAN is positively correlated with the bird's ability to discriminate among male songs (Hamilton et al. 1997). In addition, preliminary evidence suggests that female lMAN volume may also correlate positively with male song repertoire size among several species of European warblers (DeVoogd et al. 1996).

2.3.3.4. Behavioral-State Modulation of Auditory Responses

A role for the AFP in song perception does not exclude a role for the VMP in song perception. Some controversy remains as to whether the VMP contributes to song perception. Throughout the song system, auditory responses of neurons are strongly modulated by behavioral state (Dave et al. 1998a; Schmidt and Konishi 1998). Surprisingly, the modulation of response is such that activity is suppressed in awake birds relative to sleeping birds. One report that systematically studied behavioral state changes found virtually no auditory activity in RA in awake birds but strong auditory responses in sleeping birds (Dave et al. 1998a). (Subsequent single-unit studies have suggested that the sleep state may be involved in song playback, perhaps associated with motor program consolidation; see Dave and Margoliash 2000.) Another study, which focused on the contribution of HVc and lMAN to RA auditory responses in anesthetized birds, reported (in a small part of the overall design) RA auditory responses in awake birds (Vicario and Yohay 1993). It is difficult to resolve these apparently conflicting observations. The Dave, Yu, and Margoliash (1998a) study used chronic recording techniques, and male birds were sufficiently unhindered that they freely sang in response to females. The auditory stimuli were presented after the females were removed from cages adjacent to those of the subjects. The Vicario and Yohay (1993) study used restrained animals in a sound-attenuation chamber and did not directly observe the animals during acoustic stimulation. It is difficult to assess whether these differences in design could result in sufficient differences in behavioral state to explain the variation in RA auditory responsiveness. A recent lesion study of RA also claimed to show an effect on conspecific song recognition (Vicario et al. 2001). The behavioral effects were small, however, and whether the locus of the effects of the lesions was within RA remains unresolved.

The failure to find auditory responses in the VMP in awake birds is a strong challenge to the theory that the VMP is involved in song perception (see Margoliash 1997). A related question is whether any song-system neurons exhibit auditory activity during the day. One study recording multi-unit activity identified an apparent loss of auditory activity in male zebra finch HVc during daytime recordings (Schmidt and Konishi 1998). In contrast, several other studies identified daytime auditory activity in HVc of canaries, white-crowned sparrows, and zebra finches (McCasland and Konishi 1981; Margoliash 1986; Yu and Margoliash 1996). A study of single HVc neurons concluded that many neurons retain auditory responsiveness during the day, albeit at consistently lower levels than observed at night. Consistent if weak auditory responses were also observed in area X in daytime recordings (Rauske and Margoliash 1999), and daytime auditory responses in nonsinging birds have also been reported for lMAN recordings (Hessler and Doupe 1999). This suggests that the AFP but not the VMP retains auditory activity during the day. If this conclusion is established, and these results are obtained in other species and under other behavioral measures of perceptual responses, they would provide strong evidence against the motor theory hypothesis of perception and strong evidence for the ontogenetic-coupling hypothesis.

2.3.4. Plasticity and Functional Constraints

The ethologically relevant functions of any representational system derive from a combination of an organism's evolutionary history and ontogenetic experience. Both of these processes can lead to plasticity in the neural system and thus can constrain the form of a representation at any given point in time. Plasticity can result from time-dependent components of ontogeny and from experience-dependent components (i.e., learning). The combination of these two components gives rise to the so-called "critical period" in development where certain experiences must occur at specific times in order for representations to develop normally (for discussions of critical periods, see Bateson 1979). Even after a normal sensory development, representational plasticity can be induced by altering experience in the periphery (e.g., by peripheral denervation, stimulus restriction, or illusion; Pons et al. 1991; Recanzone 1998) or by central lesions (Nudo et al. 1996). These effects have been observed in many sensory modalities, and although usually examined in relation to cortical representations (e.g., Jones 2000), they have also been observed in more peripheral structures (Gao and Suga 1998; Gold and Knudsen 2000). Moreover, representational plasticity is observed not only in response to abnormal experience. Simply training a monkey to perform a frequency-discrimination task increases the cortical representation (number of cells), sharpness of tuning, and response latency for the behaviorally relevant frequencies (Recanzone et al. 1993). Likewise, in humans and bats, tonotopic responses in the auditory cortex can be modulated by simple conditioning procedures (Morris et al. 1998; Gao and

Suga 2000). These plasticity effects appear to have important differences, depending on the specializations of particular regions (Sakai and Suga 2001). Based on such data, we hypothesize that the representations of complex acoustic events, such as vocal communication signals, are also mediated by experience. Thus, it may be that, at any one time, the representational system is "tuned" to a subset of conspecific vocalizations related to relevant tasks and contexts consistent with individual experience in a manner similar to that observed for synthetic stimuli (e.g., Kilgard and Merzenich 1998; Kilgard et al. 2001). Under such conditions, uninformed choices of natural vocalizations presented during an experiment may be incongruous with the representational state. One must be aware of the behavioral relevance associated with each vocalization and attempt to match the functional significance (i.e., the induced plasticity) of specific vocalizations to the experience of specific animals.

The behavioral relevance of different vocalizations is also shaped by evolutionary history, and this provides an additional source for constraints on the representational system. However, these effects are not always intuitively obvious. For example, vocal-communication based recognition can take many forms, depending on the species under consideration. One commonly observed form of recognition is between heterospecific and conspecific signals (i.e., species recognition). Among the midshipman, *Porichthys notatus*, a species of nocturnally active fish, for example, females appear to use male acoustic signals to localize prospective mates (McKibben and Bass 1998). The same is true for many anurans (Capranica 1965; see Feng and Ratnam 2000) as well as birds and mammals (Searcy and Yasukawa 1996). Based on such widespread observations, one general function of the perceptual system may be to differentiate between heterospecific and conspecific signals. However, in many cases, animals are able to make much finer discriminations between conspecific signals, and to the extent that behaviorally relevant classifications are made on the basis of intraspecific acoustic variation, species recognition can be expected as an indirect by-product of an auditory system tuned to other information. This leads to the hypothesis that *inter*species recognition is related to the structure of the ascending auditory system. It may not be necessary to postulate any special mechanisms or "templates" to account for innate predispositions for recognition of conspecific vocalizations (cf. Konishi 1978; Marler and Sherman 1983). Perhaps the most dramatic and compelling evidence for this comes from chick/quail chimera studies, where perceptual predispositions were associated with midbrain structures that could be transferred across species (Long et al. 2001). Because auditory feature detectors have mostly been described in forebrain pathways, for example in birds and primates, we propose the hypothesis that *intra*species recognition is related in these species to the structure of the forebrain auditory system. In Section 3, we develop intraspecies recognition as a model for perceptual and cognitive mechanisms of vocal communication.

2.3.3.1. Categorization and Classification of Conspecific Vocalizations

One way that experience-dependent plasticity can influence the structure and form of perceptual representations is through the emergence of categorical boundaries between various sets of stimuli. At the behavioral level, categorical perception is observed as a nonlinear relationship between stimulus variation along some dimension and a corresponding behavior. Changes along a stimulus dimension that span the boundaries between categories are easy to detect, whereas equal-magnitude changes within a category are more difficult to detect. Such behaviors suggest the obvious hypothesis that similarly categorical (i.e., nonlinear) neural responses should be observed when particular neurons are exposed to fixed levels of variation both within and across category boundaries. In general, the complexity of most natural stimuli suggests that categorical neural responses are likely to arise at higher levels within forebrain hierarchies. An example of this is the recent data showing differential responses of neurons in the prefrontal cortex of monkeys to different categories of visual objects (Freedman et al. 2001). However, in cases where categorical, perceptual behavior corresponds closely to relevant dimensions of peripheral tunings (e.g., frequency), categorical responses may first arise in peripheral structures.

Not all categories of stimuli will necessarily meet the behavioral definitions required for strict demonstrations of categorical perception. More general category formation can be described at varying levels of abstraction (Herrnstein 1990). For instance, some categories of stimuli may simply reflect rote collections of arbitrary objects, whereas exemplars in other categories may share common physical features or be the predicate of an abstract rule. For instance, classes of predators, food items, or potential mates may form natural categories. These general forms of categorization appear to be nearly ubiquitous among vertebrate species (Herrnstein 1990), and some forms may occur among invertebrates as well (e.g., Wyttenbach et al. 1996). Moreover, the fact that learning influences most categorical behavior suggests an amazingly high degree of plasticity in the neural processes underlying these capabilities.

Behavioral tests are required to determine the location of category boundaries within natural stimulus sets. Here, the use of vocalization repertoires holds a distinct advantage because much of the behavioral work on communication systems has been directed at the determination of behaviorally relevant classes of stimuli. The use of repertoires of natural stimuli has been developed most prominently in the analysis of bird and primate auditory systems, where neuronal response variation across different categories of vocalizations has been tested. For example, in European starlings, a species of songbird, neurons throughout the field L (analogous to the primary auditory cortex in mammals), the caudal medial neostriatum (NCM), and caudal ventral hyperstriatum (cHV) respond to conspecific vocalizations, and the specificity of the response properties shows increasing complexity between field L and NCM/cHV (Leppelsack and Vogt 1976; Müller and Leppelsack 1985; see Section 3). In general, neurons responsive to specific, behaviorally significant features of vocalizations form a plausible substrate for categorical representations (see Section 2.1). Very few studies, however, have explored the neuronal responses to natural variation within and between categories of stimuli (Freedman et al. 2001). This work is only now beginning in primates and birds (Gentner and Margoliash 2001; Tian et al. 2001; Wang and Kadia 2001). This is an area ripe for future research. In Section 3, we provide a logical basis for understanding these neural mechanisms in the context of vocal recognition.

3. Emerging Model Systems for the Neuroethology of Vocal Recognition

The preceding sections discuss a number of ways in which neuroethological approaches using vocal communication signals can constrain both the input to and state of the representational system as well as its output. The biological plausibility of these constraints derives directly from the functional (i.e., adaptive) role served by vocal communication signals. Thus, the extent to which these various constraints can be invoked for any one system depends on our knowledge of how that organism uses particular signals under natural conditions. Ideally, one wants to know how behaviors are driven by acoustic variation in a given communication signal and then adapt the critical components of such behaviors to laboratory procedures that are amenable to physiological preparations. This requires sophisticated knowledge of both behavior and basic auditory physiology, and to date only a few systems meet these requirements. In this section, we review the central auditory physiology and vocal-recognition behavior in two communication systems, those of songbirds and primates. We concentrate on vocal recognition because it captures many features of communication systems described above (e.g., categorization) and is widespread among many taxa, making the results amenable to comparative analyses.

3.1. Perception of Vocal Signals

Variation in communication signals can occur in the spectro-temporal properties of the signal itself and also in the spatial-temporal distribution of signal sources. Together, this variation leads to at least two general classes of receiver behavior. The first class derives from the fact that not all acoustic events are of equal interest, so animals must be able to dissociate appropriate target signals from irrelevant/background noise, including nontarget conspecific vocalizations. This has been studied in the context of the so-called cocktail party effect (Cherry and Taylor 1954) and more generally in terms of auditory stream segregation (Bregman 1990). Despite its obvious importance, only a relatively few studies have addressed this phenomenon in nonhumans (Fay 1998; MacDougall-Shackleton et al. 1998b). Only a single study has examined stream segregation using natural stimuli (Hulse et al. 1997). It may be that the acoustic parameters governing stream segregation of acoustic communication signals vary dramatically from those involved in the segregation of pure tone sequences. In any case, the basic ability is likely to be widespread, and initial data suggest that such processing occurs at or before the level of the primary auditory cortex (Fishman et al. 2001). Recent reviews cover both stream segregation and the closely related topic of auditory spatial localization in nonhuman animals (Feng and Ratnam 2000; Klump 2000), and we do not address them further. Most research using conspecific communication signals assumes that the test subject has successfully extracted the target signal by presenting stimuli in isolation. This also assumes independence of segregation and subsequent classification behaviors.

Once an auditory object is formed, a second general class of behavior emerges as these objects or events are organized into behaviorally relevant classes. For example, females might rely on male vocalizations to help choose a mate and therefore are likely, under appropriate conditions, to distinguish between heterospecific and conspecific male vocalizations. The presence of such distinctions, or class boundaries, implies discrimination among multiple auditory objects along with an associative link between the resulting internal representation and some behavioral response. At the behavioral level, these processes are collectively referred to as recognition.

According to this definition, recognition can take many forms, depending on the specific boundaries between classes of vocalizations. Often these acoustic boundaries correspond to other behaviorally relevant distinctions (e.g., species, sex, kin, and individual). That is, they are not arbitrary but rather reflect the ecology of the particular animal under consideration. Although not all forms of recognition behavior are likely to be mediated by the same neural mechanisms, there are likely to be shared features across species, particularly when relevant classification requires discrimination among subsets of conspecific vocalizations. Recognition based on intraspecific acoustic variation is widespread (Boughman and Moss, Chapter 4), and several of the most recent examples from different taxa are given below.

3.2. Intraspecific Recognition Behavior

One common distinction within species is in the degree of relatedness between individuals. Vocal recognition often follows these lines. For example, king penguin chicks, *Aptenodytes patagonicus*, and emperor penguin chicks, *A. forsteriis*, can each recognize the calls of their parents (Jouventin et al. 1999; Aubin et al. 2000), in the latter case by using harmonic interference patterns generated by the simultaneous production of two sounds in different halves of the syrinx (the vocal-production organ in birds). Adult king penguins can also recognize their mate's calls (Lengagne et al. 2000), as can spectacled parrotlets, *Forpus conspicillatus* (Wanker et al. 1998), and several species of songbirds (Lind et al. 1997; O'Loghlen and Beecher 1997, 1999; Beguin et al. 1998).

Vocalization-based kin recognition is also apparent among a variety of mammalian species, including some populations of gray seals, *Halichoerus grypus* (McCulloch and Boness 2000), and Northern fur seals, *Callorhinus ursinus*, where mother-offspring recognition is maintained for many years beyond the breeding season (Insley 2000). In addition, bottlenose dolphins, *Tursiops truncatus*, have individually distinctive "signature" whistles long thought to function in recognition. Recent data confirm this by showing that mothers can recognize the whistles of their independent offspring and that independent offspring can recognize the whistles of their mother (Sayigh et al. 1999).

Female African elephants, *Loxodonta africana*, appear to have extensive networks of vocal recognition, distinguishing the infrasonic calls of female family, bond group, and even more distant kin from those of females outside these categories (McComb et al. 2000). Female spotted hyenas, *Crocuta crocuta*, can also recognize specific vocalizations of their own pups (Holekamp et al. 1999). Female greater spear-nosed bats, *Phyllostomus hastatus*, give screech calls whose acoustic structure varies between groups from different caves, and individuals appear to discriminate among the calls from different caves, although the capacity for individual vocal recognition remains unknown (Boughman and Wilkinson 1998).

3.3. *The Songbird Model*

Various forms of intraspecific vocal recognition have been observed in nearly every species of songbird studied to date (see Stoddard 1996) and have been examined more extensively here than in any other group of animals. In general, vocal recognition in songbirds provides for the association of specific songs with specific singers or locations and thereby serves as a basis for decisions in more elaborate social behaviors such as female choice (Wiley et al. 1991; Lind et al. 1997), female preference (O'Loghlen and Beecher 1997), and kin recognition among communally breeding birds (reviewed by Beecher 1991). Another complex social behavior in which vocal recognition plays an important role is territoriality, where it functions in both the manipulation and maintenance of territorial boundaries (Peek 1972; Falls and Brooks 1975; Falls 1982; Godard 1991) and thus may have indirect effects on reproductive success (Hiebert et al. 1989).

3.3.1. Field Studies

The function of male song in maintaining and establishing songbird territories is well-known. For example, removing a male songbird's ability to sing has dramatic effects on his success at holding a territory (Peek 1972, Smith and Reid 1979; McDonald 1989), and simply broadcasting a conspecific song from an unoccupied territory leads to significantly lower rates of settlement in that territory than in others' territories where control sounds or no sounds are played (Krebs et al. 1978; Yasukawa 1981; Falls 1987). Moreover, territory residents often respond weakly, or not at all, to a neighbor singing from a familiar location but more strongly to a stranger singing from that same location. Using these facts along with a variety of clever song-playback techniques in the field, a very large number of studies have demonstrated that males in many (at least 23) songbird species are capable of discriminating among neighbors and strangers on the basis of song alone. Furthermore, for several (at least 8) species, listeners are capable of recognizing individual singers on the basis of their songs (see Stoddard 1996).

3.3.2. Signal Variation

There are several ways that singer identity could be represented in the acoustic variation of male song. In the simplest case, individual males might sing a unique song or sets of songs (i.e., repertoires), and recognition would follow by the association of specific songs with specific singers. This strategy appears to be used by song sparrows, *Melospiza melodia* (Beecher et al. 1994), and European starlings (Gentner et al. 2000). One obvious feature of this strategy, potentially worth exploring, is that it may be heavily constrained by the memory capacity of the recognition system. Although this question deserves further attention, initial results suggest that the capacity of these systems is, in fact, quite high (see Gentner et al. 2000) and, at least for song sparrows, exceeds the number of exemplars that an individual is likely to face at a single time in the wild (Stoddard et al. 1992).

A second recognition strategy employed relies on morphological differences in the acoustics of shared song types. In both white-throated sparrows and field sparrows, the songs of neighboring territorial males share several acoustic features but vary slightly in frequency. Neighbors rely on these subtle frequency differences to recognize one another (Brooks and Falls 1975; Nelson 1989). Related to this is a third possible strategy for recognition. If the morphology of the vocal-production apparatus varies slightly between individuals, then this variation might impart unique spectral features, or so-called "voice characteristics," to all of an individual's vocalizations. The use of voice characteristics has been suggested for great tits, *Parus major* (Weary and Krebs 1992), but does not appear to be a relevant cue for either song sparrows (Beecher et al. 1994) or starlings (Gentner et al. 2000). Finally, vocal recognition might also rely on the sequence in which multiple song types are sung. That is, different males may share song types but sing them in individually distinctive temporal patterns. The role of this final cue has not been extensively studied in songbirds, but there is some evidence to suggest that European starlings are sensitive to the sequence of motifs within familiar song bouts (Gentner and Hulse 1998).

For species in which males sing multiple songs, the four mechanisms outlined above may not be mutually exclusive. There is no *a priori* reason to believe that vocal recognition in a single species relies on individual variation coded in only a single dimension, nor is there any reason to suspect that all species of songbirds use the same recognition strategies. Given the approximately 4,500 different species of songbirds—each singing acoustically distinct songs and the occurrence of vocal recognition in a wide range of behavioral contexts, it is likely that vocal-recognition information is coded at multiple levels throughout a songbird's repertoire (Braaten 2000).

3.3.3. Laboratory Studies

Given the likely diversity of vocal-recognition behaviors across songbird species, it is reasonable to consider whether there are corresponding peripheral perceptual specializations among songbirds that in theory might provide an "open channel" of communication within a species while limiting confusion across species. For instance, different species might concentrate the spectral energy with their songs in defined spectral bands. This hypothesis is supported by several observations of species-specific advantages during operant discriminations of multiple conspecific and heterospecific songs in several different species (Sinnott 1980; Okanoya and Dooling 1990; Cynx and Nottebohm 1992; Dooling et al. 1992). However, the overwhelming data from psychophysical studies of hearing in birds indicate that most basic sensory processing capabilities (e.g., frequency sensitivity) are conserved across songbird species (Dooling et al. 2000). Thus, it appears that biases for the discrimination of species-specific vocalizations, and hence mechanisms for vocal recognition, result from evolutionary or ontogenic changes in central processing structures. This inference is consistent with the more general assumption that the cognitive processes underlying vocal recognition take the neural representation of acoustically complex signals (i.e., song) as their input. Recent laboratory studies of European starlings have addressed these questions by determining more precisely the form of the acoustic signal controlling recognition in this species.

Male starlings tend to present their songs in long episodes of continuous singing referred to as bouts. Song bouts, in turn, are composed of much smaller acoustic units referred to as motifs (Adret-Hausberger and Jenkins 1988; Eens et al. 1991), which in turn are composed of still smaller units called notes. (This usage of "motif" for starling songs is at slight variance with motifs as defined for zebra finch songs in Section 2.3.3.1.) Notes can be broadly classified by the presence of continuous energy in their spectrographic representations, and although several notes may occur in a given motif, their pattern is usually highly stereotyped between successive renditions of the same motif. One can thus consider starling song as a sequence of motifs, where each motif is an acoustically complex event. The number of unique motifs that a male starling can sing (i.e., his repertoire size) can be quite large, and consequently different song bouts from the same male are not necessarily composed of the same set of motifs. This broad acoustical variation in their song provides several potential cues that starlings might use when learning to recognize the songs of an individual conspecific and while maintaining that recognition over time. One straightforward recognition mechanism is the association of specific motifs with specific singers. Although some sharing of motifs does occur among captive males (Hausberger and Cousillas 1995; Hausberger 1997), the motif repertoires of different males living in the wild are generally unique (Adret-Hausberger and Jenkins 1988; Eens et al. 1989, 1991; Chaiken et al. 1993; Gentner and Hulse 1998). Thus, learning which males sing which motifs can provide a discriminative cue for song classification.

Data from operant studies in starlings support the idea that recognition is based at the level of the motif. Starlings trained to recognize individual conspecifics by one set of song bouts can readily generalize correct recognition to novel song bouts from the same singers (Gentner and Hulse 1998; Gentner et al. 2000; Fig. 7.4A, B). However, when these novel song bouts have no motifs in common with the training songs, and when song exposure outside of the operant apparatus is restricted, recognition falls to chance (Gentner et al. 2000; Fig. 7.4C). Likewise, starlings trained to discriminate among pairs of motifs will reverse the discrimination when transferred to the same motif sung by the opposite individual and perform at chance when transferred to novel motifs sung by the training singers (Gentner 1999). This failure to generalize correct recognition to songs composed of novel motifs, or to single novel motifs, is inconsistent with the use of individually invariant source and/or filter properties (voice characteristics) for vocal recognition.

The data suggest that starlings learn to recognize the songs of individual conspecifics by attending to information contained at (or below) the level of the motif and by then associating distinct sets of motifs with individual singers. If this is true, then once recognition is learned, it should be possible to control it systematically by varying the proportions of motifs in a given bout that come from two "vocally familiar" males. That is, recognition behavior ought to track the statistical distribution of motifs from two vocally familiar males rather than the presence or absence of single motifs from either male. Recent data confirm this prediction (Gentner and Hulse 2000; Fig. 7.4) and thereby suggest that when starlings are compelled to

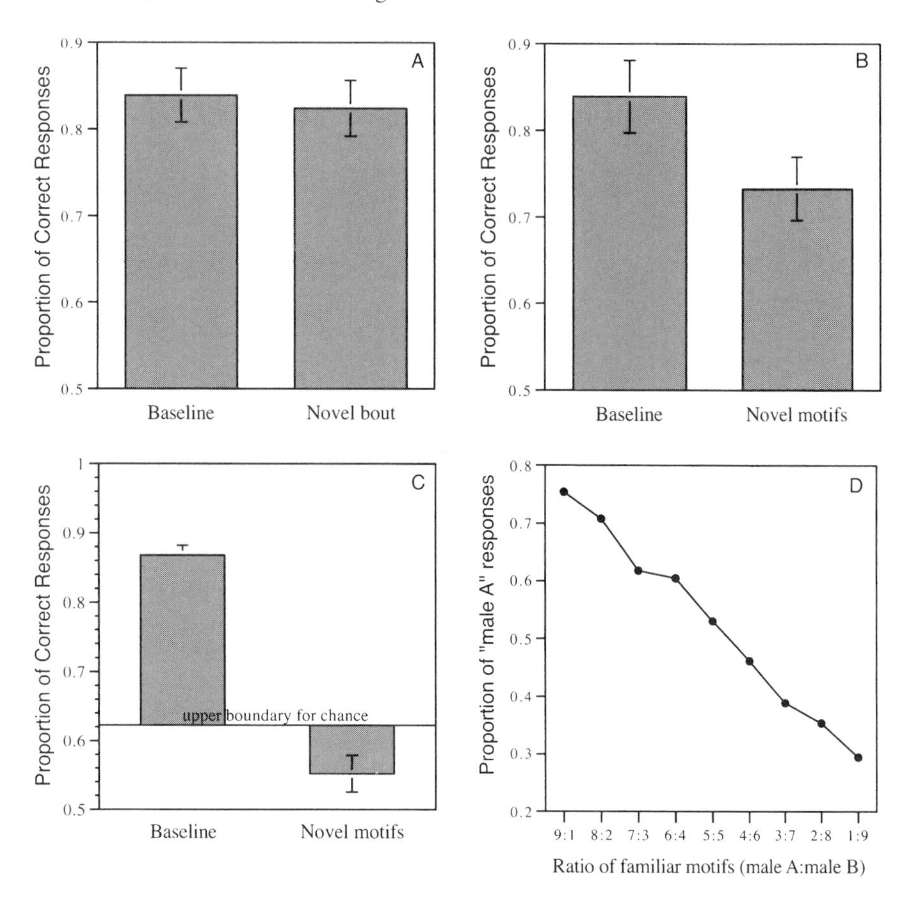

FIGURE 7.4. Vocal-recognition behavior in European starlings. (**A**) Mean (±SEM) proportion of correct responses given during asymptotic performance on an operant recognition task ("Baseline") and during initial transfer to novel songs containing familiar motifs ("Novel bout"). (**B** and **C**) Mean (±SEM) proportion of correct responses during transfer from the baseline training to novel songs from the same singers composed of "Novel motifs." Data in (**B**) show the transfer when the subjects were exposed to the training and test songs outside of the operant apparatus, whereas the data in (**C**) show the results for the same transfer after controlling for this experience. Note that in (**C**) recognition of the novel motifs falls below chance. (**D**) Data showing the close (and approximately linear) relationship between the statistical distribution of familiar motifs from two different singers and song recognition.

classify conspecific songs, they do so by memorizing large numbers of unique song components (i.e., motifs) and then organizing subsets of these motifs into separate classes. As a cognitive strategy, classifying songs according to their component structure represents a parsimonious method of dealing with these complex acoustic signals. Because individual starlings tend to possess unique motif repertoires, disjoint sets of motifs will generally correspond to individual identity. Therefore, attending to the motif structure captures a significant portion of the individual variation in the signal.

The behavioral data inform a number of hypotheses regarding the neural mechanisms of vocal recognition in starlings. First, the representational architecture of forebrain auditory regions should reflect the segmentation of song at the level of the motif. When similarity/difference computations are invoked, as they must be for recognition to proceed, the input to such computations should be some representation of the motif. Determination of the precise acoustic information corresponding to such a representation has yet to be accomplished but is amenable to behavioral study in the laboratory. Investigations of these representations at the neural level are likely to tell a great deal about the manner in which complex auditory objects are represented and processed. Second, the behavioral strategy employed by starlings for vocal recognition suggests that the upper bound on the capacity for accurate recognition is constrained by the memory capacity of the system for specific motifs. If the representation of a motif is coded by dynamic temporal and spatial patterns of connectivity among local feature detectors, then these memory constraints may derive directly from the perceptual mechanisms for coding complex stimuli.

3.3.4. Songbird Auditory System: Anatomy

The basic plan of the passerine auditory system follows a general reptile-bird pattern of connections (Ulinski and Margoliash 1990; Carr 1992; Carr and Code 2000). In birds, the auditory nerve projects to two cochlear nuclei, the nucleus magnocellularis and the nucleus angularis. These nuclei project in turn to second-order olivary nuclei, to the lemniscal nuclei, and contralaterally to the central nucleus of the nucleus mesencephalicus lateralis dorsalis (MLd), the avian analog of the inferior colliculus. The central nucleus of the MLd projects to a major target in the medial portion of the dorsal thalamus, the nucleus ovoidalis (Ov), and to a lesser extent to the subjacent nucleus semiluminaris parovoidalis (SPO) (Karten 1968) and a region surrounding the Ov-SPO complex referred to as the "shell" (Durand et al. 1992). There are a wealth of connections and patterns of connections in the more peripheral auditory structures not reviewed here. In addition to the primary projections to the auditory telencephalon, auditory fibers also project from the thalamus to the hypothalamus, providing one possible route for interaction between auditory, neuromodulatory (Li and Sakaguchi 1997), and endocrine systems (Durand et al. 1992).

Nissl and Golgi preparations of male zebra finch brains (Bonke et al. 1979b) confirm the general pattern of organization observed in other species (Karten 1968) and demonstrate that the caudal medial portion of the avian telencephalon is composed of five cytoarchitectonic subregions—L1, L2a, L2b, L3 and L—called the field L complex (Fortune and

Margoliash 1992). The field L complex is the primary telencephalic target for auditory information arriving via several parallel pathways from the Ov complex in the thalamus (Fig. 7.3). Neurons in the shell of Ov also project to the caudomedial portion of the neostriatum (NCM) (Durand et al. 1992; Vates et al. 1996). The subregions of field L are densely interconnected and project to the NCM and reciprocally to the lateral portion of the caudal ventral hyperstriatum (clHV). The NCM and clHV share reciprocal connections with the caudal medial portion of HV (cmHV). Figure 7.3 provides a schematic for this complicated pattern in songbirds.

Neurons in clHV, L1, and L3 also project to the neostriatum immediately ventral to HVc, referred to as the "shelf" (Kelley and Nottebohm 1979). Neurons in the shelf are auditory (Scheich et al. 1979a; Müller and Leppelsack 1985), and it has been proposed that the shelf is a source of auditory input to HVc. The projections of the shelf into HVc are extremely sparse, however (L. Katz, cited in Margoliash 1987; Vates et al. 1996). One proposition is that dendrites in caudal HVc, which extend into the shelf, receive auditory input from shelf axons. These hypotheses are long-standing but have never been confirmed, in part because the small size of the structures and their physical proximity to each other hinders independent manipulation of the shelf and HVc. Auditory information may also enter the VMP via the clHV-NIf-HVc pathway (Fig. 7.3). BOS selective auditory responses and correlated activity in NIf have recently been demonstrated (Janata and Margoliash 1999). NIf activity may be necessary for HVc auditory responses (Boco and Margoliash 2001). NIf receives input from clHV and possibly also from field L via the dendritic projections of a distinct class of NIf neurons lying along the dorsorostral border of NIf with L1 (Fortune and Margoliash 1995). Identification of sources of auditory input into the VMP is of particular interest because it provides a neural substrate for hypotheses about the origin of the well-known BOS selective responses in these structures. The auditory inputs to HVc also provide the basis for speculations that expand the functions of AFP structures beyond their classic role in juvenile song learning and into adult song perception (see Section 2.3.3).

The song system is associated with a series of cytoarchitectonically indistinct structures. These structures, such as the shelf ventral to HVc and medial MAN, are physically close to cytoarchitectonically distinct song-system nuclei and have patterns of connections similar to song-system nuclei. Components of this indistinct system have also been observed in several species of nonpasserine birds (Brauth et al. 1987; Brauth and McHale 1988; Korzeniewska and Güntürkün 1990; Fortune and Margoliash 1995; see also Margoliash et al. 1994). Although the functions of this pathway remain unknown, one possibility is that the song system arose as a series of specializations of the indistinct pathway in relation to selection pressures associated with vocal learning. If this is true, then the indistinct pathway in oscines and other birds may mediate more general aspects of vocal behavior, such as calling behavior in relation to complex social interactions (Margoliash et al. 1994), or other reward/reinforcement contingencies involving vocal perception in adults. Such speculations are consistent with emerging homologies between the AFP and cortico-basal ganglia loops in mammals (Luo and Perkel 1999).

3.3.5. Songbird Auditory System: Physiology

In European starlings, Ov is tonotopically organized, with best frequencies decreasing ventrally (Bigalke-Kunz et al. 1987). Neurons throughout the auditory telencephalon also show tonotopic organization (Leppelsack and Schwartzkopff 1972; Rubsamen and Dorrscheidt 1986), although in much more complex patterns. In starlings, roughly 11 different regions can be identified on the basis of the direction of the tonotopic gradient and tuning curve bandwidth (Haüsler 1997; Capsius and Lepplesack 1999), and similar patterns are observed in zebra finches (Gehr et al. 1999). These tonotopically defined regions appear to respect anatomically defined regions of the field L complex.

Relatively few studies have examined responses in the telencephalic regions using complex acoustic stimuli. Neurons in L1 and L3 have lower response rates to tone bursts than those in L2 and show greater selectivity to species-specific vocalizations (Leppelsack and Vogt 1976; Bonke et al. 1979a; Müller and Leppelsack 1985; Theunissen and Doupe 1998). This selectivity is borne out by the complexity of the spatial-temporal receptive fields (STRFs) for many neurons within field L. Indeed, more reliable estimates of the STRF are derived from responses to conspecific vocalizations than tone pips (Theunissen et al. 2000; cf. Schäfer et al. 1992). This general pattern of increasing response selectivity from field L2 to the higher-order areas continues into NCM and cHV (Müller and Leppelsack 1985), suggesting that these regions are involved in the extraction of complex features. Early data from white-crowned sparrows are consistent with this in showing a small subset of neurons in the NCM that are selective for specific directions of FM in a common trill element of conspecific song (Leppelsack 1983). Recent preliminary data (Grace and Theunissen 2000; Gentner and Margoliash 2001) support the idea that cHV in particular is involved in the extraction and/or representation of complex features in showing highly selective responses in this region to specific features in behaviorally relevant conspecific songs.

Neurons in NCM are broadly responsive to conspecific stimuli and respond to the repeated presentation of conspecific song in a stimulus-specific manner (Chew et al. 1995; Stripling et al. 1997). The repeated presentation of a single conspecific song elicits a rapid modulation in the initial firing rate of NCM neurons (Stripling et al. 1997). If the same song is repeated on the order of 200 times, this initial modulation of the firing rate is no longer observed when that same song is presented on subsequent trials.

This is true even though the initial response modulation can still be observed for other conspecific songs (Chew et al. 1995, 1996; Stripling et al. 1997). These stimulus-specific changes in the response properties of NCM neurons have led to the hypothesis that NCM may play an important role in individual vocal recognition (Chew et al. 1996). Consistent with this idea is the fact that many neurons in NCM (and cHV) show a rapid up-regulation of the immediate early gene (IEG) *zenk* in response to the presentation of conspecific songs (Mello et al. 1992) that is tuned to the acoustics of particular conspecific song syllables (Ribiero et al. 1998). Interestingly, the genomic response also habituates to the repeated presentation of the same conspecific song (Mello et al. 1995) and is elevated during specific components of the vocal-recognition task described above in starlings (Gentner et al. 1999). The mammalian homolog to *zenk* is required for expression of late LTP and long-term memories in mice (Jones et al. 2001). This suggests that *zenk* expression in NCM and cHV may be related to learning about conspecific songs and implicates these structures in concomitant processes.

3.4. The Nonhuman Primate Model

3.4.1. Vocal-Recognition Behavior

Various forms of vocal recognition are also prominent among many species of primates. In humans, *Homo sapiens*, the ability to recognize the sex and the identity of a talker is anecdotally obvious. Acoustic cues to a talker's sex are present in both the fundamental frequency (reflecting larynx size) and vocal tract length (reflecting body size). Recent data on speech perception suggest that acoustic cues to individual recognition within sexes are due to supralaryngeal vocal tract filtering caused by anatomical variation between talkers (Bachorowski and Owren 1999). Individual vocal recognition (and kin recognition) has also been demonstrated in rhesus macaques, *Macaca mulatta* (Rendall et al. 1996), and acoustic cues related to vocal tract filtering have been suggested as the basis for this behavior (Owren et al. 1997; Rendall et al. 1998). Vocal recognition is also apparent in the referential call systems of many primates (e.g., Seyfarth et al. 1980a, 1980b; Hauser 1998; Rendall et al. 1999; Fitch and Hauser, Chapter 3). In these cases, rather than having signals that are associated with specific individuals or groups of individuals, particular calls are used to refer, for example, to different classes of predators, food types, or various other behavioral events. This aspect of primate vocal communication has been reviewed elsewhere (Cheney and Seyfarth 1990; Ghazanfar and Hauser 1999), and we do not address it further except to point out that such behaviors provide another set of natural categories for vocalizations that is amenable to future neuroethological study.

A third class of vocal-recognition behavior has been studied extensively in Japanese macaques, following the observation that several subtypes of the "coo" vocalization in this species can be defined on the basis of acoustic variation (Green 1975). Much of this work has focused on two particular subtypes of coo vocalizations, the smooth-early-high (SEH) and smooth-late-high (SLH), so-called because of the relative position of the peak of one frequency component that sweeps up and then down over the course of the call. Although Japanese macaque mothers can discriminate the coos of their young from others (Pereira 1986), the role of individual recognition cues in the coo vocalizations has not been well-studied. Nevertheless, the SEH and SLH call types function in different behavioral contexts (Green 1975), and Japanese macaques possess a species-specific bias for discriminations involving these coos when the relevant variation is in the relative timing of the FM peak (Zoloth et al. 1979).

Although initial data suggested that the SEH and SLH calls are perceived categorically (May et al. 1989), more recent data from the field indicate that many adult female coo vocalizations have FM peaks within the ambiguous zone between "early" and "late" prototypes (Owren and Casale 1994). The categorical boundaries determined in laboratory tests do not coincide with natural variation in the distribution of calls in the field. Nonetheless, all data to date show clear evidence that the coo calls are perceived as perceptually distinct classes (if not categorically), and several studies now substantiate the notion that the relative position of the FM peak within the coo is the most salient cue to discrimination among different coos (May et al. 1988; Le Prell and Moody 2000). Amplitude cues also appear to function in call discrimination (Le Prell and Moody 1997). Thus, the stimulus dimensions involved in "real-world" classification and/or categorization of coo calls may involve a more complex stimulus space than that suggested by the original categorical perception studies.

Interestingly, lesions to the left, but not the right, superior temporal gyrus impair discrimination between the SEH and SLH coos by Japanese macaques, suggesting a human language-like hemispheric dominance (Heffner and Heffner 1984). Similarly, behavioral tests among free-ranging rhesus macaques, *Macaca mulatta*, indicate a left hemisphere dominance for the processing of conspecific compared to heterospecific vocalizations that is present in adults but not infants (Hauser and Andersson 1994). Apart from their comparative value in understanding the evolutionary origin of human cognition, the lateralization of specific cognitive functions can in theory allow for identification of at least some of the brain regions mediating these behaviors. Comparisons of activation in each hemisphere during tasks that call on lateralized functions, using either fMRI or immediate early gene-expression techniques, should yield regions of differential activity in the appropriate hemisphere worthy of closer study.

3.4.2. Primate Auditory System: Cortical Anatomy and Physiology

The mammalian auditory cortex extends across the superior temporal plane and onto the adjacent superior temporal gyrus. The primary auditory cortex in primates can be divided into separate core and belt areas on the basis of

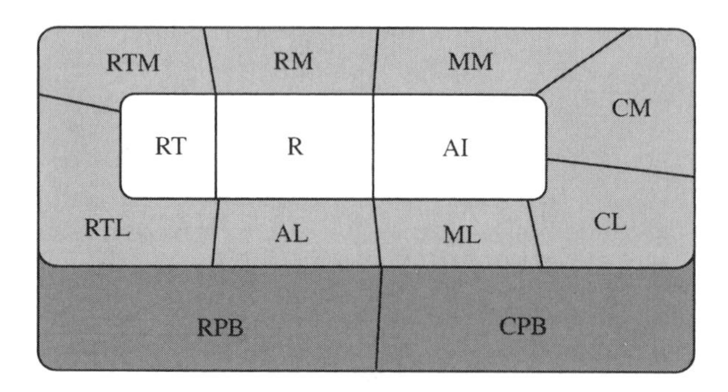

FIGURE 7.5. Schematic layout of primate auditory cortex. Areas shown in white represent the core regions of the auditory cortex, those shown in light gray the lateral belt, and those in dark gray the para-belt. Most of the areas in the core and lateral-belt area are interconnected. AI: primary auditory cortex; R: rostral area; RT: rostrotemporal area; CL: caudolateral area; CM: caudomedial area; MM: middle medial area; RM: rostromedial area; RTM: medial rostrotemporal area; RTL: lateral rostrotemporal area; AL: anterolateral area; ML: middle lateral area; RPB: rostral para-belt; CPB: caudal para-belt. (Based on Kaas and Hackett 2000).

cytoarchitecture and connectivity, with a third para-belt region positioned laterally (Fig. 7.5). Moving rostral to caudal, the core is composed of three distinct regions: the rostral temporal field RT, the rostral field R, and most caudally, AI. Both AI and R have well-defined cochlear representations, with caudorostral tonotopic gradients of best frequencies running from high to low in AI and low to high in R. The tonotopic organization of RT is less well-studied, but it appears that best frequencies are arranged from high to low as one moves rostrally (reviewed in Kaas and Hackett 2000). All three of these core areas are densely interconnected ipsilaterally and project colosally to homotopic regions in the opposite hemisphere. All three regions receive dense inputs from the medial geniculate complex, the principal nucleus of the thalamus. In the common marmoset, *Calithrix jacchus*, subpopulations of AI cells show selective responses to conspecific vocalizations as compared with synthetic variations with the same spectral but different temporal characteristics (Wang et al. 1995). This suggests that, at least in marmosets, AI cells may be sensitive to more than narrow-band frequency parameters.

Immediately surrounding the core is an area containing approximately four to eight auditory regions (Rauschecker 1998; Kaas and Hackett 2000), each with distinct cochlear representations, referred to as the belt (Fig. 7.5). Most of the thalamic input into these regions arises in the dorsal and medial division of the medial geniculate complex. It has been argued that the distribution of these thalamic projections and those to the core areas reflects the functional separation of spatial-localization and pattern-recognition mechanisms (Romanski et al. 1999; Rauschecker and Tian 2000) analogous to the so-called "what" and "where" pathways in visual processing. Although neurons in belt regions can be driven using tonal stimuli, and there are cochleotopic representations in several regions, many neurons seem to prefer specific bandwidths for frequency-centered sound bursts, independent of intensity (Rauschecker et al. 1995). The bandwidth tuning of cells in the caudolateral (CL), mediolateral (ML), and anterior later (AL) belt regions varies along an axis that is orthogonal to the cochleotopic organization.

An additional auditory processing region just lateral to the belt, and referred to as the "para-belt", has been identified histologically (Hackett et al. 1998), but neither its anatomical extent nor physiological response properties have been fully examined. In contrast to the dense reciprocal connections among the core and belt areas, the para-belt does not have direct access to information in the core but rather appears to be driven primarily by input from the belt (Kaas and Hackett 2000). The nature of this input is not fully resolved. However, both the general pattern of connections among the core, belt, and para-belt regions in primates and the change in tuning properties across these regions resemble that among field L, NCM, and cmHV in songbirds described above. Lateral-belt neurons are tuned for both direction and rate of frequency modulation (Rauschecker 1997), and, at least in macaques, many of these cells also respond vigorously to conspecific calls or components thereof (Rauschecker et al. 1995). Although responses to species-specific calls can also be observed in AI, the selectivity is significantly less pronounced than that in the lateral-belt areas (see Rauschecker 1998). Moreover, early recordings in squirrel monkeys, *Saimiri sciureus*, showing large percentages of cells responsive to conspecific calls and subpopulations therein with modest nonlinear response properties (Wollberg and Newman 1972; Newman and Wollberg 1973; Winter and Funkenstein 1973) were mostly done in the superior temporal gyrus and therefore were likely to lie within the lateral belt or para-belt. Thus, there appears to be a general pattern of increasingly complex receptive field properties that coincides with the transition from primary auditory cortex to postsynaptic regions. This can be observed in humans as well, where the dorsolateral superior temporal gyrus and planum temporale (putative belt) are more strongly activated by FM tones than noise, and the superior temporal sulcus (putative para-belt) is more activated by various speech stimuli, including pseudowords and reversed speech, than by FM tones and other nonvocal sounds (Binder et al. 1994, 2000; Belin et al. 2000).

4. Summary and Conclusions

We have discussed organizational schemes and mechanisms that underlie the coding of complex stimuli, including feature detectors and combination sensitivity, which are likely to provide the basis for representations of acoustically complex vocal communication signals. Although we have attempted to maintain a focus on perceptual mechanism, the nature of vocal communication systems necessitates an integrative approach to production and perception. Our discussion of motor-sensory linkages in the well-studied oscine song system addresses these concerns from both empirical and theoretical perspectives.

Much of the auditory cortex in mammals and auditory forebrain in birds is mapped tonotopically, but a number of additional acoustic stimulus parameter mappings have been described in birds and mammals, including pitch periodicity, frequency, intensity, spatial location, duration, amplitude, and frequency modulation (see Ehret 1997; Schulze and Langner 1997). The structure of these mappings may change significantly in awake, alert animals (Evans and Whitfield 1964; Pfingst et al. 1977; Dave et al. 1998a; Schmidt and Konishi 1998; Capsius and Leppelsack 1999; cf. Recanzone et al. 2000). Patterns of organization in numerous other systems—including barn owls, bats, electric fish, songbirds, and others—suggest that a breakdown in cochleotopic organization of responses often coincides with mappings for more complex stimulus parameters. This may represent the convergence of multiple acoustically simpler maps onto some regions (i.e., a stimulus reconstruction) or the extraction of encoded information along dimensions that have not been represented in prior mappings (i.e., emergent properties). The neuroethological approach provides a logical and theoretically sound basis for investigation of such emergent properties in complex signals.

The neuroethological approach to vocal communication is also informative of the likely functional outputs of a perceptual system. These outputs form the basis for higher cognition. We have described how vocal-recognition behavior, and more explicitly the organization of vocal communication signals into behaviorally relevant classes or categories, can be used to develop hypotheses about the perceptual and cognitive demands placed on the central nervous system.

Principles in neuroethology have been difficult to identify. Most reviews of neuroethological research are organized around a series of case examples. This can incorrectly reinforce the conclusion that principles of neuroethological research are not forthcoming. In contrast, for example, systems neuroscience has identified a number of features of CNS organization, such as mappings, lateral interactions, population dynamics, and network reconfiguration, which help to organize research and can be considered principles. Neuroethological principles do exist, such as those described in this paper, but they have been difficult to demonstrate because of the scope of work required for a comparative analysis of brain and behavior. Only now, after some 40 years of research, and only in the most extensively studied behaviors, such as vocal communication, are these principles emerging. Somewhat akin to neurobiology, neuroethology aims to explain how molecular, cellular, and systems-level phenomena result in behaviors such as perception and cognition. But by the very nature of its underlying definitions, neuroethology aims to do more, for it provides the nexus for integration of the multiple levels of analysis articulated by Tinbergen's classic four questions. Ultimate and proximate explanations are incomplete until they are united. The task for neuroethology is enormous, but the principles and patterns of organization we have described for vocal communication are examples of how successful this approach can be and how much promise future research holds.

Acknowledgments. We thank Andrea Megela-Simmons for valuable comments on the manuscript. This work was supported by grants from the NIH to DM (MH59831 and MH60276) and TQG (DC00389).

References

Adret P (1993) Operant conditioning, song learning and imprinting to taped song in the zebra finch. Anim Behav 46:149–159.

Adret-Hausberger M, Jenkins PF (1988) Complex organization of the warbling song in starlings. Behaviour 107:138–156.

Alexander RD (1962) Evolutionary change in cricket acoustical communication. Evolution 16:443–467.

Anderson SE, Dave AS, Margoliash D (1996) Template-based automatic recognition of birdsong syllables from continuous recordings. J Acoust Soc Am 100: 1209–1219.

Aubin T, Jouventin P, Hildebrand C (2000) Penguins use the two-voice system to recognize each other. Proc R Soc Lond B Biol Sci 267:1081–1087.

Bachorowski JA, Owren MJ (1999) Acoustic correlates of talker sex and individual talker identity are present in a short vowel segment produced in running speech. J Acoust Soc Am 106:1054–1063.

Bankes SC, Margoliash D (1993) Parametric modeling of the temporal dynamics of neuronal responses using connectionist architectures. J Neurophysiol 69:980–991.

Barlow HB (1972) Single units and sensation: A neuron doctrine for perceptual psychology? Perception 1:371–394.

Barlow HB (1995) The neuron doctrine in perception. In: Gazzaniga M (ed) The Cognitive Neurosciences. Cambridge, MA: MIT Press, pp. 415–435.

Bateson P (1979) How do sensitive periods arise and what are they for? Anim Behav 27:470–486.

Beecher MD (1991) Success and failures of parent–offspring recognition in animals. In: Hepper PG (ed) Kin Recognition. Cambridge, UK: Cambridge University Press.

Beecher MD, Campbell SE, Burt J (1994) Song perception in the song sparrow: Birds classify by song type but not by singer. Anim Behav 47:1343–1351.

Beer C (1982) Conceptual issues in the study of communication. In: Kroodsma DE, Miller EH (eds) Acoustic Communication in Birds, Vol 2. Song Learning and Its Consequences. New York: Academic Press, pp. 279–310.

Beguin N, LeBoucher G, Kreutzer M (1998) Sexual preferences for mate song in female canaries (*Serinus canaria*). Behaviour 135:1185–1196.

Belin P, Zatorre RJ, Lafaille P, Ahad P, Pike B (2000) Voice-selective areas in human auditory cortex. Nature 403:309–312.

Bentley DR, Hoy RR (1972) Genetic control of the neuronal network generating cricket song patterns. Anim Behav 20:478–492.

Best CT, McRoberts GW, Sithole NM (1988) Examination of the perceptual reorganization for speech contrasts: Zulu click discrimination by English-speaking adults and infants. J Exp Psychol Hum Percept Perform 14:345–360.

Bigalke-Kunz B, Rubsamen R, Dorrscheidt G (1987) Tonotopic organization and functional characterization of the auditory thalamus in a songbird, the European starling. J Comp Physiol A 161:255–265.

Binder JR, Frost JA, Hammeke TA, Bellgowan PS, Springer JA, Kaufman JN, Possing ET (2000) Human temporal lobe activation by speech and nonspeech sounds. Cereb Cortex 10:512–528.

Binder JR, Rao SM, Hammeke TA, Yetkin FZ, Jesmanowicz A, Bandettini PA, Wong EC, Estkowski LD, Goldstein MD, Haughton VM, Hyde JS (1994) Functional magnetic resonance imaging of human auditory cortex. Ann Neurol 35: 662–672.

Boco T, Margoliash D (2001) NIf is a major source of auditory and spontaneous drive to HVc. Soc Neurol Abstr 27:381.2

Bodnar DA, Bass AH (1997) Temporal coding of concurrent acoustic signals in auditory midbrain. J Neurosci 17:7553–7564.

Bodnar DA, Bass AH (1999) Midbrain combinatorial code for temporal and spectral information in concurrent acoustic signals. J Neurophysiol 81:552–563.

Bonke D, Scheich H, Langner G (1979a) Responsiveness of units in the auditory neostriatum of the Guinea fowl (Numida meleagris) to species-specific calls and synthetic stimuli. I. Tonotopy and functional zones. J Comp Physiol 132:243–255.

Bonke DA, Bonke D, Scheich H (1979b) Connectivity of the auditory forebrain nuclei in the guinea fowl (*Numida meleagris*). Cell Tissue Res 200:101–121.

Bottjer SW, Johnson F (1997) Circuits, hormones, and learning: Vocal behavior in songbirds. J Neurobiol 33:602–618.

Bottjer SW, Miesner EA, Arnold AP (1984) Forebrain lesions disrupt development but not maintenance of song in passerine birds. Science 224:901–903.

Boughman JW, Wilkinson GS (1998) Greater spear-nosed bats discriminate group mates by vocalizations. Anim Behav 55:1717–1732.

Braaten RF (2000) Multiple levels of representation of song by European starlings (*Sturnus vulgaris*): Open-ended categorization of starling song types and differential forgetting of song categories and exemplars. J Comp Psychol 114:61–72.

Bradbury JW, Vehrencamp SL (1998) Principles of Animal Communication. Sunderland, MA: Sinauer.

Bradbury JW, Vehrencamp SL (2000) Economic models of animal communication. Anim Behav 59:259–268.

Brainard MS, Doupe AJ (2000) Interruption of a basal ganglia-forebrain circuit prevents plasticity of learned vocalizations. Nature 404:762–766.

Brainard MS, Doupe AJ (2001) Postlearning consolidation of birdsong: Stabilizing effects of age and anterior forebrain lesions. J Neurosci 21:2501–2517.

Brauth SE, McHale CM (1988) Auditory pathways in the budgerigar. II. Intratelencephalic pathways. Brain Behav Evol 32:193–207.

Brauth SE, McHale CM, Brasher CA, Dooling RJ (1987) Auditory pathways in the budgerigar. I. Thalamo-telencephalic projections. Brain Behav Evol 30:174–199.
Bregman AS (1990) Auditory Scene Analysis: The Perceptual Organization of Sound. Cambridge, MA: MIT Press.
Brenowitz EA (1991) Altered perception of species-specific song by female birds after lesions of a forebrain nucleus. Science 251:303–305.
Brenowitz EA, Margoliash D, Nordeen K (eds) (1997) The neurobiology of birdsong. J Neurobiol 33:495–516.
Brooks RJ, Falls JB (1975) Individual recognition by song in white-throated sparrows. III. Song features used in individual recognition. Can J Zool 53:1749–1761.
Brosch M, Schreiner CE (2000) Sequence sensitivity of neurons in cat primary auditory cortex. Cereb Cortex 10:1155–1167.
Bruns V, Schmieszek E (1980) Cochlear innervation in the greater horseshoe bat: Demonstration of an acoustic fovea. Hear Res 3:27–43.
Burt JM, Lent KL, Beecher MD, Brenowitz EA (2000) Lesions of the anterior forebrain song control pathway in female canaries affect song perception in an operant task. J Neurobiol 42:1–13.
Capranica RR (1965) The evoked vocal response of the bullfrog. Research Monograph No. 33. Cambridge, MA: MIT Press.
Capranica RR (1978) Auditory processing in anurans. Fed Proc Fed Am Soc Exp Biol 37:2324–2328.
Capsius B, Leppelsack H-J (1999) Response patterns and their relationship to frequency analysis in auditory forebrain centers of a songbird. Hear Res 136:91–99.
Carr CE (1992) Evolution of the central auditory system in reptiles and birds. In: Webster DB, Fay RR, Popper AN (eds) The Evolutionary Biology of Hearing. New York: Springer-Verlag, pp. 511–544.
Carr CE, Code RA (2000) The central auditory system of reptiles and birds. In: Dooling RJ, Fay RR, and Popper AN (eds) Comparative Hearing: Birds and Reptiles. New York: Springer-Verlag, pp. 197–248.
Casseday JH, Covey E (1995) Mechanisms for analysis of auditory temporal patterns in the brainstem of echolocating bats. In: Covey E, Hawkins HL, Port RF (eds) Neural Representations of Temporal Patterns. New York: Plenum Press, pp. 25–51.
Casseday JH, Ehrlich D, Covey E (1994) Neural tuning for sound duration: Role of inhibitory mechanisms in the inferior colliculus. Science 264:847–850.
Casseday JH, Covey E, Grothe B (1997) Neural selectivity and tuning for sinusoidal frequency modulations in the inferior colliculus of the big brown bat, *Eptesicus fuscus*. J Neurophysiol 77:1595–1605.
Chaiken M, Böhner J, Marler P (1993) Song acquisition in European starlings, *Sturnus vulgaris*: A comparison of the songs of live-tutored, tape-tutored, untutored, and wild-caught males. Anim Behav 46:1079–1090.
Cheney DL, Seyfarth RM (1990) How Monkeys See the World. Chicago: University of Chicago Press.
Cherry EC, Taylor WK (1954) Some further experiments on the recognition of speech, with one and two ears. J Acoust Soc Am 26:554–559.
Chew SJ, Mello CV, Nottebohm F, Jarvis E, Vicario DS (1995) Decrements in auditory responses to a repeated conspecific song are long lasting and require two periods of protein synthesis in the songbird forebrain. Proc Natl Acad Sci USA 92:3406–3410.

Chew SJ, Vicario DS, Nottebohm F (1996) A large-capacity memory system that recognizes the calls and songs of conspecifics. Proc Natl Acad Sci USA 93:1950–1955.

Covey E (2000) Neural population coding and auditory temporal pattern analysis. Physiol Behav 69:211–220.

Crawford JD (1997) Feature-detecting auditory neurons in the brain of a sound-producing fish. J Comp Physiol A 180:439–450.

Cynx J (1990) Experimental determination of a unit of song production in the zebra finch (*Taeniopygia guttata*). J Comp Psychol 104:3–10.

Cynx J, Nottebohm F (1992) Role of gender, season, and familiarity in discrimination of conspecific song by zebra finches (*Taeniopygia guttata*). Proc Natl Acad Sci USA 89:1368–1371.

Dabelsteen T, Pedersen SB (1993) Song-based species discrimination and behaviour assessment by female blackbirds, *Turdus merula*. Anim Behav 45:759–771.

Damasio A, Tranel D, Damasio H (1990) Face agnosia and the neural substrate of memory. Annu Rev Neurosci 13:89–109.

Dave AS, Margoliash D (2000) Song replay during sleep and computational rules for sensorimotor vocal learning. Science 290:812–816.

Dave AS, Yu AC, Margoliash D (1998a) Behavioral state modulation of auditory activity in a vocal motor system. Science 282:2250–2254.

Dave A, Yu AC, Gilpin JJ, Margoliash D (1998b) Methods for chronic neuronal ensemble recordings in singing birds. In: Nicolelis MAL (ed) Methods for Neuronal Ensemble Recordings. Boca Raton: CRC Press, pp. 101–120.

Dawkins R, Krebs JR (1978) Animal signals: Information or manipulation? In: Krebs JR, Davies NB (eds) Behavioural Ecology. Oxford: Blackwell Scientific Publications, pp. 282–309.

deCharms RC, Zador A (2000) Neural representation and the cortical code. Annu Rev Neurosci 23:613–647.

Del Negro C, Gahr M, Leboucher G, Kreutzer M (1998) The selectivity of sexual responses to song displays: Effects of partial chemical lesion of the HVC in female canaries. Behav Brain Res 96:151–159.

Del Negro C, Kreutzer M, Gahr M (2000) Sexually stimulating signals of canary (*Serinus canaria*) songs: Evidence for a female-specific auditory representation in the HVc nucleus during the breeding season. Behav Neurol 114:526–542.

DeVoogd TJ, Cardin JA, Szekely T, Büki J, Newman SW (1996) Relative volume of lMAN in female warbler species varies with the number of songs produced by conspecific males. Soc Neurosci Abstr 22:1401.

Dickinson A (1980) Contemporary Animal Learning Theory. New York: Cambridge University Press.

Dooling RJ, Searcy MH (1980) Early perceptual selectivity in the swamp sparrow. Dev Psychobiol 13:499–506.

Dooling RJ, Brown SD, Klump GM, Okanoya K (1992) Auditory perception of conspecific and heterospecific vocalizations in birds: Evidence for special processes. J Comp Psychol 1:20–28.

Dooling RJ, Lohr B, Dent ML (2000) Hearing in birds and reptiles. In: Dooling RJ, Fay RR, Popper AN (eds) Comparative Hearing: Birds and Reptiles. New York: Springer-Verlag, pp. 308–359.

Doupe AJ (1997) Song- and order-selective neurons in the songbird anterior forebrain and their emergence during vocal development. J Neurosci 17:1147–1167.

Doupe AJ, Kuhl PK (1999) Birdsong and human speech: Common themes and mechanisms. Annu Rev Neurosci 22:567–631.
Durand S, Tepper J, Cheng M (1992) The shell region of the nucleus ovoidalis: A subdivision of the avian auditory thalamus. J Comp Neurol 323:495–518.
Eaton RC (1983) Is the Mauthner cell a vertebrate command neuron? In: Ewert JP, Capranica RR, Ingle DJ (eds) Advances in Vertebrate Neuroethology. New York: Plenum Press, pp. 629–636.
Eens M, Pinxten M, Verheyen RF (1989) Temporal and sequential organization of song bouts in the European starling. Ardea 77:75–86.
Eens M, Pinxten M, Verheyen RF (1991) Organization of song in the European starling: Species-specificity and individual differences. Belg J Zool 121:257–278.
Ehret G (1997) The auditory cortex. J Comp Physiol A 181:547–557.
Eichenbaum HB, Davis JL (eds) (1998) Neuronal Ensembles: Strategies for Recording and Decoding. New York: John Wiley and Sons.
Evans FE, Whitfield IC (1964) Classification of unit responses in the auditory cortex of the unanesthetized and unrestrained cat. J Physiol (Lond) 171:476–493.
Ewert JP, Burghagen H, Schürg-Pfeiffer E (1983a) Neuroethological analysis of the innate releasing mechanism for prey-catching in toads. In: Ewert JP, Capranica RR, Ingle DJ (eds) Advances in Vertebrate Neuroethology. New York: Plenum Press, pp. 413–475.
Ewert JP, Capranica RR, Ingle DJ (1983b) Advances in Vertebrate Neuroethology. New York: Plenum Press.
Falls JB (1982) Individual recognition by sound in birds. In: Kroodsma DE, Miller EH (eds) Acoustic communication in birds. New York: Academic Press, pp. 237–278.
Falls JB (1987) Does song deter territorial intrusion in White-throated Sparrows (*Zonotrichia albicollis*)? Can J Zool 66:206–211.
Falls JB, Brooks RJ (1975) Individual recognition by song in white-throated sparrows. II: Effects of location. Can J Zool 53:1412–1420.
Farries MA, Perkel DJ (2000) Electrophysiological properties of avian basal ganglia neurons recorded in vitro. J Neurophysiol 84:2502–2513.
Fay RR (1998) Auditory stream segregation in goldfish (*Carassius auratus*). Hear Res 120:69–76.
Feng AS, Ratnam R (2000) Neural basis of hearing in real-world situations. Annu Rev Psychol 51:699–725.
Feng AS, Schellart NAM (1999) Central auditory processing in fish and amphibians. In: Popper AN, Fay RR (eds) Comparative Hearing: Fish and Amphibians. New York: Springer-Verlag, pp. 218–268.
Fishman YI, Reser DH, Arezzo JC, Steinschneider M (2001) Neural correlates of auditory stream segregation in primary auditory cortex of the awake monkey. Hear Res 151:167–187.
Fitch WT (2000) The evolution of speech: A comparative review. Trends Cogn Sci 4:258–267.
Fitzpatrick DC, Suga N, Olsen JF (1998) Distribution of response types across entire hemispheres of the mustached bat's auditory cortex. J Comp Neurol 391:353–365.
Fodor JA (1983) The Modularity of Mind. Cambridge, MA: MIT Press.
Fortune ES, Margoliash D (1992) Cytoarchitectonic organization and morphology of cells in the field L complex in male zebra finches (*Taeniopygia guttata*). J Comp Neurol 325:388–404.

Fortune ES, Margoliash D (1995) Parallel pathways and convergence onto HVc and adjacent neostriatum of adult zebra finches (*Taeniopygia guttata*). J Comp Neurol 360:413–441.
Fowler CA (1996) Listeners do hear sounds, not tongues. J Acoust Soc Am 99:1730–1741.
Freedman DJ, Riesenhuber M, Poggio T, Miller EK (2001) Categorical representation of visual stimuli in the primate prefrontal cortex. Science 291:312–316.
Fujita I, Tanaka K, Ito M, Cheng K (1992) Columns for visual features of objects in monkey inferotemporal cortex. Nature 360:343–346.
Fuzessery ZM, Feng AS (1982) Frequency selectivity in the anuran auditory midbrain: Single unit responses to single and multiple tone stimulation. J Comp Physiol 146:471–484.
Fuzessery ZM, Feng AS (1983) Mating call selectivity in the thalamus and midbrain of the leopard frog (*Rana p. pipiens*): Single and multiunit analyses. J Comp Physiol 150:333–344.
Gallistel CR (1990) The Organization of Learning. Cambridge, MA: MIT Press.
Gao E, Suga N (1998) Experience-dependent corticofugal adjustment of midbrain frequency map in bat auditory system. Proc Natl Acad Sci USA 95:12663–12670.
Gao E, Suga N (2000) Experience-dependent plasticity in the auditory cortex and the inferior colliculus of bats: Role of the corticofugal system. Proc Natl Acad Sci USA 97:8081–8086.
Gehr DD, Capsius B, Grabner P, Gahr M, Leppelsack HJ (1999) Functional organization of the field-L-complex of adult male zebra finches. Neuroreport 10:375–380.
Gehr DD, Komiya H, Eggermont JJ (2000) Neuronal responses in cat primary auditory cortex to natural and altered species-specific calls. Hear Res 150:27–42.
Gentner TQ (1999) Behavioral and neurobiological mechanisms of song perception among European starlings. Unpublished Ph.D. Thesis, Johns Hopkins University, Baltimore, MD.
Gentner TQ, Hulse SH (1998) Perceptual mechanisms for individual recognition in European starlings (*Sturnus vulgaris*). Anim Behav 56:579–594.
Gentner TQ, Hulse SH (2000) Perceptual classification based on the component structure of song in European starlings. J Acoust Soc Am 107:3369–3381.
Gentner TQ, Margoliash D (2001) Behaviorally relevant selectivity of auditory neurons in the cmHV of European starlings. Sixth Int Congr Neuroethol Abstr:146.
Gentner TQ, Hulse SH, Ball GF (1999) IEG ZENK expression in songbirds during individual vocal recognition. Soc Neurosci Abstr 25:624.
Gentner TQ, Hulse SH, Bentley GE, Ball GF (2000) Individual vocal recognition and the effect of partial lesions to HVc on discrimination, learning, and categorization of conspecific song in adult songbirds. J Neurobiol 42:117–133.
Gentner TQ, Hulse SH, Duffy D, Ball GF (2001) Response biases in auditory forebrain regions of female songbirds following exposure to sexually relevant variation in male song. J Neurobiol 46:48–58.
Gerhardt HC (1982) Sound pattern recognition in some North American treefrogs (*Anura: Hylidae*): Implications for mate choice. Am Zool 22:581–595.
Gerhardt HC, Doherty JA (1988) Acoustic communication in the gray treefrog, *Hyla versicolor*: Evolutionary and neurobiological implications. J Comp Physiol A 162:261–278.

Ghazanfar AA, Hauser MD (1999) The neuroethology of primate vocal communication: Substrates for the evolution of speech. Trends Cogn Sci 3:377–384.
Godard R (1991) Long-term memory for individual neighbors in a migratory songbird. Nature 350:228–229.
Gold JI, Knudsen EI (2000) A site of auditory experience-dependent plasticity in the neural representation of auditory space in the barn owl's inferior colliculus. J Neurosci 20:3469–3486.
Grace JA, Theunissen FE (2000) Processing of natural and synthetic sounds in the avian auditory forebrain. Soc Neurosci Abstr 26:2030.
Green S (1975) Variation of vocal pattern with social situation in the Japanese monkey (*Macaca fuscata*): A field study. In: Rosenblum LA (ed) Primate Behavior. New York: Academic Press, pp. 1–102.
Gross CG (1992) Representation of visual stimuli in inferior temporal cortex. Philos Trans R Soc Lond B Biol Sci 335:3–10.
Gurney ME (1981) Hormonal control of cell form and number in the zebra finch song system. J Neurosci 1:658–673.
Hackett TA, Stepniewska I, Kaas JH (1998) Subdivisions of auditory cortex and ipsilateral cortical connections of the parabelt auditory cortex in macaque monkeys. J Comp Neurol 394:475–495.
Hamilton KS, King AP, Sengelaub DR, West MJ (1997) A brain of her own: A neural correlate of song assessment in a female songbird. Neurobiol Learn Mem 68: 325–332.
Hausberger M (1997) Social influences on song acquisition and sharing in the European starling (*Sturnus vulgaris*). In: Snowden C, Hausberger M (eds) Social Influences on Vocal Development. Cambridge, UK: Cambridge University Press.
Hausberger M, Cousillas H (1995) Categorization in birdsong: From behavioural to neuronal responses. Behav Process 35:83–91.
Hauser MD (1998) Functional referents and acoustic similarity: Field playback experiments with rhesus monkeys Anim Behav 56:1309–1310.
Hauser MD, Andersson K (1994) Left hemisphere dominance for processing of vocalizations in adult, but not infant rhesus monkeys: Field experiments. Proc Natl Acad Sci USA 91:3946–3947.
Haüsler U (1997) Measurement of short-time spatial activity patterns during auditory stimulation in the starling. In Syka J (ed) Acoustical Signal Processing in the Central Auditory System. New York: Plenum Press, pp. 85–91.
Hebb DO (1949) The Organization of Behavior: A Neuropsychological Theory. New York: Wiley.
Heffner HE, Heffner RS (1984) Temporal lobe lesions and perception of species-specific vocalizations by macaques. Science 226:75–76.
Heiligenberg WF (1977) Principles of Electrolocation and Jamming Avoidance in Electric Fish: A Neuroethological Approach. Berlin: Springer-Verlag.
Heiligenberg WF (1991) Neural Nets in Electric Fish. Cambridge, MA: MIT Press.
Herrnstein RJ (1990) Levels of stimulus control: A functional approach. Cognition 37:133–166.
Hessler NA, Doupe AJ (1999) Singing-related neural activity in a dorsal forebrain-basal ganglia circuit of adult zebra finches. J Neurosci 19:10461–10481.
Hiebert SM, Stoddard PK, Arcese P (1989) Repertoire size, territory acquisition and reproductive success in the song sparrow. Anim Behav 37:266–273.

Holekamp KE, Boydston EE, Szykman M, Graham I, Nutt KJ, Birch S, Piskiel A, Singh M (1999) Vocal recognition in the spotted hyaena and its possible implications regarding the evolution of intelligence. Anim Behav 58:383–395.

Hosino T, Okanoya K (2000) Lesion of a higher-order song nucleus disrupts phrase level complexity in Bengalese finches. Neuroreport 11:2091–2095.

Hoy RR, Hahn J, Paul RC (1977) Hybrid cricket auditory behavior: Evidence for genetic coupling in animal communication. Science 195:82–84.

Hulse SH, MacDougall-Shackleton SA, Wisniewski AB (1997) Auditory scene analysis by songbirds: Stream segregation of birdsong by European starlings (*Sturnus vulgaris*). J Comp Psychol 111:3–13.

Insley SJ (2000) Long-term vocal recognition in the northern fur seal. Nature 406:404–405.

Janata P, Margoliash D (1999) Gradual emergence of song selectivity in sensorimotor structures of the male zebra finch song system. J Neurosci 19:5108–5118.

Jarvis ED, Nottebohm F (1997) Motor-driven gene expression. Proc Natl Acad Sci USA 94:4097–4102.

Jones EG (2000) Cortical and subcortical contributions to activity-dependent plasticity in primate somatosensory cortex. Annu Rev Neurosci 23:1–37.

Jones MW, Errington ML, French PJ, Fine A, Bliss TV, Garel S, Charnay P, Bozon B, Laroche S, Davis S (2001) A requirement for the immediate early gene Zif268 in the expression of late LTP and long-term memories. Nat Neurosci 4:289–296.

Jouventin P, Aubin T, Lengagne T (1999) Finding a parent in a king penguin colony: The acoustic system of individual recognition. Anim Behav 57:1175–1183.

Kaas JH, Hackett TA (2000) Subdivisions of auditory cortex and processing streams in primates. Proc Natl Acad Sci USA 97:11793–11799.

Kanwal JS, Fitzpatrick DC, Suga N (1999) Facilitatory and inhibitory frequency tuning of combination-sensitive neurons in the primary auditory cortex of mustached bats. J Neurophysiol 82:2327–2345.

Karten HJ (1968) The ascending auditory pathway in the pigeon (*Columba liva*). II. Telencephalic projections of the nucleus ovoidalis thalami. Brain Res 11:134–153.

Katz LC, Gurney ME (1981) Auditory responses in the zebra finch's motor system for song. Brain Res 211:192–197.

Kelley D, Nottebohm F (1979) Projections of a telencephalic auditory nucleus—field L—in the canary. J Comp Neurol 183:455–470.

Kilgard MP, Merzenich MM (1998) Plasticity of temporal information processing in the primary auditory cortex. Nat Neurosci 1:727–731.

Kilgard MP, Pandya PK, Vazquez J, Gehi A, Schreiner CE, Merzenich MM (2001) Sensory input directs spatial and temporal plasticity in primary auditory cortex. J Neurophysiol 86:326–338.

Klein DJ, Depireux DA, Simon JZ, Shamma SA (2000) Robust spectrotemporal reverse correlation for the auditory system: Optimizing stimulus design. J Comput Neurosci 9:85–111.

Kluender KR, Diehl RL, Killeen PR (1987) Japanese quail can learn phonetic categories. Science 237:1195–1197.

Klump G (2000) Sound localization in birds. In: Dooling RJ, Fay RR, Popper AN (eds) Comparative Hearing: Birds and Reptiles. New York: Springer-Verlag, pp. 249–307.

Knudsen EI (1985) Experience alters the spatial tuning of auditory units in the optic tectum during a sensitive period in the barn owl. J Neurosci 5:3094–3109.
Kobayashi K, Uno H, Okanoya K (2001) Partial lesions in the anterior forebrain pathway affect song production in adult Bengalese finches. Neuroreport 12:353–358.
Konishi M (1970) Evolution of design features in the coding of species-specificity. Am Zool 10:67–72.
Konishi M (1978) Auditory environment and vocal development in birds. In: Walk RD, Pick HLJ (eds) Perception and Experience. New York: Plenum, pp. 105–118.
Korzeniewska E, Güntürkün O (1990) Sensory properties and afferents of the N. dorsolateralis posterior thalami of the pigeon. J Comp Neurol 292:457–479.
Krebs J, Ashcroft R, Webber M (1978) Song repertoires and territory defense in the great tit. Nature 271:539–542.
Kuhl PK, Miller JD (1975) Speech perception by the chinchilla: Voiced-voiceless distinction in alveolar plosive consonants. Science 190:69–72.
Kupfermann I, Weiss KR (1978) The command neuron concept. Behav Brain Sci 1:3–39.
Kuwabara N, Suga N (1993) Delay lines and amplitude selectivity are created in subthalamic auditory nuclei: The brachium of the inferior colliculus of the mustached bat. J Neurophysiol 69:1713–1724.
Langner G, Bonke D, Scheich H (1981) Neuronal discrimination of natural and synthetic vowels in field L of trained mynah birds. Exp Brain Res 43:11–24.
Lasky RE, Syrdal-Lasky A, Klein RE (1975) VOT discrimination by four to six and a half month old infants from Spanish environments. J Exp Child Psychol 20:215–225.
Lee DD, Seung HS (1999) Learning the parts of objects by non-negative matrix factorization. Nature 401:788–791.
Lehrman DW (1953) A critique of Konrad Lorenz's theory of instinctive behavior. Q Rev Biol 28:337–369.
Lengagne T, Aubin T, Jouventin P, Lauga J (2000) Perceptual salience of individually distinctive features in the calls of adult penguins. J Acoust Soc Am 107:508–516.
Leppelsack HJ (1983) Analysis of song in the auditory pathway of song birds. In: Evert JP, Capranica BR, Ingle DJ (eds) Advances in Vertebrate Neuroethology. New York: Plenum Press, pp. 783–799.
Leppelsack HJ, Schwartzkopff J (1972) Properties of acoustic neurons in the caudal neostriatum of birds. J Comp Physiol A 80:137–140.
Leppelsack HJ, Vogt M (1976) Response to auditory neurons in the forebrain of a song bird to stimulation with species-specific sounds. J Comp Physiol A 107:263–274.
Le Prell CG, Moody DB (1997) Perceptual salience of acoustic features of Japanese monkey coo calls. J Comp Psychol 111:261–274.
Le Prell CG, Moody DB (2000) Factors influencing the salience of temporal cues in the discrimination of synthetic Japanese monkey (*Macaca fuscata*) coo calls. J Exp Psychol Anim Behav Process 26:261–273.
Leroy SA, Wenstrup JJ (2000) Spectral integration in the inferior colliculus of the mustached bat. J Neurosci 20:8533–8541.
Lettvin JY, Maturana HR, McCulloch WS, Pitts WH (1959) What the frog's eye tells the frog's brain. Proc Inst Radio Eng NY 47:1940–1951.

Lewicki MS (1996) Intracellular characterization of song-specific neurons in the zebra finch auditory forebrain. J Neurosci 16:5855–5863.

Lewicki MS, Konishi M (1995) Mechanisms underlying the sensitivity of songbird forebrain neurons to temporal order. Proc Natl Acad Sci USA 92:5582–5586.

Li R, Sakaguchi H (1997) Cholinergic innervation of the song control nuclei by the ventral paleostriatum in the zebra finch: a double-labeling study with retrograde fluorescent tracers and choline acetyltransferase immunohistochemistry. Brain Res 763:239–246.

Liberman AM, Mattingly IG (1985) The motor theory of speech perception revised. Cognition 21:1–36.

Liberman AM, Cooper FS, Shankweiler DP, Studdert-Kennedy M (1967) Perception of the speech code. Psychol Rev 74:431–461.

Lind H, Dabelsteen T, McGregor PK (1997) Female great tits can identify mates by song. Anim Behav 52:667–671.

Lindblom B (1991) The status of phonetic gestures. In: Mattingly IG, Studdert-Kennedy M (eds) Modularity and the Motor Theory of Speech Perception. Hillsdale, NJ: Lawrence Erlbaum, pp. 7–24.

Logan JS, Lively SE, Pisoni DB (1989) Training Japanese listeners to identify /r/ an /l/. J Acoust Soc Am 85:137–138.

Long KD, Kennedy G, Balaban E (2001) Transferring an inborn auditory perceptual predisposition with interspecies brain transplants. Proc Natl Acad Sci USA, 98:5862–5867.

Lotto AJ, Kluender KR, Holt LL (1997) Perceptual compensation for coarticulation by Japanese quail (*Coturnix coturnix japonica*). J Acoust Soc Am 102:1134–1140.

Luo M, Perkel DJ (1999) Long-range GABAergic projection in a circuit essential for vocal learning. J Comp Neurol 403:68–84.

MacDougall-Shackleton SA, Hulse SH, Ball GF (1998a) Neural bases of song preferences in female zebra finches (*Taeniopygia guttata*). Neuroreport 9:3047–3052.

MacDougall-Shackleton SA, Hulse SH, Gentner TQ, White W (1998b) Auditory scene analysis by European starlings (*Sturnus vulgaris*): Perceptual segregation of tone sequences. J Acoust Soc Am 103:3581–3587.

Manley GA (1990) Peripheral Hearing Mechanisms in Reptiles and Birds. Berlin: Springer-Verlag.

Manley JA, Müller-Preuss P (1978) Response variability in the mammalian auditory cortex: An objection to feature detection? Fed Proc Fed Am Soc Exp Biol 37:2355–2359.

Margoliash D (1983) Acoustic parameters underlying the responses of song-specific neurons in the white-crowned sparrow. J Neurosci 3:1039–1057.

Margoliash D (1986) Preference for autogenous song by auditory neurons in a song system nucleus of the white-crowned sparrow. J Neurosci 6:1643–1661.

Margoliash D (1987) Neural plasticity in birdsong learning. In: Rauschecker JP, Marler P (eds) Imprinting and Cortical Plasticity. New York: Wiley, pp. 23–54.

Margoliash D (1997) Functional organization of forebrain pathways for song production and perception. J Neurobiol 33:671–693.

Margoliash D (2002) Offline learning and the role of autogenous speech: New suggestions from birdsong research. Speech Communication, in press.

Margoliash D, Fortune ES (1992) Temporal and harmonic combination-sensitive neurons in the zebra finch's HVc. J Neurosci 12:4309–4326.

Margoliash D, Konishi M (1985) Auditory representation of autogenous song in the song-system of white-crowned sparrows. Proc Natl Acad Sci USA 82:5997–6000.

Margoliash D, Fortune ES, Sutter ML, Yu AC, Wren-Hardin BD, Dave A (1994) Distributed representation in the song system of oscines: Evolutionary implications and functional consequences. Brain Behav Evol 44:247–264.

Marler P, Sherman V (1983) Song structure without auditory feedback: Emendations of the auditory template hypothesis. J Neurosci 3:517–531.

Marr D (1982) Vision. San Francisco: Freeman.

Martin KA (1994) A brief history of the "feature detector." Cereb Cortex 4:1–7.

May B, Moody DB, Stebbins WC (1988) The significant features of Japanese macaque coo sounds: A psychophysical study. Anim Behav 36:1432–1444.

May B, Moody DB, Stebbins WC (1989) Categorical perception of conspecific communication sounds by Japanese macaques, *Macaca fuscata*. J Acoust Soc Am 85:837–847.

McCasland JS (1987) Neuronal control of bird song production. J Neurosci 7:23–39.

McCasland JS, Konishi M (1981) Interaction between auditory and motor activities in an avian song control nucleus. Proc Natl Acad Sci USA 78:7815–7819.

McComb K, Moss C, Sayialel S, Baker L (2000) Unusually extensive networks of vocal recognition in African elephants. Anim Behav 59:1103–1109.

McCulloch S, Boness DJ (2000) Mother-pup vocal recognition in the gray seal (*Halichoerus grypus*) of Sable Island, Nova Scotia, Canada. J Zool 251:449–455.

McDonald MV (1989) Function of song in Scott's seaside sparrow, *Ammodramus maritimus peninsulae*. Anim Behav 38:468–485.

McKibben JR, Bass AH (1998) Behavioral assessment of acoustic parameters relevant to signal recognition and preference in a vocal fish. J Acoust Soc Am 104:3520–3533.

Mello CV, Nottebohm F, Clayton D (1995) Repeated exposure to one song leads to a rapid and persistent decline in an immediate early gene's response to that song in zebra finch telencephalon. J Neurosci 15:6919–6925.

Mello CV, Vicario DS, Clayton DF (1992) Song presentation induces gene expression in the songbird forebrain. Proc Natl Acad Sci USA 89:6818–6822.

Miller PJO, Bain DE (2000) Within-pod variation in the sound production of a pod of killer whales, *Orcinus orca*. Anim Behav 60:617–628.

Mittman DH, Wenstrup JJ (1995) Combination-sensitive neurons in the inferior colliculus. Hear Res 90:185–191.

Moiseff A, Konishi M (1981) Neuronal and behavioral sensitivity to binaural time differences in the owl. J Neurosci 1:40–48.

Mooney R (2000) Different subthreshold mechanisms underlie song selectivity in identified HVc neurons of the zebra finch. J Neurosci 20:5420–5436.

Morris JS, Friston KJ, Dolan RJ (1998) Experience-dependent modulation of tonotopic neural responses in human auditory cortex. Proc R Soc Lond B Biol Sci 265:649–657.

Mudry KM, Constantine-Paton M, Capranica RR (1977) Auditory sensitivity of the diencephalon of the leopard frog (*Rana p. pipiens*). J Comp Physiol 114:1–13.

Müller CM, Leppelsack HJ (1985) Feature extraction and tonotopic organization in the avian forebrain. Exp Brain Res 59:587–599.

Narins PM, Capranica RR (1976) Sexual differences in the auditory system of the tree frog *Eleutherodactylus coqui*. Science 192:378–380.

Nearey TM (1997) Speech perception as pattern recognition. J Acoust Soc Am 101:3241–3254.
Nelken I, Prut Y, Vaadia E, Abeles M (1994a) Population responses to multi-frequency sounds in the cat auditory cortex: One- and two-parameter families of sounds. Hear Res 72:206–222.
Nelken I, Prut Y, Vaadia E, Abeles M (1994b) Population responses to multi-frequency sounds in the cat auditory cortex: Four-tone complexes. Hear Res 72:223–236.
Nelson DA (1989) Song frequency as a cue for recognition of species and individuals in the field sparrow (*Spizella pusilla*). J Comp Psychol 103:171–176.
Newman JD, Wollberg Z (1973) Multiple coding of species-specific vocalizations in the auditory cortex of squirrel monkeys. Brain Res 54:287–304.
Nicolelis MAL (ed) (2001) Advances in Neural Population Coding. Progress in Brain Research, Vol 130. Amsterdam: Elsevier.
Nieder A, Klump GM (1999) Adjustable frequency selectivity of auditory forebrain neurons recorded in a freely moving songbird via radiotelemetry. Hear Res 127:41–54.
Nordeen KW, Nordeen EJ (1993) Long-term maintenance of song in adult zebra finches is not affected by lesions of a forebrain region involved in song learning. Behav Neural Biol 59:79–82.
Nottebohm F, Arnold AP (1976) Sexual dimorphism in vocal control areas of the songbird brain. Science 194:211–213.
Nottebohm F, Stokes TM, Leonard CM (1976) Central control of song in the canary, *Serinus canarius*. J Comp Neurol 165:457–486.
Nudo RJ, Wise BM, SiFuentes F, Milliken GW (1996) Neural substrates for the effects of rehabilitative training on motor recovery after ischemic infarct. Science 272:1791–1794.
Ohlemiller KK, Kanwal JS, Suga N (1996) Facilitative responses to species-specific calls in cortical FM-FM neurons of the mustached bat. Neuroreport 7:1749–1755.
Okanoya K, Dooling RJ (1990) Song syllable perception in song sparrows (*Melospiza melodia*) and swamp sparrows (*Melospiza georgiana*): An approach from animal psychophysics. Bull Psychon Soc 28:221–224.
Okanoya K, Yamaguchi A (1997) Adult Bengalese finches (Lonchura striata var. domestica) require real-time auditory feedback to produce normal song syntax. J Neurobiol 33:343–356.
Okanoya K, Tsumaki S, Honda E (2000) Perception of temporal properties in self-generated songs by Bengalese finches (*Lonchura striata var. domestica*). J Comp Psychol 114:239–245.
O'Loghlen AL, Beecher MD (1997) Sexual preferences for mate song types in female song sparrows. Anim Behav 53:835–841.
O'Loghlen AL, Beecher MD (1999) Mate, neighbor and stranger songs: A female song sparrow perspective. Anim Behav 58:13–20.
Olsen JF (1994) Medial geniculate neurons in the squirrel monkey sensitive to inter-component delays that categorize species-specific calls. Assoc Res Otolaryngol Abstr 21:84.
Olsen JF, Suga N (1991a) Combination-sensitive neurons in the medial geniculate body of the mustached bat: Encoding of relative velocity information. J Neurophysiol 65:1254–1274.

Olsen JF, Suga N (1991b) Combination-sensitive neurons in the medial geniculate body of the mustached bat: Encoding of target range information. J Neurophysiol 65:1275–1296.

O'Neill WE (1985) Responses to pure tones and linear FM components of the CF-FM biosonar signal by single units in the inferior colliculus of the mustached bat. J Comp Physiol A 157:797–815.

O'Neill WE (1995) The bat auditory cortex. In: Popper AN, Fay RR (eds) Hearing by Bats. New York: Springer-Verlag, pp. 416–480.

O'Neill WE, Suga N (1979) Target range-sensitive neurons in the auditory cortex of the mustache bat. Science 203:69–73.

Owings DH, Morton ES (1998) Animal Vocal Communication: A New Approach. Cambridge, UK: Cambridge University Press.

Owren MJ, Casale TM (1994) Variations in fundamental frequency peak position in Japanese macaque (*Macaca fuscata*) coo calls. J Comp Psychol 108:291–297.

Owren MJ, Seyfarth RM, Cheney DL (1997) The acoustic features of vowel-like grunt calls in chacma baboons (*Papio cyncephalus ursinus*): Implications for production processes and functions. J Acoust Soc Am 101:2951–2963.

Peek FW (1972) An experimental study of the territorial function of vocal and visual display in the male red-winged blackbird (*Agelaius phoeniceus*). Anim Behav 20:112–118.

Pereira MR (1986) Maternal recognition of juvenile offspring coo vocalizations in Japanese macaques. Anim Behav 34:935–937.

Perkel DJ, Farries MA (2000) Complementary "bottom-up" and "top-down" approaches to basal ganglia function. Curr Opin Neurobiol 10:725–731.

Pfingst BE, O'Connor TA, Miller JM (1977) Response plasticity of neurons in auditory cortex of the rhesus monkey. Exp Brain Res 29:393–404.

Pons TP, Garraghty PE, Mishkin M (1991) Serial and parallel processing of tactual information in somatosensory cortex of rhesus monkeys. J Neurophysiol 68:518–527.

Popper AN, Fay RR (eds) (1995) Hearing by Bats. New York: Springer-Verlag.

Portfors CV, Wenstrup JJ (1999) Delay-tuned neurons in the inferior colliculus of the mustached bat: Implications for analyses of target distance. J Neurophysiol 82:1326–1338.

Pytte CL, Suthers RA (1999) A bird's own song contributes to conspecific song perception. Neuroreport 10:1773–1778.

Rauschecker JP (1997) Processing of complex sounds in the auditory cortex of cat, monkey, and man. Acta Otolaryngol Suppl 532:34–38.

Rauschecker JP (1998) Cortical processing of complex sounds. Curr Opin Neurobiol 8:516–521.

Rauschecker JP, Tian B (2000) Mechanisms and streams for processing of "what" and "where" in auditory cortex. Proc Natl Acad Sci USA 97:11800–11806.

Rauschecker JP, Tian B, Hauser M (1995) Processing of complex sounds in the macaque nonprimary auditory cortex. Science 268:111–114.

Rauske PL, Margoliash D (1999) Does behavioral state modulate sensorimotor properties in HVc? Soc Neurosci Abstr 25:624.

Recanzone GH (1998) Rapidly induced auditory plasticity: The ventriloquism aftereffect. Proc Natl Acad Sci USA 95:869–875.

Recanzone GH, Schreiner CE, Merzenich MM (1993) Plasticity in the frequency representation of primary auditory cortex following discrimination training in adult owl monkeys. J Neurosci 13:87–103.
Recanzone GH, Guard DC, Phan ML (2000) Frequency and intensity response properties of single neurons in the auditory cortex of the behaving macaque monkey. J Neurophysiol 83:2315–2331.
Rendall D, Rodman PS, Emond RE (1996) Vocal recognition of individuals and kin in free-ranging rhesus monkeys. Anim Behav 51:1007–1015.
Rendall D, Owren MJ, Rodman PS (1998) The role of vocal tract filtering in identity cueing in rhesus monkey (*Macaca mulatta*) vocalizations. J Acoust Soc Am 103:602–614.
Rendall D, Seyfarth RM, Cheney DL, Owren MJ (1999) The meaning and function of grunt variants in baboons. Anim Behav 57:583–592.
Ribeiro S, Cecchi GA, Magnasco MO, Mello CV (1998) Toward a song code: Evidence for a syllabic representation in the canary brain. Neuron 21:359–371.
Rieke F, Warland D, Van Steveninck RDR, Bialek W (1996) Spikes, Exploring the Neural Code. Cambridge, MA: MIT Press.
Riquimaroux H, Gaioni SJ, Suga N (1991) Cortical computational maps control auditory perception. Science 251:565–568.
Roeder KD (1966) Auditory system of noctuid moths. Science 154:1515–1521.
Rolls ET (1992) Neurophysiological mechanisms underlying face processing within and beyond the temporal cortical visual areas. Philos Trans R Soc Lond B Biol Sci 335:11–20.
Romanski LM, Tian B, Fritz J, Mishkin M, Goldman-Rakic PS, Rauschecker JP (1999) Dual streams of auditory afferents target multiple domains in the primate prefrontal cortex. Nat Neurosci 2:1131–1136.
Rose G, Capranica RR (1983) Temporal selectivity in the central auditory system of the leopard frog. Science 219:1087–1089.
Rose GJ, Kawasaki M, Heiligenberg W (1988) "Recognition units" at the top of a neuronal hierarchy? Prepacemaker neurons in *Eigenmannia* code the sign of frequency differences unambiguously. J Comp Physiol A 162:759–772.
Rubsamen R, Dorrscheidt GJ (1986) Tonotopic organization of the auditory forebrain in a songbird, the European starling. J Comp Physiol A 158:639–646.
Ryan MJ (1998) Sexual selection, receiver biases, and the evolution of sex differences. Science 281:1999–2003.
Sakai M, Suga N (2001) Plasticity of the cochleotopic (frequency) map in specialized and nonspecialized auditory cortices. Proc Natl Acad Sci USA 98:3507–3512.
Sayigh LS, Tyack PL, Wells RS, Solow AR, Scott MD, Irvine AB (1999) Individual recognition in wild bottlenose dolphins: A field test using playback experiments. Anim Behav 57:41–50.
Schäfer M, Rubsamen R, Dorrscheidt GJ, Knipschild M (1992) Setting complex tasks to single units in the avian auditory forebrain. II. Do we really need natural stimuli to describe neuronal response characteristics? Hear Res 57:231–244.
Scharff C, Nottebohm F (1991) A comparative study of the behavioral deficits following lesions of various parts of the zebra finch song system: Implications for vocal learning. J Neurosci 11:2896–2913.
Scharff C, Nottebohm F, Cynx J (1998) Conspecific and heterospecific song discrimination in male zebra finches with lesions in the anterior forebrain pathway. J Neurobiol 36:81–90.

Scheich H, Bonke BA, Bonke D, Langer HJ (1979a) Functional organization of some auditory nuclei in the guinea fowl demonstrated by the 2-deoxyglucose technique. Cell Tissue Res 204:17–27.

Scheich H, Langner G, Bonke D (1979b) Responsiveness of units in the auditory neostriatum of the Guinea fowl (*Numida meleagris*) to species-specific calls and synthetic stimuli. II. Discrimination of iambus-like calls. J Comp Physiol 132:257–276.

Schmidt MF, Konishi M (1998) Gating of auditory responses in the vocal control system of awake songbirds. Nat Neurosci 1:513–518.

Schulze H, Langner G (1997) Periodicity coding in the primary auditory cortex of the Mongolian gerbil (*Meriones unguiculatus*): Two different coding strategies for pitch and rhythm? J Comp Physiol A 181:651–663.

Schwarz DW, Tomlinson RW (1990) Spectral response patterns of auditory cortex neurons to harmonic complex tones in alert monkey (*Macaca mulatta*). J Neurophysiol 64:282–298.

Searcy WA (1996) Sound-pressure levels and song preferences in female red-winged blackbirds (*Agelaius phoeniceus*) (Aves, Emberizidae). Ethology 102:187–196.

Searcy WA, Yasukawa K (1996) Song and female choice. In: Kroodsma DE, Miller EH (eds) Ecology and Evolution of Acoustic Communication in Birds. Ithaca, NY: Cornell University Press, pp. 454–473.

Seyfarth RM, Cheney DL, Marler P (1980a) Monkey responses to three different alarm calls: Evidence of predator classification and semantic communication. Science 210:801–803.

Seyfarth RM, Cheney DL, Marler P (1980b) Vervet monkey alarm calls: Semantic communication in a free-ranging primate. Anim Behav 28:1070–1094.

Shea SD, Rauske PL, Margoliash D (2001) Identification of HVc projection neurons in extracellular records by antidromic stimulation. Soc Neurosci Abstr 27:381.6.

Simoncelli E, Olshausen A (2001) Natural image statistics and neural representation. Annu Rev Neurosci 24:1193–1215.

Singer W, Gray CM (1995) Visual feature integration and the temporal correlation hypothesis. Annu Rev Neurosci 18:555–586.

Sinnott JM (1980) Species-specific coding in bird song. J Acoust Soc Am 68:494–497.

Smith DG, Reid FA (1979) Roles of the song repertoire in red-winged blackbirds. Behav Ecol Sociobiol 5:279–290.

Smith WJ (1997) The behavior of communicating, after twenty years. In: Owings DH, Beecher MD, Owings DH, Thompson NS (eds) Perspectives in Ethology, Vol 12: Communication. New York: Plenum Press, pp. 7–53.

Sohrabji F, Nordeen EJ, Nordeen KW (1990) Selective impairment of song learning following lesions of a forebrain nucleus in the juvenile zebra finch. Behav Neural Biol 53:51–63.

Solis MM, Brainard MS, Hessler NA, Doupe AJ (2000) Song selectivity and sensorimotor signals in vocal learning and production. Proc Natl Acad Sci USA, 97:11836–11842.

Stoddard PK (1996) Vocal recognition of neighbors by territorial passerines. In: Kroodsma DE, Miller EH (eds) Ecology and Evolution of Acoustic Communication in Birds. Ithaca, NY: Cornell University Press, pp. 356–374.

Stoddard PK, Beecher MD, Loesche P, Campbell SE (1992) Memory does not constrain individual recognition in a bird with song repertoires. Behaviour 122:274–287.

Stripling R, Volman SF, Clayton DF (1997) Response modulation in the zebra finch neostriatum: Relationship to nuclear gene expression. J Neurosci 17:3883–3893.
Suga N, Horikawa J (1986) Multiple axes for representation of echo delays in the auditory cortex of the mustached bat. J Neurophysiol 55:776–805.
Suga N, O'Neill WE, Manabe T (1978) Cortical neurons sensitive to combinations of information-bearing elements of biosonar signals in the mustache bat. Science 200:778–781.
Suga N, O'Neill WE, Manabe T (1979) Harmonic-sensitive neurons in the auditory cortex of the mustache bat. Science 203:270–274.
Suga N, Zhang Y, Yan J (1997) Sharpening of frequency tuning by inhibition in the thalamic auditory nucleus of the mustached bat. J Neurophysiol 77:2098–2114.
Suga N, Yan J, Zhang Y (1998) The processing of species-specific sounds by the ascending and descending auditory systems. In: Poon PWF, Brugge JF (eds) Central Auditory Processing and Neural Modeling. New York: Plenum Press.
Sullivan WEI, Konishi M (1984) Segregation of stimulus phase and intensity coding in the cochlear nucleus of the barn owl. J Neurosci 4:1787–1799.
Sutter ML, Schreiner CE (1991) Physiology and topography of neurons with multipeaked tuning curves in cat primary auditory cortex. J Neurophysiol 65:1207–1226.
Takahashi T, Moiseff A, Konishi M (1984) Time and intensity cues are processes independently in the auditory system of the owl. J Neurosci 4:1781–1786.
Taniguchi I, Niwa H, Wong D, Suga N (1986) Response properties of FM-FM combination-sensitive neurons in the auditory cortex of the mustached bat. J Comp Physiol A 159:331–337.
Theunissen FE, Doupe AJ (1998) Temporal and spectral sensitivity of complex auditory neurons in the nucleus HVc of male zebra finches. J Neurosci 18:3786–3802.
Theunissen FE, Sen K, Doupe AJ (2000) Spectral-temporal receptive fields of non-linear auditory neurons obtained using natural sounds. J Neurosci 20:2315–2331.
Tian B, Reser D, Durham A, Kustov A, Rauschecker JP (2001) Functional specialization in rhesus monkey auditory cortex. Science 292:290–293.
Tinbergen N (1963) On aims and methods of ethology. Z Tierpsychol 20:410–433.
Tsunoda K, Yamane Y, Nishizaki M, Tanifuji M (2001) Complex objects are represented in macaque inferotemporal cortex by the combination of feature columns. Nat Neurosci 4:832–838.
Tsuzuki K, Suga N (1988) Combination-sensitive neurons in the ventroanterior area of the auditory cortex of the mustached bat. J Neurophysiol 60:1908–1923.
Ulinski PS (1984) Design features in vertebrate sensory systems. Am Zool 24:717–731.
Ulinski PS, Margoliash D (1990) Neurobiology of the reptile-bird transition. In: Jones EG, Peters A (eds) Cerebral Cortex. New York: Plenum, pp. 217–265.
Vallet E, Kreutzer M (1995) Female canaries are sexually responsive to special song phrases. Anim Behav 49:1603–1610.
van Dijk P, Wit HP, Segenhout JM (1997) Dissecting the frog inner ear with Gaussian noise. II. Temperature dependence of inner ear function. Hear Res 114:243–251.
Vates GE, Broome BM, Mello CV, Nottebohm F (1996) Auditory pathways of caudal telencephalon and their relation to the song system of adult male zebra finches (*Taenopygia guttata*). J Comp Neurol 366:613–642.

Venkatachalam S, Fee MS, Kleinfeld D (1999) Ultra-miniature headstage with 6-channel drive and vacuum-assisted micro-wire implantation for chronic recording from the neocortex. J Neurosci Methods 90:37–46.
Vicario DS, Yohay KH (1993) Song-selective auditory input to a forebrain vocal control nucleus in the zebra finch. J Neurobiol 24:488–505.
Vicario DS, Naqvi NH, Raksin JN (2001) Behavioral discrimination of sexually dimorphic calls by male zebra finches requires an intact vocal motor pathway. J Neurobiol 47:109–120.
Volman S (1993) Development of neural selectivity for birdsong during vocal learning. J Neurosci 13:4737–4747.
von der Heydt R, Peterhans E, Baumgartner G (1984) Illusory contours and cortical neuron responses. Science 224:1260–1262.
von Helversen O, von Helversen D (1994) Forces driving coevolution of song and song recognition in grasshoppers. Fortschr Zool 39:253–284.
Vu ET, Mazurek ME, Kuo Y-C (1994) Identification of a forebrain motor programming network for the learned song of zebra finches. J Neurosci 14:6924–6934.
Vu ET, Kuo Y, Chance FS (1995) Effects of lesioning nucleus interfacialis on adult zebra finch song. Soc Neurosci Abstr 21:964.
Wang X, Kadia SC (2001) Differential representation of species-specific primate vocalizations in the auditory cortices of marmoset and cat. J Neurophysiol 86:2616–2620.
Wang X, Merzenich MM, Beitel R, Schreiner CE (1995) Representation of a species-specific vocalization in the primary auditory cortex of the common marmoset: Temporal and spectral characteristics. J Neurophysiol 74:2685–2706.
Wang Y, Fujita I, Murayama Y (2000) Neuronal mechanisms of selectivity for object features revealed by blocking inhibition in inferotemporal cortex. Nat Neurosci 3:807–813.
Wanker R, Apcin J, Jennerjahn B, Waibel B (1998) Discrimination of different social companions in spectacled parrotlets (*Forpus conspicillatus*): Evidence for individual vocal recognition. Behav Ecol Sociobiol 43:197–202.
Weary DM, Krebs JR (1992) Great tits classify songs by individual voice characteristics. Anim Behav 43:283–287.
Wenstrup JJ (1999) Frequency organization and responses to complex sounds in the medial geniculate body of the mustached bat. J Neurophysiol 82:2528–2544.
Wenstrup JJ, Grose CD (1995) Inputs to combination-sensitive neurons in the medial geniculate body of the mustached bat: The missing fundamental. J Neurosci 15:4693–4711.
Wenstrup JJ, Leroy SA (2001) Spectral integration in the inferior colliculus: Role of glycinergic inhibition in response facilitation. J Neurosci 21:RC124.
Wenstrup JJ, Mittmann DH, Grose CD (1999) Inputs to combination-sensitive neurons of the inferior colliculus. J Comp Neurol 409:509–528.
Werker JF, Tees RC (1984) Cross language speech perception: Evidence for perceptual reorganization during the first year of life. Infant Behav Dev 7:49–63.
Wiley HB, Hatchwell BJ, Davies NB (1991) Recognition of individual males' songs by female Dunnocks: A mechanism increasing the number of copulatory partners and reproductive success. Ethology 88:145–153.
Williams H, Mehta N (1999) Changes in adult zebra finch song require a forebrain nucleus that is not necessary for song production. J Neurobiol 39:14–28.

Williams H, Nottebohm F (1985) Auditory responses in avian vocal motor neurons: A motor theory for song perception in birds. Science 229:279–282.

Williams H, Vicario DS (1993) Temporal patterning of song production: Participation of nucleus uvaeformis of the thalamus. J Neurobiol 24:903–912.

Winter P, Funkenstein HH (1973) The effects of species-specific vocalization on the discharge of auditory cortical cells in the awake squirrel monkey (*Saimiri sciureus*). Exp Brain Res 18:489–504.

Wollberg Z, Newman JD (1972) Auditory cortex of squirrel monkey: Response patterns of single cells to species-specific vocalizations. Science 175:212–214.

Wyttenbach RA, May ML, Hoy RR (1996) Categorical perception of sound frequency by crickets. Science 273:1542–1544.

Yan J, Suga N (1999) Corticofugal amplification of facilitative auditory responses of subcortical combination-sensitive neurons in the mustached bat. J Neurophysiol 81:817–824.

Yasukawa K (1981) Song repertoires in the red-winged blackbird (*Agelaius phoeniceus*): A test of the Beau Geste hypothesis. Anim Behav 29:114–125.

Yu A, Margoliash D (1996) Temporal hierarchical control of singing in birds. Science 273:1871–1875.

Zhang Y, Suga N (2000) Modulation of responses and frequency tuning of thalamic and collicular neurons by cortical activation in mustached bats. J Neurophysiol 84:325–333.

Zhang Y, Suga N, Yan J (1997) Corticofugal modulation of frequency processing in bat auditory system. Nature 387:900–903.

Zoloth SR, Peterson MR, Beecher MD, Green SG, Marler P, Moody DB, Stebbins WC (1979) Species-specific perceptual processing of vocal sounds by monkeys. Science 204:870–873.

Index

This index combines both subject and species. In most cases, species are listed by the scientific name. Common names are provided where available, but these are cross-listed to the scientific names.